Essentials of *In Vivo* Biomedical Imaging

Essentials of In Vivo Biomedical Imaging

Edited by
Simon R. Cherry
Ramsey D. Badawi
Jinyi Qi

CRC Press is an imprint of the
Taylor & Francis Group, an **informa** business

CRC Press
Taylor & Francis Group
6000 Broken Sound Parkway NW, Suite 300
Boca Raton, FL 33487-2742

Printed and bound in India by Replika Press Pvt. Ltd.

Printed on acid-free paper
Version Date: 20141118

International Standard Book Number-13: 978-1-4398-9874-1 (Pack - Book and Ebook)

Library of Congress Cataloging-in-Publication Data

Essentials of in vivo biomedical imaging / [edited by] Simon R. Cherry, Ramsey D. Badawi, Jinyi Qi.
p. ; cm.
Includes bibliographical references and index.
ISBN 978-1-4398-9874-1 (alk. paper)
I. Cherry, Simon R., editor. II. Badawi, Ramsey D., editor. III. Qi, Jinyi, 1970- , author.
[DNLM: 1. Diagnostic Imaging. WN 180]

RC78.7.D53
616.07'54--dc23 2014042964

Visit the Taylor & Francis Web site at
http://www.taylorandfrancis.com

and the CRC Press Web site at
http://www.crcpress.com

Contents

Preface

In vivo biomedical imaging technologies provide a noninvasive window into the structure and function of the living body and have become widely adopted in biomedical research, spanning preclinical studies in animal models through clinical research in human subjects. The technologies and methods of biomedical imaging are used in many disciplines and across many disease areas, and also are increasingly employed by industry in the development and validation of new therapeutic interventions. There is hardly an area of biomedical research in which imaging has not become an essential part of the experimental toolbox. *In vivo* imaging has unique strengths, which include the ability to noninvasively and nondestructively survey large volumes of tissue (whole organs and often the entire body) and the ability to visualize and quantify changes (often over time) in tissue morphology and function in normal health, in disease, and in response to treatment. Since most imaging techniques are also highly translational, this provides a unified experimental platform for moving across species, from preclinical studies in small or large animal disease models to clinical research studies in humans.

Users of these imaging technologies in biomedical research come from a staggering array of backgrounds, including cancer biology, neuroscience, immunology, chemistry, biochemistry, material science, nutrition, veterinary and human medicine, toxicology, drug development, and many more. While there are many excellent textbooks focused on clinical medical imaging as practiced daily throughout the world, there are few books that approach *in vivo* imaging technologies from the perspective of a scientist or physician-scientist using, or interested in using, these techniques in their research. It is for these scientists that this book is written, with the hope of providing a reference source that can help answer the following often-asked questions: Can imaging address this question? Which technique should I use? How does it work? What information does it provide? What are its strengths and limitations? What applications is it best suited for? How can I analyze the data? Through attempting to address these questions, our goal is to help scientists choose appropriate *in vivo* imaging technologies and methods and use them as effectively as possible in their research.

The book is written by leading authorities in the field and with the understanding that readers will come to this book with a wide variety of training and expertise. While material is presented at some depth, using appropriate mathematics, physics, and engineering when necessary for those who really want to dig into a particular imaging technique, it also is a book for the more casual user of imaging to dip into. Large fractions of the text are accessible to researchers independent of their specific scientific background, where the emphasis is on explaining what each imaging technology can measure, describing major methods and approaches, and giving examples demonstrating the rich repertoire of modern biomedical imaging to address a wide range of morphological, functional, metabolic, and molecular parameters in a safe and noninvasive manner. We hope you will gain as much pleasure and insight from reading this book as we have had in editing it.

Editors

Simon R. Cherry, PhD, is a distinguished professor in the Departments of Biomedical Engineering and Radiology, as well as director of the Center for Molecular and Genomic Imaging, at the University of California, Davis. He earned a PhD in medical physics in 1989 from the Institute of Cancer Research, London. Dr. Cherry's research interests focus around radiotracer imaging, optical imaging, and hybrid multimodality imaging systems, focusing on the development of new technologies, instrumentation, and systems. Dr. Cherry has over 25 years of experience in the field of biomedical imaging and has authored more than 200 publications, including the textbook *Physics in Nuclear Medicine*. He is a fellow of the Institute for Electrical and Electronic Engineers (IEEE), the Biomedical Engineering Society, and the Institute of Physics in Engineering and Medicine.

Ramsey D. Badawi, PhD, is an associate professor in the Departments of Radiology and Biomedical Engineering at the University of California, Davis (UC Davis). He currently serves as chief of the Division of Nuclear Medicine and holds the molecular imaging endowed chair in the Department of Radiology. Dr. Badawi earned a bachelor's degree in physics in 1987 and a master's in astronomy in 1988 from the University of Sussex, UK. He entered the field of medical imaging in 1991, when he joined St. Thomas' Hospital in London. He earned a PhD in positron emission tomography (PET) physics at the University of London in 1998. Subsequently, he worked at the University of Washington, Seattle, and at the Dana Farber Cancer Institute in Boston prior to joining UC Davis in 2004. Dr. Badawi's current research interests include PET and multimodality imaging instrumentation, image processing, and imaging in clinical trials.

Jinyi Qi, PhD, is a professor in the Department of Biomedical Engineering at the University of California, Davis (UC Davis). He earned a PhD in electrical engineering from the University of Southern California (USC) in 1998. Prior to joining the faculty of UC Davis, he was a research scientist in the Department of Functional Imaging at the Lawrence Berkeley National Laboratory. Dr. Qi is an associate editor of *IEEE Transactions of Medical Imaging*. He was elected as a fellow of the American Institute for Medical and Biological Engineering in 2011, and a fellow of the IEEE in 2013. Dr. Qi's research interests include statistical image reconstruction, medical image processing, image quality evaluation, and imaging system optimization.

Contributors

Jeff R. Anderson
MR Core Facilities
Department of Translational Imaging
Houston Methodist Research Institute
Houston, Texas, USA

Ramsey D. Badawi
Department of Radiology
University of California, Davis Medical Center
Sacramento, California, USA

John M. Boone
Department of Radiology
University of California, Davis Medical Center
Sacramento, California, USA

Simon R. Cherry
Department of Biomedical Engineering
Center for Molecular and Genomic Imaging
University of California, Davis
Davis, California, USA

Joel R. Garbow
Biomedical MR Laboratory
Mallinckrodt Institute of Radiology
Washington University
Saint Louis, Missouri, USA

Vasilis Ntziachristos
Institute for Biological and Medical Imaging
Helmholtz Zentrum München
München, Germany

Jinyi Qi
Department of Biomedical Engineering
University of California, Davis
Davis, California, USA

K. Kirk Shung
Department of Biomedical Engineering
University of Southern California
Los Angeles, California, USA

Adrian Taruttis
Institute for Biological and Medical Imaging
Helmholtz Zentrum München
München, Germany

Wen-Yih I. Tseng
Center for Optoelectronic Biomedicine
National Taiwan University
Taipei, Taiwan (ROC)

Hsiao-Ming Wu
School of Medicine
Fu-Jen Catholic University
Taipei, Taiwan (ROC)

Kai Yang
Department of Radiological Sciences
University of Oklahoma Health Sciences Center
Oklahoma City, Oklahoma, USA

Pat B. Zanzonico
Department of Medical Physics
Memorial Sloan Kettering Cancer Center
New York, New York, USA

List of Abbreviations and Acronyms

3DRP	three-dimensional reprojection
%ID/g	% of injected dose per gram
A-mode	amplitude mode
a-Si	amorphous silicon
A/D	analog to digital
ACD	annihilation coincidence detection
ACF	attenuation correction factor
AD	Alzheimer's disease
ADC	analog-to-digital converter (electronics)
ADC	apparent diffusion coefficient (magnetic resonance imaging)
AIF	arterial input function
APD	avalanche photodiode
ARFI	acoustic radiation force imaging
ART	algebraic reconstruction technique
ASL	arterial spin labeling
AUC	area under the curve
B-mode	brightness mode
BGO	bismuth germanate
BOLD	blood oxygenation level dependence
BP	binding potential
CBF	cerebral blood flow
CBV	cerebral blood volume
CCD	charge-coupled device
C-D	contrast detail
CHO	channelized Hotelling observer
CMOS	complementary metal oxide semiconductor
CMRG	cerebral metabolic rate of glucose
CMRO	cerebral metabolic rate of oxygen
COR	center of rotation

CR	computed radiography
CSI	chemical shift imaging
CT	computed tomography
CW	continuous wave
CZT	cadmium zinc telluride
DCE	dynamic contrast enhanced
DECT	dual-energy computed tomography
DNP	dynamic nuclear polarization
DOI	depth of interaction
dreMR	delta relaxation enhanced magnetic resonance
DSA	digital subtraction angiography
DSC	dynamic susceptibility contrast
DSCT	dual-source computed tomography
DTI	diffusion tensor imaging
DV	distribution volume
DVR	distribution volume ratio
DWI	diffusion-weighted imaging
ECG	electrocardiogram
EES	extravascular extracellular space
ESSE	effective scatter source estimation
FBP	filtered backprojection
fcMRI	functional connectivity magnetic resonance imaging
FDDNP	2-(1-{6-[(2-[F-18]fluoroethyl)(methyl)amino]-2-naphthyl}ethylidene) malononitrile
FDG	2-deoxy-2-[^{18}F]fluoro-D-glucose (^{18}F-fluorodeoxyglucose)
FDM	finite difference method
FEM	finite element method
FFT	fast Fourier transform
FID	free induction decay
FITC	fluorescein isothiocyanate
FLIM	fluorescence lifetime imaging microscopy
fMRI	functional magnetic resonance imaging
FMT	fluorescence molecular tomography
FORE	Fourier rebinning
FPF	false-positive fraction
FRET	fluorescence resonance energy transfer
FT	Fourier transform
FWHM	full width at half maximum
HER2	human epidermal growth factor receptor 2
HIFU	high-intensity focused ultrasound
HSP90	heat shock protein 90
HVL	half-value layer
IAUC	initial area under the curve
ICG	indocyanine green
IP	imaging plate
I_{SA}	spatial average intensity
I_{SP}	spatial peak intensity
I_{SPTA}	spatial peak temporal average intensity

I_{SPTP} spatial peak temporal peak intensity
I_{TA} temporal average intensity
I_{TP} temporal peak intensity
IVUS intravascular ultrasound
LAD left anterior descending
LCD liquid crystal display
LOR line of response
LSF line spread function
LSO lutetium oxyorthosilicate
LV left ventricle
LYSO lutetium yttrium oxyorthosilicate
MDCT multidetector computed tomography
MIBI methoxyisobutylisonitrile
MLEM maximum-likelihood expectation maximization
MPR myocardial perfusion ratio
MR magnetic resonance
MRA magnetic resonance angiography
MRE magnetic resonance elastography
MRG metabolic rate of glucose
MRI magnetic resonance imaging
MSRB multislice rebinning
MTBI mild traumatic brain injury
MTF modulation transfer function
MTT mean transit time
NMR nuclear magnetic resonance
NPS noise power spectrum
NSF nephrogenic systemic fibrosis
OPO optical parametric oscillator
OSEM ordered-subset expectation maximization
PET positron emission tomography
PIB Pittsburgh compound B
PMT photomultiplier tube
PSF point spread function
PSP photostimulable phosphor
PSPMT position-sensitive photomultiplier tube
PVDF polyvinylidene difluoride
PW pulsed wave
PZT lead zirconate titanate
QDE quantum detection efficiency
RAM random access memory
RAMLA row-action maximization likelihood algorithm
rCMRglc regional cerebral metabolic rate of glucose
RF radiofrequency
ROC receiver operating characteristic
ROI region of interest
SAR specific absorption rate
SI signal intensity

SiPM	silicon photomultiplier
S/N	signal to noise
SPECT	single-photon emission computed tomography
SPIO	superparamagnetic iron oxide
SPM	statistical parametric mapping
SSRB	single-slice rebinning
SUV	standardized uptake value
SWIFT	sweep imaging with Fourier transform
TAC	time-activity curve
TE	echo time
TEW	triple-energy window
TFT	thin-film transistor
TGC	time gain compensation
TOF	time of flight
TPF	true-positive fraction
TR	repetition time
T_r	repetition period
VOI	volume of interest

1

Overview

Simon R. Cherry, Ramsey D. Badawi, and Jinyi Qi

1.1 INTRODUCTION

Imaging has become an indispensable tool in the practice of modern medicine and biology. It's uses span clinical diagnostics, monitoring response to therapy, drug and biomarker development, and the scientific study of the body in health and disease across the entire life span, from fetus to cadaver. Several billion imaging procedures are performed for diagnostic purposes annually across the world. In the research setting, biomedical imaging is a highly translational experimental platform, providing assays and measurements that often can move seamlessly across species, from rodent to larger animal models and into the human.

The field of biomedical imaging also is, by necessity, highly multidisciplinary. Broadly speaking, physicists are involved in inventing new technologies, chemists in designing new contrast agents, mathematicians and computer scientists in developing advanced analysis and visualization tools, and engineers in designing and implementing high-performance imaging systems. The end users are biomedical researchers and clinicians who ultimately apply the technologies and methods in innovative ways to address a dizzying array of questions related to human health and disease intervention. But increasingly, encouraged by interdisciplinary training programs such as those found in many biomedical engineering departments, we see a new breed of imaging scientist—scientists whose expertise cuts across two or more of these areas and who are equally comfortable working in the physical, engineering, or biomedical sciences.

This book is designed with this new generation of interdisciplinary biomedical scientists in mind and is aimed at providing both an introductory text for those starting to explore or apply imaging techniques as well as a reference text to dip into, as needed, for the more advanced students and practitioners. The book is targeted at those using imaging in biomedical research rather than clinical practice. This distinguishes the book from the many outstanding texts on clinical medical imaging, as the range of techniques and applications used in research is far broader, and there also tends to be a stronger emphasis on quantification. Nonetheless, we hope the text will also be of interest to clinical practitioners. It is likely that some of today's research imaging methods foreshadow future clinical uses of imaging.

The book focuses on those technologies and methods that image at the macro tissue/organ scale, that is to say, methods that can examine large volumes of tissue (e.g., an entire organ) or even the entire body in one acquisition. This includes x-ray computed tomography (CT), ultrasound, magnetic resonance imaging (MRI), nuclear imaging (positron emission tomography [PET] and single photon emission computed tomography [SPECT]), and optical imaging (including bioluminescence, fluorescence, and photoacoustic imaging). This book does not concern itself with the various "microscopies" (e.g., confocal and multiphoton microscopy, or electron microscopy) or the use of some of the techniques described in this book at the cellular or subcellular level in excised specimens. Rather, the focus is on noninvasive and nondestructive *in vivo* imaging, at the tissue, organ, or whole organism level, capturing, in many cases, the complex anatomic interconnections or the myriad of signaling and communication pathways that characterize the biology of the intact organism and often are critical for accurate diagnosis of disease and subsequent treatment.

1.2 IMAGE CHARACTERISTICS

A major theme of this book is to communicate an understanding of the basic imaging properties of each technique. Each imaging modality has certain strengths and weaknesses based on its underlying physics (or, in some cases, chemistry), and it is useful to ask questions such as "how good is this image?", "how can I make the image better?", and "is this image better than that image"? While image characteristics can be quantified in a number of different ways, the answer to which image is "best" can only be given when the imaging task at hand is clearly defined. An image generally is used to allow the researcher or physician to detect or quantify the object (or some property of the object) of interest, and the image attributes that permit this will vary depending on the specific question or task. For example, one needs different attributes to detect a very small structural abnormality in the gray matter in the cerebral cortex than one does to quantify the level of a specific receptor being expressed on the surface of the cells in a tumor. This is one reason why a wide range of imaging modalities and methods exist. Each is designed to address different questions, based on its different capabilities.

Nonetheless, we can broadly describe certain characteristics that generally are desirable in an image. The most intuitive of these is high *spatial resolution*—the ability to resolve fine detail and see small structures inside the body. However, equally critical, in our ability to "see" something, is image *contrast*. If all tissues produced the same intensity in the image, we could not distinguish them however good the spatial resolution was. Contrast depends on the physics behind how the signal is generated and is often enhanced through the administration of contrast agents to the subject. In some modalities (e.g., imaging radioactivity inside the body with PET or SPECT), there is essentially no signal or contrast unless a contrast agent (in this case, a radiolabeled substance or "radiotracer") is introduced into the body.

Every imaging modality also has sources of *noise*. Noise may be in the form of statistical fluctuations in the number of information carriers (e.g., photons) detected or electronic noise that comes from the imaging system and its components. Whether a specific signal can be detected often depends quite strongly on the *contrast-to-noise ratio* of the image. Thus, the ability to detect an object generally can be improved either by increasing the contrast of the object in the image or decreasing the noise level.

Another key factor is the *sensitivity* of an imaging modality. This term is typically used in the context of injected contrast agents (although it also can apply to endogenous biomolecules) and is related to the concentration of an agent or biomolecule that needs to be present in a tissue of interest to produce a detectable change in the image intensity. This is most critical for imaging relatively low-abundance targets inside the body (for example, a cell-surface receptor) because the amount of the injected agent should be low enough that it does not cause any pharmacological or toxicological effect yet must still be sufficient to produce a big enough change in the imaging signal so that it may be visualized or quantified. Thus, for imaging of many molecular/metabolic pathways and targets, techniques that have high sensitivity are often a prerequisite.

The body is not static, tissues move (respiration, the beating heart, blood pulsing through the vessels, etc.), and therefore, how fast an image can be acquired, the *temporal resolution*, also can be of importance. In most cases, there are significant trade-offs in acquiring images very fast, involving giving up some combination of spatial resolution, the volume of tissue being imaged (the *field of view* of the imaging device), and increased noise levels. To overcome this, many imaging modalities can use techniques known as *gating*, in which respiratory and cardiac motion are monitored using external sensors (or directly from the images themselves), and images for specific phases of the respiratory and/or cardiac cycles can be averaged over time to reduce image noise while reducing blurring of the images due to the physiological motion. In other instances, physiological motion is actually used as the basis for signal or contrast. For example, in diffusion-weighted MRI, the diffusive motion of water molecules can be used to gain insights on the cellularity and organization structure of tissues. Only ultrasound and x-ray fluoroscopy can truly be classified as real-time imaging techniques, where images are displayed as they are actually acquired, at rates of many frames per second.

There also are important safety considerations that come into play. Some techniques use ionizing radiation (e.g., x-ray CT, PET, SPECT), and therefore, radiation dose must always be considered in the context of risk and benefit. Even for modalities that do not use ionizing radiation, there are limits for power deposition in the body that must be observed to prevent tissue damage (both ultrasound and light at high intensities can be used for treatment via heating effects rather than imaging). Lastly, in practice and application, there also are considerations of cost and accessibility that will drive decisions regarding which imaging modality to choose and which technique to apply.

These key characteristics apply to all the imaging techniques discussed and are highlighted, where appropriate, in each of the chapters. The fact that each modality has somewhat distinct sets of characteristics is one reason why each modality makes its own individual contributions to biomedical research. It is also the reason that images from different modalities often are combined (e.g., a high-sensitivity image of a molecular target overlaid on a high-resolution structural image of the anatomy), either through software image registration or, increasingly now, through integrated hybrid imaging scanners (e.g., PET/CT scanners).

1.3 HISTORICAL PERSPECTIVE

Although the light microscope had been around since the early 1600s, it was the discovery of x-rays by Wilhelm Roentgen in late 1895 that ushered in the era of biomedical imaging and revolutionized clinical diagnostics. Until that point, the only way to see deep inside the human body was by postmortem dissection. Diagnosis could only be based on external signs, patients' descriptions of their symptoms, and an examination of bodily fluids such as blood and urine. The penetrating nature of x-rays changed that picture with astonishing speed, with initial clinical use of x-ray imaging (albeit with a poor appreciation of the issues related to radiation dose) occurring within a year or so of the discovery. The phenomenon of radioactivity was described just a year later by Henri Bequerel, and Marie Curie's pioneering work in discovering and separating new naturally occurring radioactive elements led to the first injection of radioisotopes into a patient in the mid 1920s. The subsequent development of particle accelerators that could produce man-made radioisotopes on demand, and electronic radiation detectors, led to early functional imaging studies of the thyroid using radioactive iodine in the 1950s. The first medical uses of ultrasound were also being developed at around the same time, adapting techniques used in military sonar and radar.

While the phenomenon and underlying physics of nuclear magnetic resonance (NMR) had been described in the 1940s, it was not until the 1970s that methods to encode the spatial location were developed, allowing NMR to evolve into the imaging method we now call MRI. The 1970s was the decade of tomography—the development of the mathematical framework that enabled cross-sectional images ("slices") to be reconstructed from a series of x-ray images obtained at different angles around the subject. This led to x-ray CT and the ability for the first time to produce an image representing a virtual section through the human body. The same mathematical principles also could be used in "emission" tomography, leading to the techniques of PET and SPECT, which produce cross-sectional images showing the distribution of a radioactive material that had been injected into a subject. This mathematics also was used to create the first MRI image and later led to the frequency and phase encoding widely used in modern MRI. In subsequent years, most imaging modalities evolved rapidly from producing a single image slice, or just a few image slices, to full volumetric imaging. New instruments could simultaneously, or in rapid succession, acquire "stacks" of contiguous image slices that made up a 3-D image volume that could be rendered into a 3-D view or computationally "sliced" into any desired image slice orientation.

In recent years, there have been many stunning improvements and advances that allow images to be taken with a far higher level of detail (better spatial resolution) and in far shorter times. Today, it is routine to acquire high-resolution volumetric images of whole organs or even large sections of the human body in acquisition times that range from a few minutes to under one second. These improvements, along with new technologies and methods to increase the signal and contrast, as well as to reduce noise, have allowed imaging modalities to look in ever higher detail within the living subject to improve our understanding of disease and disease treatment.

In other developments, new methods for generating native tissue contrast have been exploited and optimized to allow better visualization of tissues. A wide range of contrast agents or "probes" are being introduced, providing highly specific image contrast and underpinning the field of molecular imaging, in which metabolic and molecular pathways can now be imaged. There also have been major advances on the algorithmic side, such as sophisticated reconstruction methods that build in models of the underlying physics and noise properties of the raw data in computing the final image volume, and robust tools for spatially registering images obtained from different modalities.

Another major trend has been the emergence of hybrid imaging devices, in which two different imaging modalities are integrated into a single device. The idea is to harness the complementary strengths of two separate imaging techniques and is motivated by the fact that different imaging modalities provide quite different information and also that many patients and research subjects undergo studies with more than one imaging technique. The most common hybrid imaging device, used widely in clinical diagnostics as well as biomedical research, is the PET/CT scanner. This device combines the high-resolution structural imaging achievable by CT with the high-sensitivity imaging of specific metabolic and molecular pathways and targets provided by PET. Knowing the anatomic location (provided by CT) of the radiotracer signal (provided by PET) often has important diagnostic consequences and assists with interpretation and quantification of research studies. SPECT/CT and PET/MRI scanners also are commercially available, and other multimodal instruments, as well as multimodal contrast agents, are being actively developed (see "New Horizons," Section 1.5).

1.4 APPLICATIONS

Biomedical imaging has touched research into virtually every organ system, every disease, and every new therapeutic strategy. We can noninvasively look at fine anatomic detail just about everywhere inside the human (or animal model) body, even in organs that are rapidly moving, such as the heart. We can map the regions of the brain that respond when a subject is given a particular task and also interrogate how different brain regions are connected to each other. We can visualize the vasculature, including the coronary arteries and the contorted and disorganized vasculature often found in tumors. We can image the delivery and kinetics of drugs and also determine whether a drug acts on its target. Merging imaging with the modern tools of molecular biology, techniques are available to image the control of gene expression (for example, the process of RNA interference or the activity of a specific gene promoter) and also to study protein–protein interactions. And with the advent of cellular therapies and nanomedicine, techniques to track cells and nanoparticles *in vivo* have been developed. Imaging also is becoming a crucial tool in the field of tissue engineering and regenerative medicine, where novel biomaterials and cellular scaffolds/grafts can be monitored noninvasively and longitudinally. Finally, there has been a trend toward integrating therapy and imaging, for example, the use of light or ultrasound at low intensities for imaging and at higher intensities to exert direct therapeutic effects or increase localized drug delivery by releasing drug cargo from a carrier. These and other approaches form the basis for the field of *theranostics* (combining therapy and diagnostics).

While the role of imaging in human medicine has been long established, development of specialized imaging systems for animal studies has led to a rapid growth of imaging in basic biomedical research and preclinical animal studies as well. This has allowed imaging to become a valuable translational tool, as imaging approaches can often be moved across species with little difficulty from a technical point of view. (Regulatory barriers, however, typically are a rate-limiting step.) Specialized imaging systems also have been developed for a range of different organs and tissues, for example, the brain, the heart, the breast, and the prostate (due to the prevalence of cancers in these organs), and the extremities.

1.5 NEW HORIZONS

There are several clear areas of current development in biomedical imaging. One has been the trend toward multimodal imaging, the idea of taking advantage of the complementary strengths of two or more imaging modalities to gain more information, either by spatially registering data sets taken at different times or by using hybrid imaging devices, such as PET/CT, PET/MRI, and SPECT or fluorescence with CT or MRI, to acquire the two data sets simultaneously or near-simultaneously, which provides both spatial and temporal registration. Typically, a high-sensitivity molecular imaging approach (such as optical or radiotracer imaging) is combined with structural (and, in some cases, functional) imaging using CT or MRI. There also are examples in which a single image is produced by exploiting two apparently distinct imaging modalities. The best known example of this is photoacoustic imaging, in which light is used as the radiation source but absorption of light in tissue or by contrast agents leads to the production of ultrasound that can be picked up using an ultrasound system.

The concept of multimodal imaging also has been extended into the realm of contrast agent design. Approaches are being developed for constructing nanoparticles that can be imaged by two or more of the following mechanisms: through their effects on the tissue relaxation time in MRI, via an increase in absorption of x-rays, through excitation by an external light source and the release of fluorescence, or through the addition of a radioactive label.

A second trend has been in developing theranostic agents, that is, contrast agents that provide diagnostic information but that also can exert a therapeutic effect. Examples include nanoparticles that can carry a drug cargo, nanoparticles that can be heated by absorption of radiation, radiolabeled antibodies, and light-activated therapeutic molecules and nanoparticles.

New methods to enhance contrast or signal also continue to be developed. For example, a number of metabolically relevant compounds can be hyperpolarized to enhance the signal level for MRI studies by several orders of magnitude. For such compounds, high-sensitivity MRI imaging over short time periods becomes feasible. A second example is the use of phase contrast in x-ray imaging and CT.

Another area of focus has been to make imaging even safer than it already is. Significant efforts are underway to reduce radiation dose still further for CT by using sophisticated reconstruction algorithms and/or by developing advanced detector technologies that can "count" each individual x-ray photon, which leads to a significant reduction in noise for a given signal level. In radiotracer imaging, PET scanner designs with much higher efficiency are being considered for whole-body imaging that could allow significant reductions in radiation dose. With all modalities, efforts continue to be made to reduce scanning time and also to find ways to reduce cost, to allow imaging techniques to be more broadly applied on a global scale.

1.6 CONTENTS

There are many books that cover the basics of clinical medical imaging; however, this one tries to span the broader use of imaging technologies from preclinical through clinical diagnostic imaging, capturing both research and clinical uses, but with a focus on the use of imaging in biomedical research. It also integrates optical imaging approaches, which are frequently ignored in medical imaging texts due to the relatively small number of clinical applications to date in humans. While the penetration of light through tissue remains an obstacle for some human applications, optical imaging is extensively used in preclinical studies in small-animal models, where light in the red part of the spectrum has sufficient penetration to access the entire body of a mouse. The flexibility of optical contrast sources allows a number of unique applications for optical imaging *in vivo*, and some of these also have promising translational prospects for future clinical applications with respect to surgical guidance, and catheter- or endoscopic-based diagnostics.

The book is organized as a series of chapters that cover each of the major imaging modalities: x-ray and x-ray CT, MRI, ultrasound, optical (including photoacoustic) imaging, and radiotracer (PET/SPECT) imaging. Each chapter focuses on the fundamentals of how signals are generated, the characteristics of the images (in terms of spatial and temporal resolution, contrast, noise), standard methods employed, and examples of applications in biomedical research. Chapter 7 contains information that is relevant for most imaging methods, regarding how imaging data may be processed, analyzed, and quantified. This is of increasing importance to the imaging practitioner, as these methods are used in quantifying a wide range of signals from the images or a time series of images and have broad applications in evaluating disease progression and response to therapy.

FURTHER READINGS

Grignon, B., Mainard, L., Delion, M., Hodez, C., Oldrini, G. Recent advances in medical imaging: Anatomical and clinical applications. *Surg Radiol Anat* 34; 675–686, 2012.

Laine, A.F. In the spotlight: Biomedical imaging. Annual articles in the journal *IEEE Reviews of Biomedical Engineering*, 2008–2013.

Mould, R.F. *A Century of X-rays and Radioactivity in Medicine*. IOP Publishing, Bristol, UK, 1993.

Pysz, M.A., Gambhir, S.S., Willmann, J.K. Molecular imaging: Current status and emerging strategies. *Clin Radiol* 65; 500–516, 2010.

Tempany, C.M., McNeil, B.J. Advances in biomedical imaging. *JAMA* 285; 562–567, 2001.

Webb, S. *From the Watching of Shadows. The Origins of Radiological Tomography*. Adam Hilger, Bristol, UK, 1990.

2

X-Ray Projection Imaging and Computed Tomography

Kai Yang and John M. Boone

2.1 INTRODUCTION

X-rays are a form of "ionizing radiation" because x-rays are energetic enough to ionize atoms and molecules during interactions. With about 10,000 times more energy than visible light photons, x-ray photons can penetrate objects including the human body. Since the first x-ray image taken in 1895 by Roentgen, x-ray imaging has become one of the most common diagnostic procedures performed in medicine. The development of modern x-ray tubes and detectors has enabled a wide range of medical imaging applications, including x-ray radiography, x-ray fluoroscopy, and x-ray computed tomography (CT). With the capability to produce cross-sectional images, x-ray CT revolutionized traditional x-ray imaging and provided an invaluable diagnostic tool. The usage of CT has rapidly increased over the past two decades. In 2011, 85.3 million x-ray CT scans were performed in the United States. In addition to their role in diagnostic medicine, x-ray methods are widely used for clinical research across a broad spectrum of disease states. X-ray projection imaging and micro-CT (high-resolution x-ray CT imaging of small volumes) have also become important tools in biomedical research studies of animal models and tissue specimens. This chapter focuses on the fundamentals of x-ray imaging and the two major classes of x-ray imaging: x-ray projection imaging and x-ray CT.

2.2 X-RAY IMAGING BASICS

2.2.1 X-Ray Production and X-Ray Spectrum

2.2.1.1 X-Ray Production

X-ray photons used for biomedical imaging are produced from a relatively complex device, the x-ray tube. The core of an x-ray tube, called the x-ray tube insert (Figure 2.1), is a vacuum sealed by a glass or metal enclosure. Within the vacuum insert, a heated filament, the *cathode*, emits electrons in a process called thermionic emission. Electrons ejected from the cathode are accelerated toward a positively charged metal *anode* by the high-voltage electric field between the cathode and anode. Being of like charge, electrons repel each other during their transit from cathode to anode. To counter this, a focusing cup produces an electric field to constrain the electron cloud and keep it focused as it travels toward the anode. These focused electrons gain kinetic energy as they are accelerated by the electric field and eventually strike the anode. The kinetic energy of the electrons is converted into x-ray photons and excess heat within the anode. The energies of the emitted photons are commonly expressed in electron volts (eV)—1 eV is defined as the kinetic energy acquired by an electron as it travels through an electrical potential difference of 1 V in a vacuum.

The efficiency of x-ray photon production is determined mainly by the atomic number of the anode/target material and the kinetic energy of the electrons, the latter being determined by the voltage applied between the anode and cathode. Typical x-ray tubes use

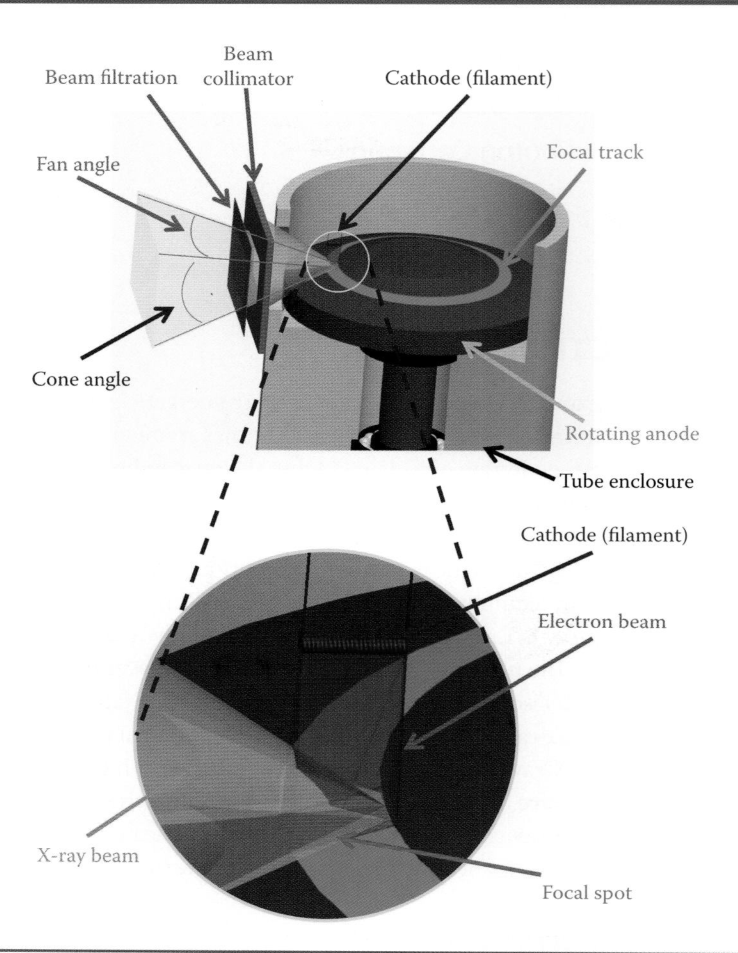

FIGURE 2.1 X-ray tube insert with rotating-anode design. (Courtesy of George W. Burkett, University of California, Davis.)

high-atomic-number elements such as tungsten (W), molybdenum (Mo), or rhodium (Rh) as the anode material. The peak potential between the anode and cathode is controlled by the x-ray generator and ranges from 20,000 to 150,000 V (20 to 150 kV) for x-ray tubes used in biomedical applications.

As shown in Figure 2.2, the area of the electron interaction site on the anode surface (called the *focal spot*) and the angle of the anode surface relative to the central ray of the x-ray beam (called the *anode angle*) determine the *effective* or *projected focal spot size*. The very shallow anode angle (normally between 7° and 20°) converts the actual focal spot area into a much smaller effective focal spot (Figure 2.2). This geometry is called the *line-focus* principle, which leads to the apparent reduction of focal spot size as it projects to the detector. Smaller focal spots produce higher-resolution images, in general. However, due to the constraints of anode heating, smaller focal spots also limit the x-ray tube power and, thus, the rate of x-ray production. Therefore, there exists a trade-off between the x-ray tube power and the minimum focal spot size. Many high-power x-ray tubes are designed with a rotating anode (at a very high speed, up to 10,000 rotations per minute) to increase heat dissipation and permit greater x-ray output. As shown in Figure 2.1, for rotating-anode

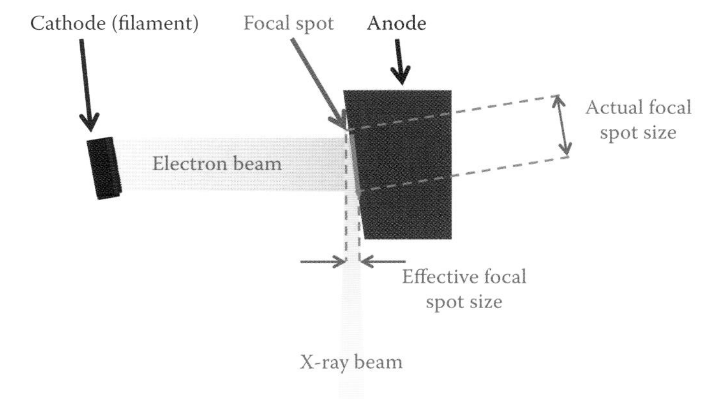

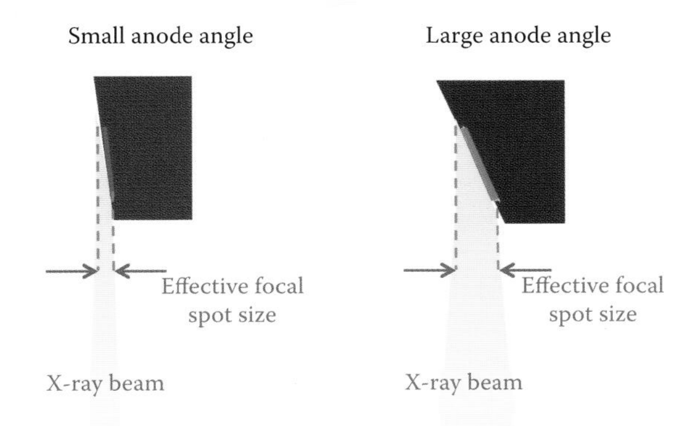

FIGURE 2.2 Line-focus principle. The effective focal spot size is much smaller than the actual focal spot size and is dependent on the anode angle.

x-ray tubes, a continuous *focal track* instead of a fixed focal spot is struck by the electrons. For clinical radiographic and fluoroscopic applications, the effective x-ray focal spots are typically 0.6 to 1.2 mm. For mammography systems, 0.1 and 0.3 mm effective focal spots are common. For biomedical applications that require very high image resolution (such as micro-CT systems), effect focal spot dimensions may be as small as 10 μm, and these tubes are called *micro-focus x-ray tubes.*

In biomedical imaging systems, normally, there are collimators (dense, metallic structures that block x-rays in specific directions) both inside and outside of the x-ray tube to limit the x-ray radiation field (Figure 2.1). In addition to this physical collimation, there is a limited solid angle (represented in Figure 2.1 by the *fan angle* and the *cone angle*) that the x-ray beam from a specific x-ray tube can cover. The maximum solid angle and collimation fundamentally limit the physical size of an object that can be imaged for a given x-ray tube-to-object distance from a single exposure. This is referred to as *coverage.*

Within the maximum solid angle, x-ray photons have a nonuniform intensity across the usable field of view. X-ray intensity is typically measured by the *x-ray photon fluence*, which is defined as the number of photons per unit area. In practice, due to the challenge of counting photons, x-ray intensity is measured using the quantity *air kerma*, which is the energy imparted to charged particles in a unit mass of dry air. The SI unit of air kerma is the *gray* (1 Gy = 1 J/kg), defined as 1 J of energy imparted in 1 kg of air. X-ray air kerma is an important parameter, which describes x-ray signal amplitude and is useful to estimate potential radiation risks and evaluate the efficiency of imaging systems. The spatial nonuniformity of x-ray intensity results from two different phenomena. Firstly, x-ray intensity from an x-ray tube is typically lower toward the anode side. This is called the *heel effect* and is due to the nonuniform attenuation of x-ray photons from the angled anode. Secondly, x-ray intensity decreases with increasing distance from the focal spot, at a rate proportional to the square of the distance. This is called the *inverse square law*. This is due to the divergent nature of the x-ray beam from a point source and the relationship between the surface area (A) and the radius (r) of a sphere ($A = 4\pi r^2$). As a given number of x-ray photons are emitted isotropically from the focal spot (the center of the sphere), they are distributed onto an increasingly larger surface area when traveling away from the source. Thus, the x-ray fluence or the intensity decreases as the square of the distance (the radius of the sphere). The inverse square law has an important impact on the design of x-ray imaging systems, especially affecting the source-to-object distance (radiation safety purpose) and the source-to-imager distance (image quality purpose).

2.2.1.2 X-Ray Spectrum

X-rays photons are produced at the anode of an x-ray tube through two different mechanisms: *Bremsstrahlung* and *characteristic* radiation. These photons have a range of energies, and hence, an *x-ray spectrum* is produced.

During Bremsstrahlung production, electrons lose their kinetic energy through interactions with the target nuclei at subatomic distances. Bremsstrahlung ("braking radiation" in German) x-ray photons have a continuous energy distribution from 0 up to the maximum kinetic energy of the accelerated electrons (Figure 2.3a). For example, an x-ray tube with an applied voltage of 100 kV produces electrons with a maximum kinetic energy of 100 keV. X-ray photons are generated at different depths within the anode, and most are absorbed within the target, while others are absorbed by the x-ray tube housing. Since the probability of absorption is higher for photons with lower energies, these processes result in a filtered Bremsstrahlung spectrum that contains a much smaller proportion of low-energy x-ray photons (Figure 2.3a).

In contrast to the continuous nature of the Bremsstrahlung spectrum, monoenergetic *characteristic* radiation can occur if the maximum electron energy exceeds the K-shell binding energy of the target materials. This phenomenon is a result of energetic electrons from the cathode colliding with orbital electrons in the anode, causing them to be ejected from the target atoms. The target atom becomes ionized and has a vacancy in one of its inner electronic shells. An outer-shell electron will then migrate to the vacancy, and this electron transition results in the release of a photon with energy equal to the difference of the binding energies between the two orbital shells. The binding energy for each shell is unique for each element, and thus, the emitted x-ray photon energies are specific to the anode material. This is why these x-ray photons are called *characteristic x-ray* photons. The x-ray spectrum generated from an x-ray tube is a combination of the filtered Bremsstrahlung spectrum and characteristic x-rays (Figure 2.3b). For a typical tungsten anode system operated at 120 kV, characteristic x-rays comprise about 10% of the photon emission. In this chapter, the x-ray spectrum is described by the symbol $\Phi(E)$, which describes photon fluence as a function of energy. In practice, it is also useful to normalize the x-ray spectrum to a given air kerma level (Figure 2.3b).

The raw x-ray spectrum from an x-ray tube still includes a very high proportion of low-energy photons. For medical imaging applications, a low-energy photon has a low

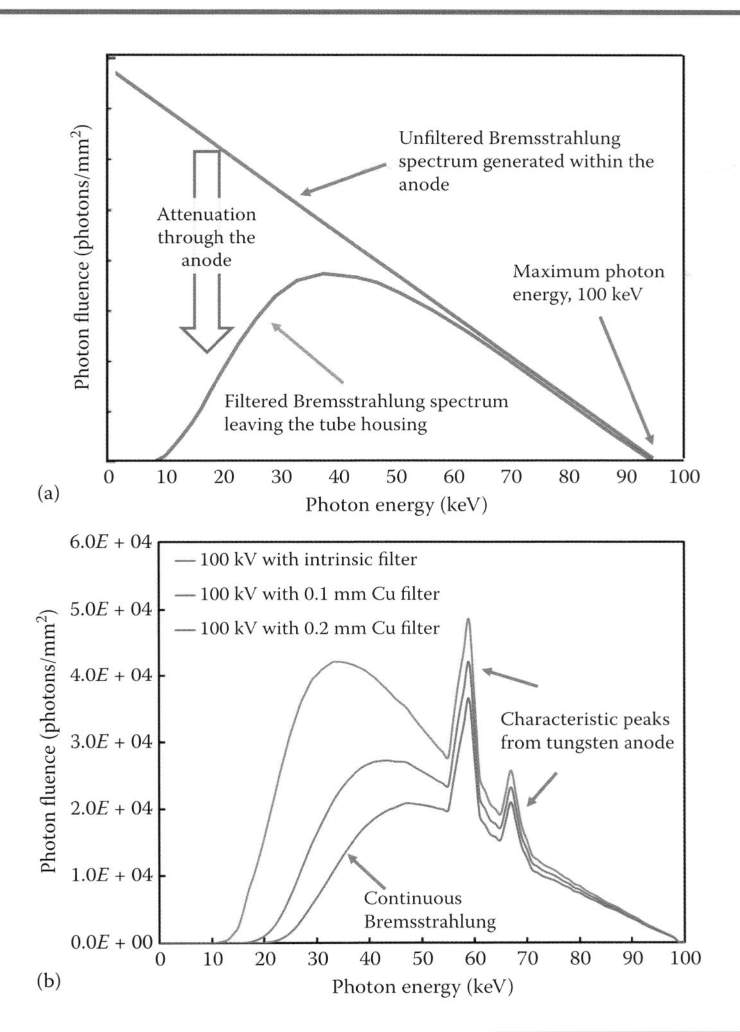

FIGURE 2.3 X-ray spectrum. (a) Bremsstrahlung spectrum at 100 kV. Due to the attenuation of the anode and tube housing, the filtered Bremsstrahlung spectrum has fewer low-energy x-ray photons compared to the unfiltered spectrum. (b) Observed x-ray spectra at 100 kV from a tungsten anode tube. The photon fluence of the spectrum with intrinsic filter is normalized to 1 mGy air kerma. The proportion of low-energy photons (which give dose but provide little information) can be reduced by adding metal filters in front of the x-ray beam.

probability of penetrating an imaging object. Therefore, this part of the spectrum imposes a significant radiation dose to biological materials and contributes little to the final image. To suppress these unwanted low-energy x-ray photons, a thin sheet of metal such as aluminum or copper is placed in the x-ray beam as a filter (Figure 2.1). The filtered x-ray spectrum has fewer low-energy photons (Figure 2.3b), and the added filters on x-ray tubes significantly reduce unnecessary radiation dose associated with imaging. A filter can tailor the raw x-ray spectrum for medical imaging at the cost of reducing x-ray tube output.

2.2.1.3 Technique Factors in X-Ray Imaging

The physical parameters selected for x-ray tube operation determine key characteristics of the x-ray beam and spectrum for a specific x-ray imaging task. The voltage applied to the x-ray tube, usually quoted in kilovolts (kV), determines the maximum energy of the x-ray photons produced. As shown in Figure 2.3b, the maximum energy is in units of keV. The kV is a loose measure of the penetration capability of the x-ray photon beam. The kV is often adjusted based on the maximum patient/sample thickness. Thicker samples require higher x-ray tube voltages. The x-ray tube current, in milliamps (mA), controls the number of electrons emitted from the cathode to the anode per unit time and, thus, the number of x-ray photons generated per unit time. The product of current (mA) and exposure time (s), abbreviated as mAs, is linearly proportional to the total number of x-ray photons generated in one exposure, that is, the total x-ray fluence. The mAs, together with kV and filtration, determines the overall radiation dose to the subject and also influences the statistical noise of the resulting x-ray image.

2.2.2 X-Ray Interaction and Detection

2.2.2.1 X-Ray Photon Interaction with Matter

There are three major interactions between x-ray photons and matter for the x-ray photon energy range used in biomedical imaging applications, *Rayleigh scattering*, *Compton scattering*, and the *photoelectric effect*. The probability of each interaction depends on the x-ray photon energy and the interaction medium.

An x-ray photon can be absorbed by an orbital electron within an atom and immediately be reemitted as a new photon in a slightly different direction without any loss of energy. This nonionizing process is called Rayleigh scattering or coherent scattering. For soft tissue, Rayleigh scattering mainly occurs at photon energies below 30 keV, such as in mammography or micro-CT of small specimens. The probability of Rayleigh scattering decreases with increasing energy and increases with increasing atomic number (Z) of the medium. Since no energy is deposited in this interaction, Rayleigh scattering does not result in any radiation dose. The detection of scattered x-ray photons reduces image contrast and increases image noise. However, outside of low-energy mammography and micro-CT applications, the probability of Rayleigh scattering is very small.

Compton scattering, also known as incoherent scattering, is the most prevalent interaction between x-ray photons and biological tissues in biomedical imaging applications with x-ray photon energies above 26 keV. The incident x-ray photon interacts with a valence electron, conveying kinetic energy and ejecting that electron. The photon is scattered from the interaction site while losing a fraction of its energy. The scattered photon energy, E_{sc}, has a simple dependency on its initial energy, E_0, and the scattering angle, θ (with respect to the incident trajectory):

$$E_{sc} = \frac{E_0}{1 + \frac{E_0}{511}(1 - \cos\theta)}, \tag{2.1}$$

where the photon energies are in units of keV. The probability of Compton scattering in soft tissue is relatively independent of the atomic number, Z, of the medium. Thus, most of the image contrast resulting from Compton scattering is dependent on the local density. In general, Compton-scattered photons can degrade image quality when detected, reducing image contrast and increasing image noise.

The photoelectric effect is an interaction that occurs between an x-ray photon and an inner-shell orbital electron, leading to the absorption of the x-ray photon. This effect can only occur when the incident photon energy is equal to or greater than the binding energy of the orbital electron. The ejected electron is called a *photoelectron*, and its initial kinetic energy is equal to the difference between the photon energy and its binding energy. The probability of photoelectric absorption per unit mass is proportional to Z^3/E^3, where Z is the atomic number of the medium and E is the x-ray photon energy. This relationship has been exploited in two key processes for biomedical x-ray imaging: (1) to generate image contrast between different materials such as bone and soft tissue and (2) to capture transmitted x-ray photons by an x-ray detector. The probability of photoelectric absorption decreases dramatically with increasing photon energy. However, the reduction is not continuous—"absorption edges" occur at the binding energies of the inner electron shells (normally, it is the innermost and most tightly bound K-shell electrons that are responsible for the absorption) of the attenuating medium. When the photon energy is equal to or just above the binding energy of one of the inner shells, photoelectric interaction becomes more energetically favorable, and there is an abrupt increase in interaction probability. The x-ray photon energy corresponding to the absorption edge increases as a function of the atomic number (Z) of the medium. The K-edges of soft tissues (C, H, O, N) are normally below 1 keV and have no significant effect for imaging. Some higher-Z materials, such as iodine ($Z = 53$) or barium ($Z = 56$), have K-edges that are in the energy range appropriate for biomedical imaging. These materials are therefore used as *contrast agents* when introduced into the subject. The greatly accentuated x-ray photon absorption by a contrast agent due to the K-edge photoelectric effect can generate very high image contrast between the agent and background tissues. This contrast-enhanced technique can provide a wide range of functional and anatomical information for *in vivo* imaging tasks. For example, iodine-based contrast agents are widely used to image the vasculature in angiography, while barium-based contrast agents are used to image the gastrointestinal tract, including the stomach and bowel.

2.2.2.2 Attenuation Coefficient and Beer's Law

When an x-ray beam passes through a medium, a fraction of the photons is removed from the beam through a combination of scattering and absorption interactions, described in Section 2.2.2.1. This removal of photons is called *attenuation* of the x-ray beam. Attenuation is the fundamental mechanism that generates x-ray image contrast and includes both photoelectric absorption and scattering interactions. If N_0 is the total number of x-ray photons incident on a thin slab of a medium with a thickness of x cm, the number of x-ray photons that are transmitted through the medium (without being attenuated), N, is given by

$$N = N_0 \cdot e^{-\mu x}, \tag{2.2}$$

where μ is called the *linear attenuation coefficient* and represents the probability that an x-ray will be removed from the beam per unit length traveled in the medium. The units of μ are typically cm^{-1}. The linear attenuation coefficient, μ, represents the total probability of attenuation from all three photon interactions described in Section 2.2.2.1 (Figure 2.4a):

$$\mu = \mu_{\text{Rayleigh}} + \mu_{\text{Compton}} + \mu_{\text{Photoelectric}} \tag{2.3}$$

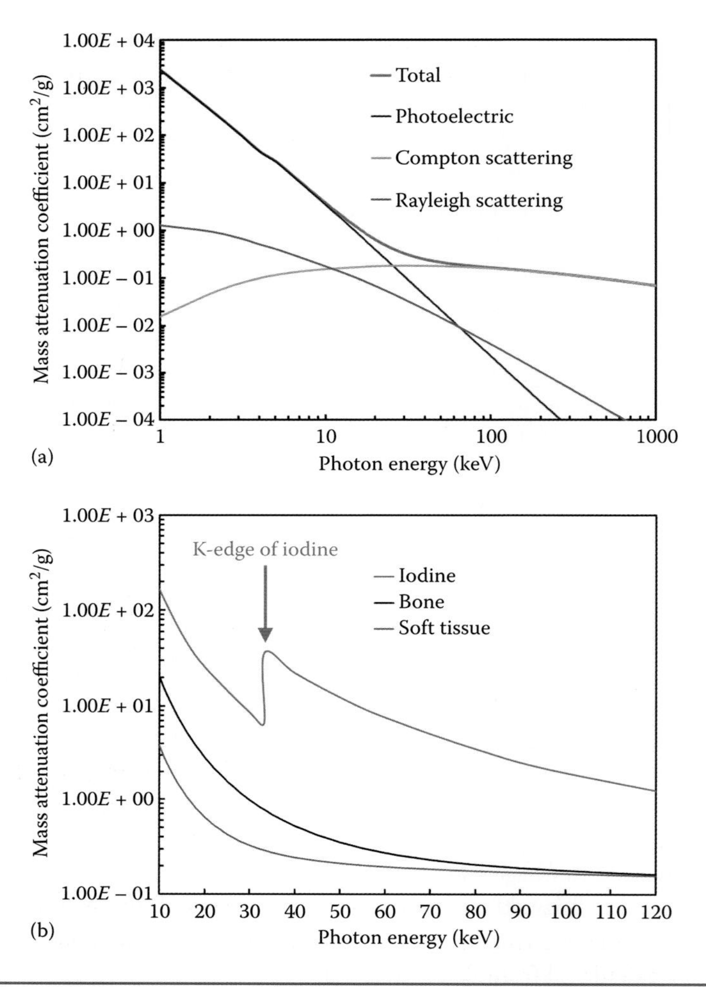

FIGURE 2.4 Attenuation coefficients. (a) Mass attenuation coefficients of soft tissue as a function of photon energy. (b) Comparison of attenuation coefficients between different materials.

Equation 2.2 is called the Beer-Lambert law. The simple relationship in Equation 2.2 only holds under the following conditions:

1. When measuring the attenuated x-ray beam, the majority of scattered x-ray photons do not reenter into the primary beam after interacting with the medium. This is the so-called *good geometry* or *narrow beam* condition for x-ray imaging.
2. The x-ray photons are of the same energy, and the medium is homogeneous. This is because the linear attenuation coefficient is a function of photon energy and the atomic number of the medium.

As described previously, x-ray beams are not comprised of monoenergetic photons, and biological tissues are not homogeneous either. Thus, Beer's law is more accurately expressed as

$$\Phi(E) = \Phi_0(E)e^{-\int_{x=0}^{x<=L} \mu(E,x)dx}, \tag{2.4}$$

where $\Phi_0(E)$ and $\Phi(E)$ are the x-ray spectra before and after attenuation, $\mu(E,x)$ is the linear attenuation coefficient at energy E and for location x in the medium, and L is the total thickness of the object.

The *mass attenuation coefficient* is a related and important parameter and is defined as

$$\text{Mass attenuation coefficient} = \frac{\mu}{\rho}, \tag{2.5}$$

where μ and ρ correspond to the linear attenuation coefficient and density for a specific material, respectively. Mass attenuation coefficients (unit, cm^2/g) are frequently used to compare the attenuation properties between different materials per unit density.

Using the mass attenuation coefficient, Equation 2.2 can also be expressed as

$$N = N_0 \cdot e^{-\left(\frac{\mu}{\rho}\right)\rho x}. \tag{2.6}$$

Beer's law is a simple function that reflects the exponential nature of x-ray photon attenuation. X-ray image contrast is fundamentally generated from x-ray photon attenuation, which is determined by the linear (or mass) attenuation coefficients of different materials. Figure 2.4b shows the comparison of mass attenuation coefficients of bone, soft tissue, and iodine. As described in Section 2.2.2.1, due to the large differences between the attenuation coefficients of iodine and biological tissues, iodine is the most commonly used contrast agent in x-ray imaging.

An important construct used in medical imaging is the *half-value layer* (HVL). From Equation 2.2, the HVL is defined as the thickness of material, L, when $N = \frac{N_0}{2}$, that is, the HVL is the thickness of the attenuating material required to attenuate the x-ray intensity (measured in terms of air kerma in units of mGy) by 50%. For a monoenergetic x-ray beam, the HVL can be calculated from Equation 2.2 as

$$\text{HVL} = \frac{\ln 2}{\mu} = \frac{0.693}{\mu}. \tag{2.7}$$

For polyenergetic x-ray beams, the HVL can be practically determined through an iterative approach by measuring the x-ray intensity with increasing thicknesses of attenuating material (typically aluminum) until the value drops by 50%. The HVL is most commonly used as an indicator for x-ray beam penetrability or *beam quality* in biomedical imaging. For a given material (e.g., Al) and the same x-ray tube kV, a higher HVL corresponds to increased penetrability of the x-ray beam (a "harder" beam), and a lower HVL indicates a "softer" x-ray beam.

2.2.2.3 X-Ray Photon Detection

After x-ray photons are transmitted through an object, they are captured and converted into an image by an x-ray detector. Radiographic film was the first widely used x-ray detector. With the development of digital technology, radiographic films have been gradually replaced by digital x-ray detectors that are composed of arrays of detector elements or *dexels* (with the exception of computed radiography [CR]; see Section 2.3.2). Each individual detector element can absorb the energy imparted by incident x-ray photons and produce measurable electrical signals (voltage or current signals) using a variety of mechanisms. For digital detectors, analog-to-digital (A/D) convertors (ADCs) convert the electrical signals into digital signals. These signals are used to form a digital x-ray image, similar to the gray-scale picture acquired on a digital photographic camera.

X-ray detectors are made from a variety of different materials, such as noble gases or solid materials. For biomedical imaging systems, most detectors are solid detectors due to their higher density and absorption efficiency. The following discussion focuses on these.

The majority of x-ray imaging detectors are designed to generate signals proportional to the integrated x-ray photon energy accumulated in each detector element, without differentiating the energy of each individual photon. This type of detector is an *energy-integrating* detector. *Photon counting* detectors, which can generate signals proportional to the energy of each individual detected x-ray photon, also are available. While photon counting detectors are widely used in nuclear imaging (see Chapter 6), they are still in the experimental stage for x-ray imaging because of the very high *photon flux* (or *fluence rate*, defined as the number of x-ray photons incident onto the detector per unit area per unit time). Photon counting detectors are discussed further in Section 2.5.2.2.

There are two types of x-ray detection mechanisms for biomedical imaging systems: direct detection and indirect detection (Figure 2.5). For direct detection, incident x-ray photons interact with the detector material through ionization, and the electrons generated are collected to produce a signal that is proportional to the accumulated energy deposited by absorbed x-ray photons in each dexel. A solid-state direct detector system is normally designed with a uniform slab of photoconductor (a material that conducts when exposed to ionizing radiation) across which an electric field is applied using two electrodes on the top and bottom. When the x-ray beam is off, almost no charge flows between the two electrodes because the photoconductor acts as an insulator. When the x-ray beam is on, electrons created by ionization move under the influence of the applied electric field, are accumulated on readout electronics, and generate an electrical signal, which is digitized. The majority of direct detection detectors for biomedical imaging of x-rays are made of amorphous selenium (a-Se).

For indirect detection, incident x-ray photons first interact with a scintillator or phosphor material that absorbs the x-rays and converts the accumulated energy into visible (or near-ultraviolet) light photons. These visible light photons are subsequently converted into an electrical signal by optical sensors to produce a signal proportional to the accumulated

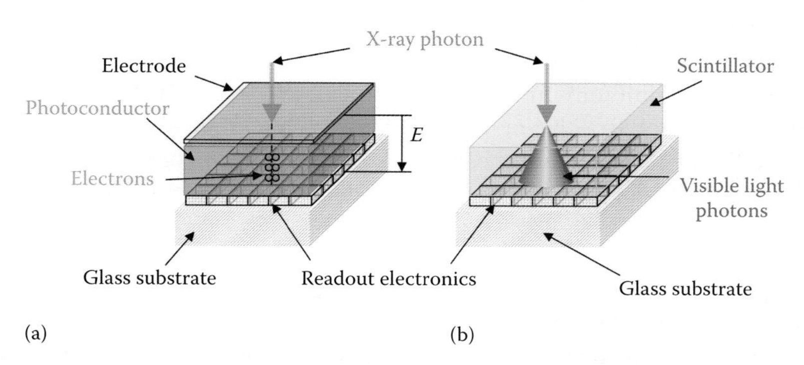

FIGURE 2.5 Direct and indirect detection detectors. Notice the key difference between the information carriers: electrons for direct detection (a) and visible light photons (which are then subsequently converted to electrons) for indirect detection (b).

energy deposited by the incident x-rays. For indirect detectors, widely used phosphor materials include thallium-doped cesium iodide (CsI:Tl), gadolinium oxysulfide (Gd_2O_2S), and calcium tungstate ($CaWO_4$). Amorphous silicon (a-Si) photodiodes are commonly used to convert the light from these materials into an electrical signal.

We have covered the fundamental concepts of x-ray detection in this section. The more detailed aspects of different x-ray detector technologies will be discussed in Sections 2.3 and 2.4.

2.2.2.4 Quantitative Metrics for Characterizing X-Ray Detectors

Despite the differences in detection mechanism and detection medium, there are several key parameters that can be used to characterize the performance of any x-ray detector. These parameters include detection efficiency, additive noise, dynamic range, and spatial resolution.

Detection efficiency is determined by the overall absorption coefficient of the detector. The *quantum detection efficiency* (QDE) is defined as

$$\mathrm{QDE} = \frac{\int_{E=0}^{E=E_{\max}} \Phi(E)\cdot\left(1-e^{-\mu(E)\cdot x}\right)\mathrm{d}E}{\int_{E=0}^{E=E_{\max}} \Phi(E)\,\mathrm{d}E}, \tag{2.8}$$

where $\Phi(E)$ is the x-ray spectrum, $\mu(E)$ is the total linear attenuation coefficient, and x is the thickness of the detector material [1]. A detector that absorbs all incident x-ray energy would have a QDE of 1; however, all practical detectors have a value less than this. As shown in Equation 2.8, QDE is directly related to the linear attenuation coefficient of the x-ray detection medium material (e.g., a phosphor for indirect detection or a photoconductor for direct detection) and its thickness. A thicker detector will absorb more x-ray photons. However, for indirect detectors, a thicker layer of scintillator will also lead to a wider spread of the scintillation light on the photodiode array, thereby degrading the image resolution. Therefore, the optimal thickness of an x-ray detector represents a task-dependent trade-off between detection efficiency and image resolution.

The *additive noise* in an electronic detector refers to the signal component that is independent of the detected x-ray fluence levels and is often thermal in origin. For a well-designed detector system, additive noise is normally constrained to be significantly below the typical signal level. Under some imaging conditions, such as for a very large or dense object (which can result in a very low x-ray photon intensity at the detector), additive noise can be comparable in amplitude to the signal level and will degrade the image quality. For an x-ray imaging system, if the signal level is several orders of magnitude higher than the additive noise level, image noise will be dominated by x-ray quantum noise, and the system is considered to be working as a *quantum-limited* detector. This will be discussed in Section 2.2.3.1.

An x-ray detector can only respond up to a certain maximum x-ray intensity level and will become saturated if the incident x-ray intensity exceeds this level. The *dynamic range* of a detector is defined as the ratio of the maximum signal level to the additive noise level. The dynamic range describes the effective signal range a detector can measure. Dynamic range is determined primarily by the signal amplification that occurs within the detector and the bit depth (quantization) of the ADC used to digitize the electronic signal.

For digital detectors, the physical dimensions of each detector element (typically called "dexel size") directly determine the maximum *spatial resolution* of the detector. However, there are factors other than dexel size in the imaging chain that can affect the overall spatial resolution of an x-ray imaging system. For example, as mentioned previously, the thickness of the scintillator layer will affect the light spread on the surface of the photodiode array and thus will influence the resulting image resolution. Other factors such as the x-ray tube focal spot size and the geometric setup of the imaging system also contribute to the overall spatial resolution of the system.

2.2.3 Intrinsic Issues Affecting X-Ray Image Quality

2.2.3.1 Limitation of Radiation Dose

As a form of ionizing radiation, x-rays can penetrate and interact within biological tissues through various mechanisms, as described previously. Potential damage can be caused to the imaging subject due to absorbed energy from x-ray photons. There exists a small risk of cancer induction when live humans and animals are exposed to x-ray radiation. *Radiation dose*, or more accurately, the *absorbed dose*, is a parameter that is defined as the energy imparted per unit mass. The SI unit of absorbed dose is the gray (1 Gy = 1 J/kg). For the purpose of biomedical imaging, radiation dose to the subject has to be as low as possible while producing an image with adequate quality for uncompromised interpretation.

In an idealized model, x-ray photons behave as individual particles traveling along straight lines through the imaging object until they impinge upon the x-ray detector. There are random statistical fluctuations in the number of detected x-ray photons at each individual detector element, and hence, the energy integrated in each dexel also experiences random fluctuation. A good analogy for this process is to observe raindrops falling on patio tiles. Each time, the number of raindrops falling on each tile is not the same and has random statistical fluctuations. If we repeat the experiment many times, the average number of drops collected at each tile can be used to predict approximately what the number will be next time, which will always fall in a range of possible numbers around this predicated or average value. Mathematically, there are two parameters that describe such a random process: the *mean* value and the *variance* (which characterizes the variability from the mean).

Bearing the same statistical property, the number of x-ray photons (or quanta) detected by a detector can be modeled as a random variable described by the Poisson distribution. One important feature of the Poisson distribution is that the mean value of the random variable is always equal to its standard deviation squared (also called variance). If we assume a simple construct of monoenergetic x-ray photons being detected through photon counting, an estimation of the *signal-to-noise ratio* (S/N) is

$$\mathrm{S/N} = \frac{\text{signal}}{\text{noise}} = \frac{N}{\sqrt{N}} = \sqrt{N}, \tag{2.9}$$

where N is the mean value of the total number of x-ray photons striking each dexel. $\sqrt{N}$ is the standard deviation (which is the square root of the variance) and is the parameter described as the "noise." For polyenergetic x-ray spectra and energy-integrating detectors, calculation of the S/N is more complex but is still proportional to $\sqrt{N}$. Equation 2.9 shows that the more x-ray photons are detected, the better the S/N for an x-ray image, due to the better overall statistical integrity of the image.

As stated previously, radiation dose to the subject is directly proportional to the total number of x-ray photons imparted to the subject. Thus, the trade-off between radiation dose and image S/N is the key issue when designing an x-ray system for imaging live subjects. This is a unique limitation for imaging modalities using ionization radiation. A simple analogy to help understand this limitation is to take a low-light photograph without using a flash. This photograph will typically be very grainy (noisy). In order to overcome this limitation and reduce unnecessary radiation dose to the subject, extensive efforts have been expended by both academic researchers and industrial developers to improve photon detection efficiency and suppress image noise.

X-ray image *contrast* is also indirectly related to radiation dose through the x-ray spectrum. An x-ray photon with higher energy has a smaller possibility of being absorbed by the subject through the photoelectric interaction and thus will be less likely to deposit its entire energy as a radiation dose. On the other hand, the reduced photoelectric absorption will produce lower image contrast between attenuating materials with different atomic numbers Z, such as between bone and soft tissue. Generally, the trade-off between radiation dose and image contrast can be adjusted by selecting a proper combination of kV and beam filtration to deliver an optimal x-ray spectrum. The size of the object to be imaged is critical in determining the appropriate x-ray tube potential (kV). For example, x-ray mammography only needs to penetrate about 4 to 6 cm of breast tissue and uses an x-ray tube potential from 24 to 32 kV. For human whole-body x-ray CT systems, tube potentials in the range of 100 to 140 kV are used to penetrate the adult subject, who can range from approximately 30 to 50 cm in thickness. As described previously, proper metal filters are also added to suppress lower-energy photons in the spectrum. Since these photons are less likely to penetrate the object, they contribute little to image contrast but can deliver significant radiation dose.

2.2.3.2 X-Ray Photon Scattering

Rayleigh scattering and Compton scattering are common interactions within tissues for x-ray photons in the energy range used for biomedical imaging (Figure 2.4a). The trajectory of scattered photons is changed from the original beam path, yet the scattered photons may still be captured at essentially random locations on the detector. These photons do not provide meaningful information for imaging but, rather, create a noisy background on the image. This causes two main issues for x-ray imaging: increased image noise and reduced dynamic range. The first issue is more prominent for systems based on digital detectors, and the second issue reduces image contrast in older screen-film radiography (where x-rays are converted to light by a phosphor screen and the light is converted into an image using photographic film).

Various approaches have been taken in x-ray imaging systems to suppress or correct for scattering. Firstly, the x-ray beam can be collimated to a very narrow slit prior to reaching the subject, and an associated collimator behind the patient, but in front of the detector, can be aligned with the slit to reduce the detection of scattered photons. This scanning slit-slot approach, however, requires a longer image acquisition time, which makes the image susceptible to subject motion. Secondly, an absorbing "grid" for scattered photons is placed in front of the detector and is used to reduce the number of scattered photons being detected. This approach requires a very accurately manufactured grid and reduces the signal strength, as a significant number of x-ray photons are absorbed by the grid. Grids work because they attenuate far more scattered photons relative to primary photons due to their geometric design. A third approach to suppress the detection of scattered radiation is to place the detector at some distance away from the object. This is known as the "air-gap" approach, and it reduces the probability of scattered photons being detected by reducing the effective solid angle to the detector. However, this approach will also reduce the size of the object that can be imaged by the system due to magnification effects. Finally, software approaches can be used to correct for the magnitude of the scattered signal in the acquired images. While this approach can theoretically restore the signal magnitude, it has no effect on the increased noise that results from the detection of scattered photons.

Different biomedical imaging systems might utilize a combination of the different approaches described in this section. For example, clinical x-ray CT scanners normally utilize narrow beam collimation, an air gap, and software correction. Mammography systems use specially designed scatter-rejection grids. Micro-CT and specimen imagers typically use power-limited x-ray tubes due to the small focal spot necessary for high-resolution applications. Thus, the air gap method is most commonly applied in these systems, with very-high-efficiency antiscatter grids currently being developed.

2.3 X-RAY PROJECTION IMAGING

2.3.1 Introduction

Once the x-ray beam generated by an x-ray tube is transmitted through an object, an x-ray "shadow" or "shadow-gram" produced by the attenuation properties of the object is recorded by the x-ray detector. This image records the spatial distribution of 3-D attenuation properties of the object integrated along a particular direction through the object. This image is called an x-ray *projection image*. Each individual picture element (*pixel*) in the projection image is a measure of the x-ray intensity after photon attenuation along a straight line between the x-ray focal spot and the corresponding detector element (Figure 2.6). Scattered radiation also contributes an unwanted bias to the signal in each pixel. The acquired x-ray projection images can be used either directly (as in x-ray radiography or fluoroscopy) or indirectly (as in x-ray CT; see Section 2.4) to provide relevant information about the subject.

X-ray radiography refers to the acquisition of stationary images of the subject, while x-ray fluoroscopy refers to the acquisition of a real-time image sequence (composed of a series of projection images) of the subject. Examples of fluoroscopy uses include monitoring the placement of vascular catheters and recording the motion of the esophagus during swallowing. This section will focus on radiographic applications.

For projection imaging, the 3-D information within the object is projected onto a 2-D image (Figure 2.6). This is a fundamental limitation of projection imaging—structure overlap within the image frequently poses challenges for image interpretation. In particular, the resulting image contrast might not be sufficient to provide accurate answers for specific clinical or biomedical research questions. Different approaches have been taken to reduce the limitations due to structure overlap and improve image contrast for specific imaging tasks, such as spectrum optimization (in mammography), dual-energy subtraction (in digital chest radiography), and the addition of contrast agents (in gastrointestinal fluoroscopy and digital subtraction angiography [DSA]). One benefit of 2-D projection imaging through a 3-D object is the enormous data compression that results—a single image contains anatomic information throughout the field of view.

There also exist x-ray imaging applications utilizing scattered photon information (typically through Compton scattering) instead of the transmitted photon information.

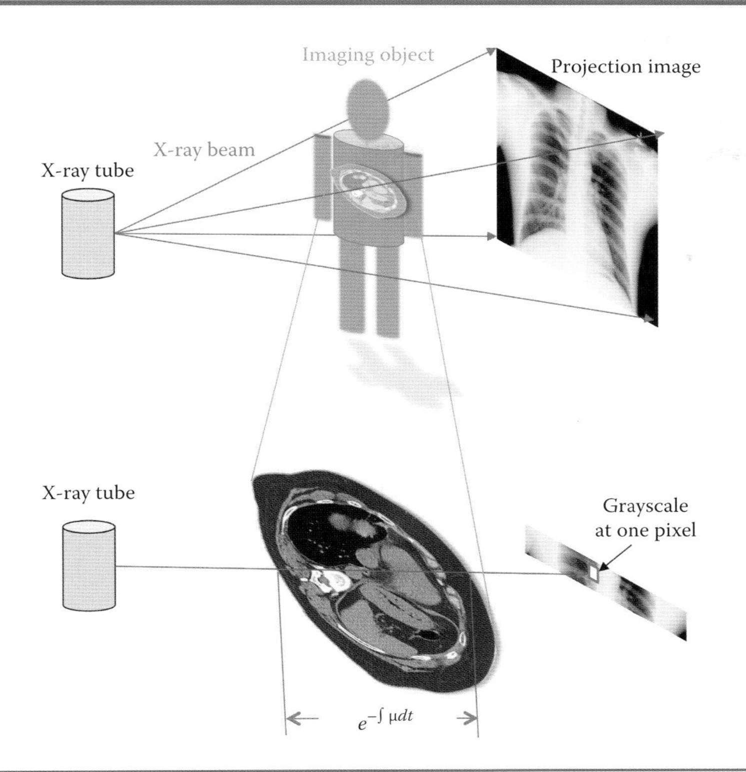

FIGURE 2.6 The concept of x-ray projection imaging.

They are primarily used in nonbiomedical applications, including industrial nondestructive detection and airport security exams [2].

2.3.1.1 Basic Geometric Principles

Projection images are always magnified with respect to the subject due to the divergent nature of x-rays emitted from a small focal spot. The magnification factor, M, is defined as

$$M = \frac{L_{\text{image}}}{L_{\text{object}}} = \frac{\text{Source to imager distance}}{\text{Source to object distance}}, \tag{2.10}$$

where L_{image} is the length of a structure measured on the projection image and L_{object} is the structure's physical length. As shown in Figure 2.7a, the magnification factor can also be determined by the ratio between the following two distances: the focal spot–to–imager distance and the focal spot–to–object distance. In practice, for biomedical imaging, the magnification factor is normally between 1.0 and 2.0. As mentioned previously, both the x-ray focal spot and the dexel size can affect the overall image resolution for x-ray imaging systems. By selecting a proper magnification factor, the x-ray focal spot size and the detector element size can be balanced to achieve optimal overall image resolution. As shown in Figure 2.7b, a relatively large magnification factor (around 2.0) can be used if the focal spot is small and the overall imaging system resolution is dominantly determined by the dexel size. On the other hand, for systems with relatively large focal spot dimensions, a lower magnification factor (close to 1.0) should be maintained to minimize focal spot blurring

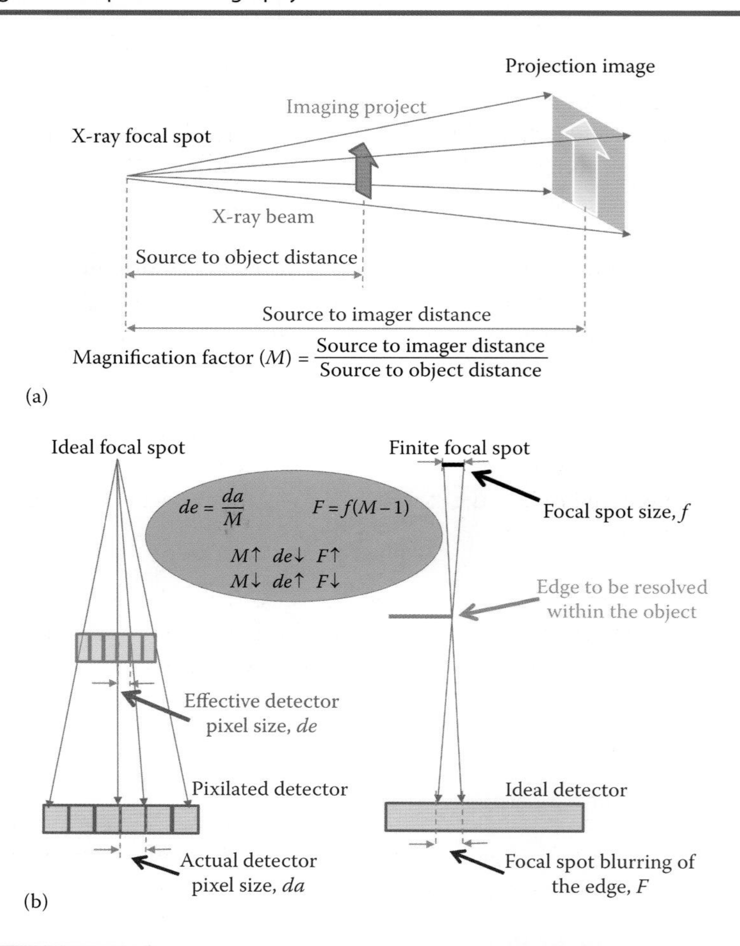

FIGURE 2.7 Geometric principles for x-ray imaging. (a) The definition of the magnification factor. (b) The competing effect between focal spot size and detector pixel size and their relationship with the magnification factor.

and improve spatial resolution. Low magnification occurs when the object is placed as close to the detector as possible. A good example of using an optimal magnification factor is digital mammography, which is a radiographic procedure with an extremely high requirement for spatial resolution.

2.3.2 Digital X-Ray Radiography and Detector Systems

As the most commonly used diagnostic imaging modality, x-ray radiography provides fast, flexible, high-quality, and cost-effective images for many clinical tasks, including dental evaluation, bone fracture evaluation, cancer detection, and disease assessment. For example, chest radiography or abdominal radiography can reveal important diagnostic information and is often the first choice for imaging in a systematic diagnostic workup. Similar to human imaging, x-ray radiography has been the primary diagnostic tool for veterinary imaging (e.g., from small pets to horses [3]) and small animal imaging in life sciences research (e.g., mice and rats [4,5]). In addition to these *in vivo* imaging applications, x-ray radiography of biological specimens extracted from live subjects (*ex vivo* imaging) is also widely utilized as a convenient analysis tool to provide clinical information pertinent to surgical procedures and for pathological evaluation.

Although screen-film radiography was used successfully for over 100 years, x-ray radiography is now digital in most clinical and research settings. Compared to traditional screen-film images, the major advantages of digital radiography include the reduced labor and cost relative to film (chemical processing and handling); the improved efficiency of image acquisition, display, transfer, and storage; the flexibility of digital image processing; and enhanced patient throughput. Digital x-ray detectors in general have much larger dynamic range, and contrast can be adjusted after exposure. Retakes due to underexposure or overexposure can be generally avoided, reducing radiation dose to patients.

Digital x-ray detector systems have been developed for radiography with either direct or indirect detection designs (as described in Section 2.2.2). Four major types of digital detectors are widely used in x-ray projection imaging applications. CR uses a storage phosphor for detection and can be considered to be a digital version of x-ray film. Light-sensitive cameras such as charged-coupled devices (CCDs) and complementary metal oxide semiconductor (CMOS) systems belong to the class of indirect x-ray detectors since they detect visible light photons induced by x-ray interaction with fluorescent screens or scintillators. Thin-film transistor (TFT)–based flat panel detectors are manufactured with either direct or indirect detection designs. There has been a recent trend of portable digital x-ray detectors (using CCD, CMOS, or TFT) emerging for many radiographic applications. The advantage of these digital detectors compared to CR is that there is no need for a separate readout process.

2.3.2.1 Computed Radiography

CR refers to a digital radiography imaging system that uses a *photostimulable phosphor* (PSP) as the detector. The PSP detector, also called a CR imaging plate (IP), is enclosed in a cassette. PSP screens typically consist of barium fluorobromide (BaFBr) and a small quantity of europium (Eu). As the IP is exposed to x-rays, the deposited energy excites electrons associated with europium atoms into a higher energy state in the conduction band. The excited electrons can be trapped locally within the PSP crystalline matrix and remain in a metastable state for days to weeks. Thus, PSP systems are called *storage phosphors*. The number of trapped electrons is proportional to the total energy deposited from x-ray exposure, and this charge distribution represents the latent image, which can be read out later to produce a digital image. To read out the latent image from a PSP screen, the cassette is placed in a CR reader, where the IP is mechanically removed from the cassette, and the plate is scanned using red laser light. Trapped electrons within the PSP gain sufficient excitation energy from the red laser light to enter the conduction band, where a large fraction will fall back to the ground energy state and, in the process, emit blue-green light. This process is called *photostimulated luminescence*. As shown in Figure 2.8, the emitted blue-green light is collected through a fiber-optic light guide by a photomultiplier tube (PMT), and the amplified electronic signal is converted into digital signals using an ADC. CR plates are widely used digital radiographic detectors and come in a range of sizes (from 18 × 24 cm to 35 × 43 cm). Without discrete dexels, the effective image pixel size from a CR plate is determined by the spot size and increment step of the readout laser and ranges from 50 × 50 μm^2 for mammography to 200 × 200 μm^2 for chest radiography. The key advantage of CR is its lower cost compared to other digital x-ray detector systems. Furthermore, the unique flexibility and portability of CR enables its wide application in portable veterinary imaging, small animal imaging, biological specimen imaging, and dental imaging systems.

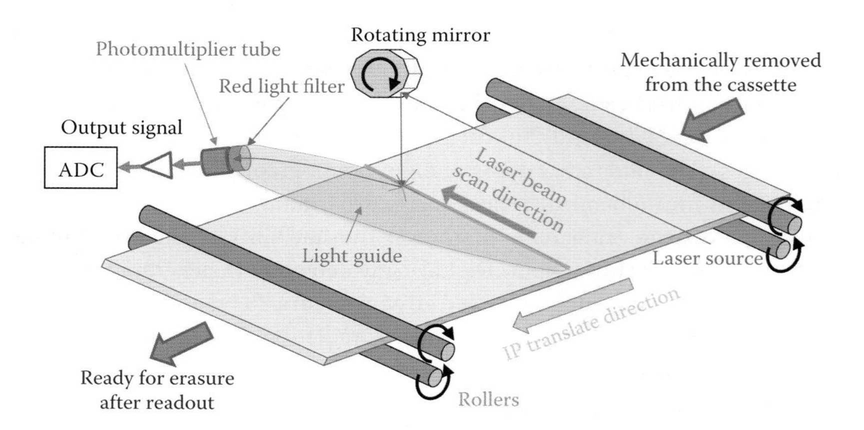

FIGURE 2.8 The readout process for a photostimulable phosphor imaging plate. Note the laser beam scan direction and the IP translation direction.

2.3.2.2 Charge-Coupled Devices

CCDs are silicon-based light sensors widely used in digital cameras and video cameras. CCD cameras are very sensitive to visible light and produce high-quality images. Visible light photons are absorbed in the silicon, releasing electrons, which are collected within each individual discrete detector element of the CCD camera. To read out the charge stored in each element after the CCD is exposed to light, the charges are shifted dexel by dexel, and the entire data matrix is digitized and read out from one corner of the CCD. The "bucket brigade" is a very good analogy for the readout dynamics of a CCD device, where charges are moved from one element to the next and are eventually amplified and digitized.

When a CCD is used as the detector for digital x-ray radiography (and fluoroscopy), it is coupled with a scintillator (normally made from CsI) or an image intensifier, which converts x-rays into visible light photons. Even though CCDs have very small detector element dimensions (on the order of $20 \times 20\ \mu m^2$) that have the potential for high spatial resolution, the key limitation is the limited detector area, typically smaller than $5 \times 5\ cm^2$. For some small sample applications, such as biospecimen or small animal imaging, this field of view is sufficient. However, for many clinical applications, visible light generated from a large scintillator screen has to be focused onto the smaller CCD chip using either optical lenses or fiber-optic coupling, in order to compensate for the limited size of the CCD. This demagnification process results in a large loss of light photons. If insufficient light photons are collected to form the image, a so-called secondary quantum sink occurs; in such a situation, the image S/N is degraded for a given radiation dose.

2.3.2.3 Complementary Metal-Oxide Semiconductor Systems

CMOS light-sensitive arrays are an alternative to CCDs for x-ray-induced light detection. Building on the concept of random access memory (RAM) used in computers, CMOS arrays are essentially composed of an array of RAM lithographed onto a crystalline silicon substrate (a process to etch integrated circuits on the surface of the crystal substrate). Each detector element can be randomly accessed or read out. The advantage of CMOS detectors over CCD is this independent readout process of each dexel. Thus, CMOS detectors can be more suitable for applications that require capturing x-ray images with high speed, such as x-ray fluoroscopy or cone-beam CT. Both CCD and CMOS detectors are based on crystalline silicon (in contrast to the a-Si used in TFT detectors). The crystalline structure of silicon has many advantages for x-ray imaging, including low noise, high reliability, low image lag, and high spatial resolution. Benefiting from the vast knowledge in crystalline silicon technology and relatively low manufacturing costs, the usage of CMOS detectors in biomedical imaging has gained momentum recently. Although CMOS detectors have the limitation of small detector area (as with CCDs), multiple CMOS detectors can be "tiled" together to form a bigger detector sufficient for most biomedical applications.

2.3.2.4 TFT-Based Flat Panel Detectors

TFT-based flat panel detectors leverage the manufacturing technology developed for flat panel liquid crystal display (LCD) systems (common consumer products like TVs and computer monitors), exploiting similar fabrication and pixel addressing schemes. In a flat panel monitor, the computer sends electrons to the monitor for display, while in an x-ray detector, electrons are stored and sent from the panel to the electronics for digitization. For TFT-based detectors, millions of discrete detector elements (with typical size ranges from $75 \times 75\ \mu m^2$ to $150 \times 150\ \mu m^2$) are assembled on a-Si substrates. Each detector element contains the effective detection area (either a storage capacitor for direct detection or a photodiode for indirect detection) and a switching transistor (the TFT). During x-ray exposure, the TFT switch is off, and each detector element will accumulate charge generated either directly by x-ray interaction within the photoconductor or indirectly by visible light photons converted using a phosphor screen (as shown in Figure 2.5). When the exposure is completed, the TFT switch is activated, and the accumulated charge stored in each detector element is read out through drain lines to produce the image data. All the TFT switches are controlled through gate lines. This unique "drain-and-gate" scheme has a much higher efficiency in data readout with a very compact circuitry design. When one gate line is opened, an entire row or column of data can be read out sequentially. Each detector element can be addressed with a combination of one gate line and one drain line. This *active matrix* design only requires $M + N$ signal lines (with M gate lines plus N drain lines) to address an $M \times N$ element array. Figure 2.9 depicts the basic structure of a TFT array.

With relatively low manufacturing costs compared to CCD and CMOS detectors, active matrix-based TFT flat panel detectors can be fabricated into large array sizes (up to $43 \times 43\ cm^2$) that are sufficient for all clinical x-ray imaging procedures. In addition to their utility in digital radiography, TFTs have been quite successful in replacing image intensifiers for digital fluoroscopy. With the recent development of wireless data transfer TFT detectors, TFTs can also replace CR for portable radiography applications.

The major drawback for TFT detectors is the limitation of the a-Si, including relatively higher readout noise and image lag, which refer to the residual signal within the detector after the previous readout. These signals mostly come from uncollected electrons (or positively charged holes) that are trapped in the defects within the a-Si substrate. Another limitation of TFTs, especially when used as an indirect detector coupled to a scintillator, is the presence of TFT electronics that cover part of the surface of each dexel (Figure 2.9). This reduces the overall detection efficiency of each detector element (the *fill factor*, the fraction of each dexel that is photosensitive), and the physical size of the electronics places limitations on how small TFT dexels can be made. Typical values of the fill factor are ~80% for ~$200 \times 200\ \mu m$ detector elements, decreasing rapidly for smaller detector elements.

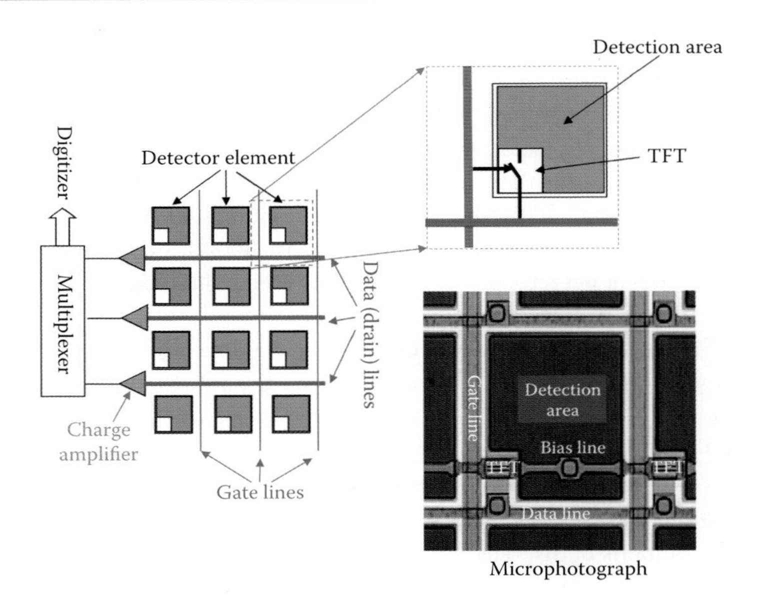

FIGURE 2.9 Structure of thin-film transistor (TFT) detectors.

2.3.3 Image Quality

In Section 2.2.3, the intrinsic factors that affect x-ray image quality were discussed, especially the trade-off between image quality and radiation dose to the subject. In this section, the terminology used to describe image quality and practical approaches to assess it are discussed. For biomedical imaging systems, the essential parameters of image quality include contrast, noise, and spatial resolution. Each of these will be briefly reviewed in Sections 2.3.3.1 through 2.3.3.3.

2.3.3.1 Contrast

Contrast is the difference in grayscale values ("image intensity") between adjacent regions in the image. For x-ray projection images, the image contrast is a consequence of differences of the attenuation properties along different x-ray beam paths through the subject. Image contrast depends on many factors, including x-ray interactions, x-ray detection, and digital image display. The x-ray spectrum, subject composition, and density in 3-D; the distribution of any introduced contrast agent; and x-ray detector characteristics fundamentally determine x-ray contrast in projection radiography. The x-ray spectrum characterizes the photon energy distribution used for imaging and can be optimized by changing the kV and tube filtration to enhance contrast between various tissue types depending on the imaging task. Size, composition, and density of the object to be imaged affect the overall image contrast and have to be carefully considered when designing an imaging system and the acquisition parameters used. For many radiographic/fluoroscopic applications, the contrast agent injection protocol (including the type, amount, and delay time between injection and image acquisition) plays a very significant role with regard to the final image contrast. As discussed in Section 2.2.2, x-ray detectors also play an important role in producing the final image contrast. For example, digital x-ray detectors have much larger dynamic range compared to older screen-film approaches, and thus, smaller differences between x-ray intensity can be detected. For digital imaging systems, the display settings such as the window and level settings can be used to enhance contrast to the final viewer, for example, a radiologist.

2.3.3.2 Noise

Noise refers to the stochastic or random fluctuations present in each image. It can be most readily observed and quantified in an image taken without an object between the x-ray tube and the detector. Assuming that the x-ray beam is homogeneous, the intensity in the image should be the same at all locations. However, in any real image, random fluctuations in the gray values (intensity) of the image are observed. The relative magnitude of this noise with respect to image contrast determines the ability of an observer to detect features in an image. Intuitively, a high level of image noise corresponds to degraded system performance and will result in a reduction in detectability for low-contrast features. As discussed in Section 2.2.3, although there are many sources contributing to x-ray image noise, a well-designed system should always be "x-ray quantum limited," which means that x-ray quantum noise is the dominant source of x-ray image noise.

Image noise is normally measured from either a uniform image or a uniform region of an image. The variance or standard deviation of pixel values in the measured image area can provide a good estimate of the noise level. Higher variance indicates a higher level of image noise. To have a complete description of the image noise for a specific imaging system, spatial frequency analysis is an important tool. This involves applying the Fourier transform to a region on the image such that the *spatial domain* information (intensity as a function of x and y location) is converted into the *spatial frequency domain*. Fine, high-resolution details (small objects) in the spatial domain correspond to signal levels in the high spatial frequency range, while slowly changing contrast (large objects) in the spatial domain corresponds to signal levels in the low spatial frequency range. The mathematical details are beyond the scope of this chapter, and interested readers are directed elsewhere for more detail [1,6]. The *noise power spectrum* (NPS) is the standard metric used to measure image noise intensity as a function of spatial frequency. The NPS can normally be determined through the Fourier transform of a uniform region of an image. Practical implementation requires that substantial averaging of the NPS be performed over many regions of interest. Figure 2.10 shows the NPS measured from a TFT flat panel detector. At higher tube currents, more x-rays are produced, and thus, noise (as measured by the NPS) is reduced.

2.3.3.3 Spatial Resolution

The spatial resolution of an imaging system describes its capability to distinctly depict two objects as they become smaller and closer to each other. Better spatial resolution indicates that smaller objects or finer details can be resolved in the image. Every imaging system has limited resolution, and a certain level of blurring occurs for every object. For x-ray imaging, since x-ray photons are generated from a finite-sized focal spot instead of an ideal point, the resulting x-ray image will be blurred, and thus, the image resolution will be reduced. As we have previously described in Section 2.3.1, magnification has been utilized to compensate for the limits of the physical detector element size. However, in general, magnification will amplify blurring and reduce spatial resolution from most focal spot sizes.

In the spatial domain, the image recorded for an infinitesimally small point object is known as the *point spread function* (PSF), and it measures the intrinsic blurring of an imaging system. The PSF is typically a 2-D function for radiographic imaging. The broader the

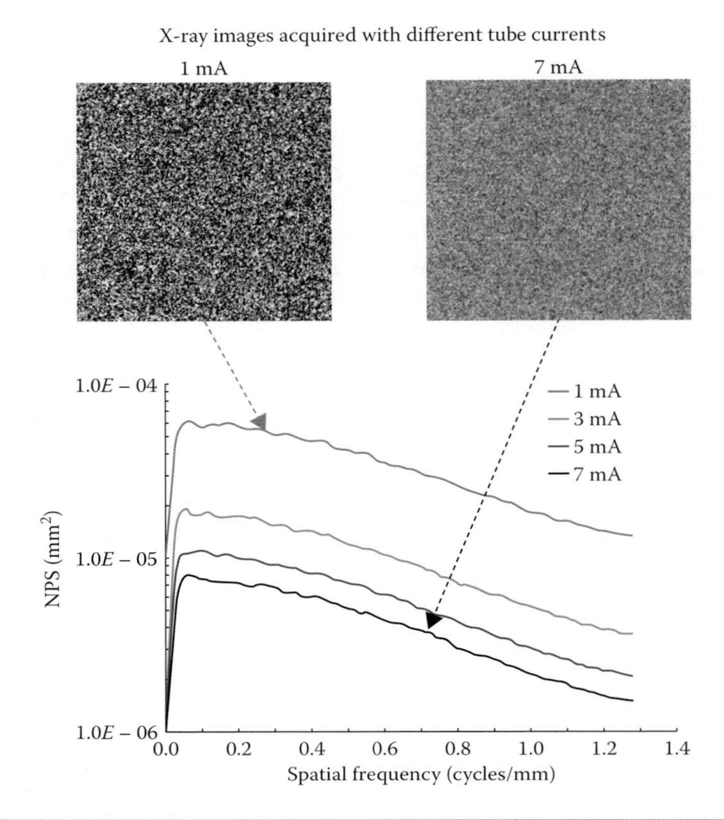

FIGURE 2.10 Noise power spectrum (NPS). NPS is a quantitative measurement of x-ray image noise in the spatial frequency domain. The image acquired with lower x-ray intensity (e.g., 1 mA) has higher noise power than the image acquired with higher x-ray intensity (e.g., 7 mA). Thus, the 1 mA image appears noisier than the 7 mA image.

PSF, the worse the image resolution is. For x-ray radiography, the 2-D PSF is relatively challenging to measure directly, and often, the 1-D *line spread function* (LSF) is measured along the two orthogonal dimensions of the x-ray detector. A very narrow slit can be imaged for x-ray projection imaging to measure the 1-D LSF.

While the LSF is a useful description of spatial resolution, it is common to describe spatial resolution in the spatial frequency domain. The *modulation transfer function* (MTF) is determined by Fourier transformation of the LSF (with normalization). Similar to the NPS, the MTF also describes the system response as a function of spatial frequency. A higher value of the MTF implies better resolving power at that particular spatial frequency. Figure 2.11 illustrates the MTF measured from a TFT flat panel detector. The magnitude of the MTF is typically lower at the higher spatial frequencies, which means that larger details (lower spatial frequencies) can be better resolved than smaller details (higher spatial frequencies) by a typical imaging system.

2.3.3.4 Detective Quantum Efficiency

Image contrast, noise, and spatial resolution are not independent of each other and all play important roles with regard to the overall image quality. In some cases, these parameters are combined to provide a sole indication of the image quality performance of an imaging

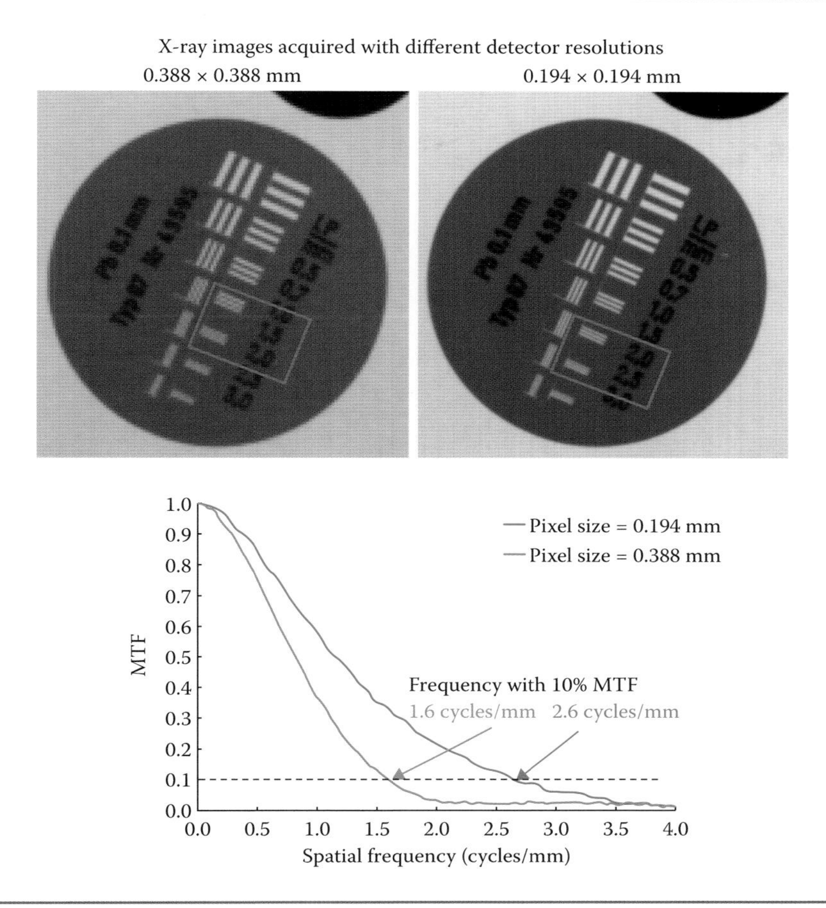

FIGURE 2.11 Modulation transfer function (MTF). MTF is a quantitative measurement of x-ray image resolution in the spatial frequency domain. The image acquired with a smaller pixel size (0.194 mm) has a higher amplitude of the MTF function (the frequency with 10% MTF occurs at 2.6 cycles/mm, shown in red) than that of the image acquired with a larger pixel size (0.388 mm), where the 10% MTF occurs at 1.6 cycles/mm (show in green). Thus, the 0.194 mm image appears sharper than the 0.388 mm image and can better reveal fine details.

system. The *detective quantum efficiency* (DQE) of an imaging system is a comprehensive and quantitative description of image quality and is especially useful in digital radiography systems. Conceptually, as a description of the efficiency of an imaging system, DQE is defined as the ratio of the output S/N squared, $(S/N_{\text{output}})^2$, to the input S/N squared, $(S/N_{\text{input}})^2$, of an imaging system:

$$\text{DQE} = \frac{(S/N_{\text{output}})^2}{(S/N_{\text{input}})^2}. \tag{2.11}$$

The DQE is also represented as a function of spatial frequency, *f*. Since the MTF describes the output "signal" component and NPS describes the output "noise" variance component, the output S/N squared can be determined in the spatial frequency domain as

$$[S/N_{\text{output}}(f)]^2 = \frac{[\text{MTF}(f)]^2}{\text{NPS}(f)}. \tag{2.12}$$

The NPS term in the Equation 2.12 is not squared because it is equivalent to the variance distribution, which is the square of the standard deviation (or the noise). As discussed in Section 2.2.3, input photon intensities for x-ray projection imaging follow a Poisson distribution, and their S/N-squared equates to the mean value of total number of photons, N (Equation 2.9). Thus, the DQE can be expressed as

$$\text{DQE}(f) = \frac{k \cdot [\text{MTF}(f)]^2}{N \cdot \text{NPS}(f)}, \tag{2.13}$$

where k is a constant to convert units [1,7]. The DQE combines system performance in terms of contrast, noise, and spatial resolution and can also be used to directly compare different systems. The DQE has become widely accepted as the standard to measure and compare the performance of x-ray radiographic systems.

2.3.4 Representative Applications of Digital Radiography

2.3.4.1 Digital Mammography

Digital mammography (Figure 2.12) is an x-ray radiographic procedure with the goal to image the breast with high resolution and low radiation dose for breast cancer screening. Since the breast is composed of soft tissues such as fibroglandular and adipose (fat) tissue, the contrast in the x-ray projection image is very limited due to the small attenuation differences between these different tissue types. To achieve sufficient image contrast while maintaining a low radiation dose level, two approaches are utilized in mammography. Firstly, the breast is always compressed between a radiation-transparent paddle and the detector enclosure. This spreads breast tissues over a larger area and reduces the breast thickness (attenuation path length). Therefore, compression provides several advantages, including scatter reduction,

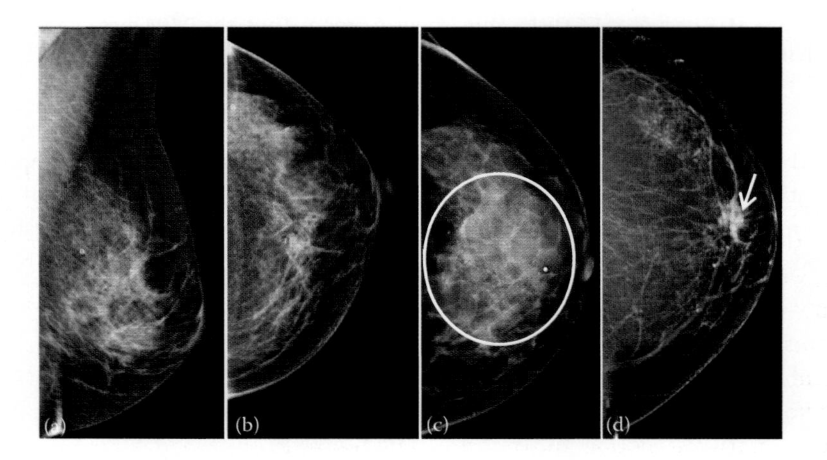

FIGURE 2.12 Example images from x-ray mammography. (a and b) Two different projection views of a breast. (c) A mammogram showing microcalcifications (within the circled area). (d) A mammogram showing a breast cancer mass (arrow).

reduction of tissue overlap, and an overall lower radiation dose to the breast. Secondly, a dedicated mammography x-ray tube is designed to produce low-kV spectra (from 24 to 35 kV). With this x-ray energy range, the ratio of image contrast to radiation dose is optimized by using an average x-ray photon energy of around 19 keV. In order to detect very small calcium specks (normally smaller than 0.2 mm, called *microcalcifications*, as shown in Figure 2.12c) that often are deposited in the presence of breast cancer cells, mammography has the highest image resolution of all clinical x-ray imaging procedures. For this purpose, mammography x-ray tubes are designed with a focal spot size combination of 0.3 and 0.1 mm, and digital detectors used for mammography are manufactured with a dexel size between 50 and 100 μm. The larger focal spot size (0.3 mm) is typically utilized to image the entire breast when it is in contact with the detector (i.e., a magnification factor close to 1). The smaller focal size (0.1 mm) is used with a special magnified view of a small region of suspicious tissues within the breast. Digital mammography is one of the most technically demanding radiographic applications. However, it also is one of the most widely used and remains the standard of care for breast cancer screening and diagnosis.

2.3.4.2 Dual-Energy Radiography

Dual-energy radiography utilizes the different energy dependencies of the attenuation coefficients of different tissue types (especially between bone and soft tissue) in order to reduce the overlapping tissue problem for projection radiography. As discussed in Section 2.2.2, the attenuation coefficients of bone and soft tissue decrease as the x-ray energy increases. More importantly, the photoelectric effect strongly depends on the atomic number (Z) of the interacting material, while the Compton scatter does not have such a dependency. Bones (composed of a large proportion of calcium, atomic number $Z = 20$) and soft tissues (with an effective atomic number $Z \approx 7.6$) produce different attenuation levels when two different effective x-ray energies are transmitted through them. Although there are different realizations of dual-energy radiography, it always involves acquiring two x-ray projection images, one at a high effective x-ray energy (e.g., 120 kV) and the other at a low effective energy (e.g., 60 kV). These two images can be weighted and logarithmically subtracted to remove the contrast associated with either the bones or the soft tissues. Dual-energy subtraction is widely used for digital chest radiography, where the ribs impede the visualization of soft tissue structures in the lungs and mediastinum. Figure 2.13 illustrates this technique to remove either bone or soft tissue contrast from the original radiographic images.

2.3.4.3 Digital Subtraction Angiography

DSA is a radiographic procedure that is used to visualize blood vessels inside a subject. It is useful in the diagnosis and treatment of vascular abnormities such as arterial and venous stenosis, vascular malformations, and vascular trauma. DSA exploits temporal differences between x-ray projection images acquired before and after the injection of highly attenuating contrast agents such as iodine. The first image captured before the injection of the contrast agent is called the *mask* and represents the anatomical background of the patient. A series of images is then captured after the contrast agent is injected into the vasculature. In the absence of patient motion, the only difference between the contrast-containing images and the mask image is the concentration of iodine agent within the vascular system. After the anatomical background is removed by logarithmic subtraction, detailed vascular structures can be visualized in isolation. Figure 2.14 shows images obtained by DSA.

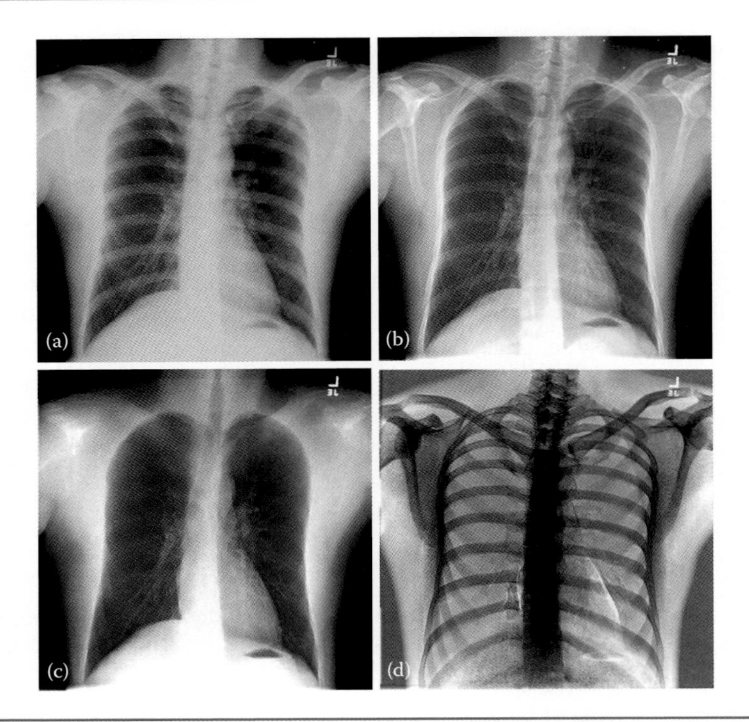

FIGURE 2.13 Example images of dual-energy subtraction radiography. Single-energy chest radiographs were acquired at (a) 60 kV and (b) 120 kV. (c) Soft-tissue-only (bone-subtracted) and (d) bone-only (soft-tissue-subtracted) images.

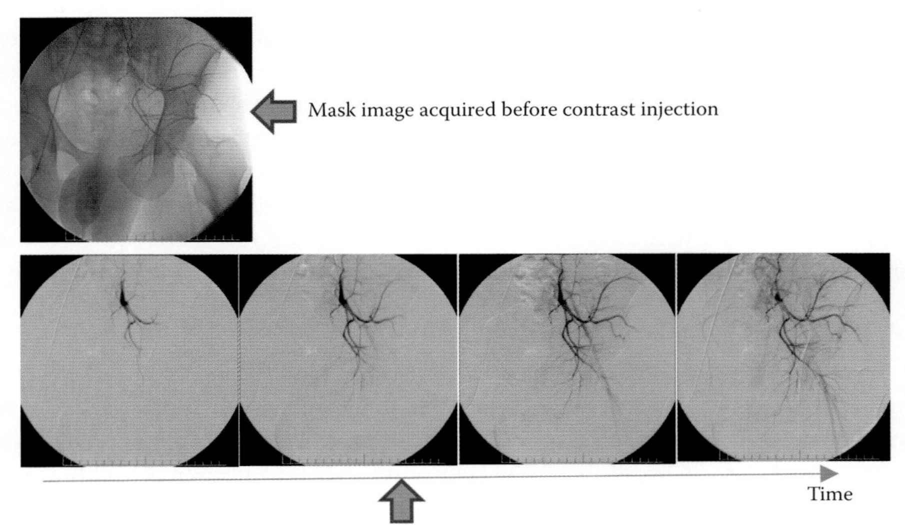

FIGURE 2.14 Example images of digital subtraction angiography. Digital subtraction angiography uses temporal subtraction following intravenous iodine-based contrast agent administration to eliminate the static anatomy of a radiographic projection acquired prior to contrast. Therefore, only the vasculature is seen in the images. Digital video processors acquire the angiographic images and can perform pixel-by-pixel subtraction between the mask image (no contrast) and the iodinated image sequence in real time.

2.4 X-RAY CT

2.4.1 INTRODUCTION

CT was first introduced for medical imaging in the early 1970s, and the technology has developed rapidly since. Tomography (*tomo* is the Greek root for *slices*) is an approach to produce cross-sectional images within an object. Utilizing the penetration and interaction properties of x-ray photons, CT can overcome the fundamental limitation of projection imaging: structure overlap. CT data acquisition bears many similarities to x-ray projection imaging; however, the x-ray tube and the detector system rotate around the object, and a series of projection images (either 1-D or 2-D) are acquired at different angles during the rotation. It is the different information present in each angular view that allows a cross-sectional CT image to be obtained through a mathematical process, called *reconstruction*, where the acquired projection images from different angles are used to compute the final image. CT has become a mature and sophisticated modality, capable of producing high-resolution and fully 3-D image data sets of the subject in just a few seconds.

As shown in Figure 2.15, a typical CT image is a reconstructed distribution of the effective linear attenuation coefficients from a very thin "slice" (normally between 0.5 and 5 mm in thickness) through the subject. In contrast to x-ray projection images, each pixel value from a CT image represents the effective linear attenuation coefficient within a very small

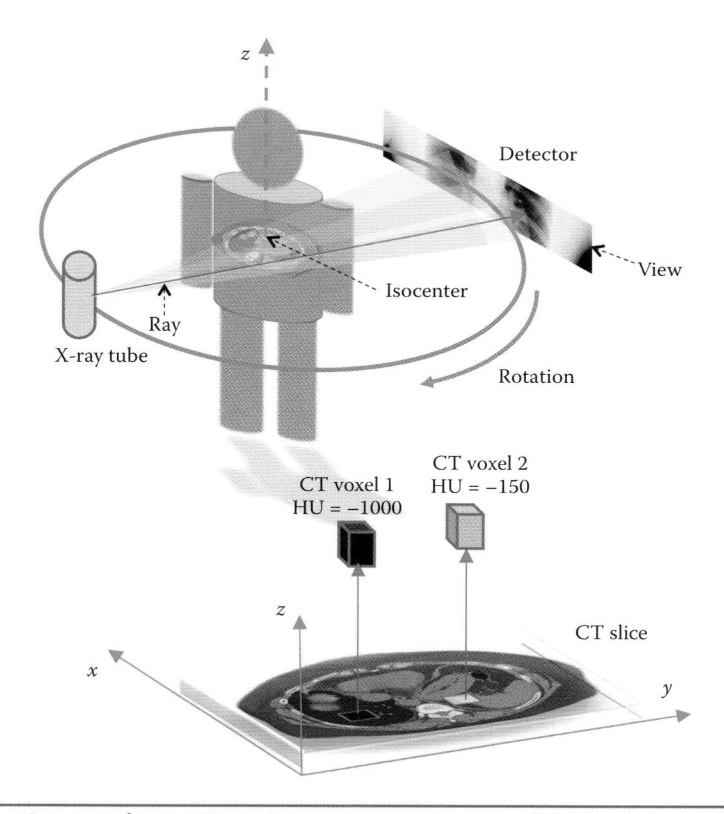

FIGURE 2.15 Concept of x-ray CT imaging. A reconstructed x-ray CT image is the distribution of the effective linear attenuation coefficients from a very thin "slice." Each pixel value from a CT image represents the effective linear attenuation coefficient within a very small volumetric element (called a *voxel*).

volume element (called a *voxel*). The measured attenuation coefficients from CT images are normally converted into integers called Hounsfield units (HUs). The HU is named after one of the inventors of CT, Sir Godfrey Hounsfield, and is defined as

$$\mathrm{HU} = \frac{\mu - \mu_{\mathrm{water}}}{\mu_{\mathrm{water}}} \times 1000, \tag{2.14}$$

where μ and μ_{water} are the effective linear attenuation coefficients of the original CT image pixel and water, respectively. Thus, water, by definition, has an HU value of 0, while air (where $\mu \approx 0$) has an HU value of −1000. Depending on the tissue density and composition, HU values for soft tissues range from around −200 to +100. Bone and artificially implanted metal objects have much higher HUs in the range from several hundred to several thousand. For the most commonly used iodine contrast agent in CT, the HUs have a very wide range, from ~10 to ~1000, since they are directly proportional to the iodine concentration and largely depend on x-ray tube potential.

Although an individual CT image is reconstructed in 2-D and only represents a very thin transverse slice within the object, a complete stack of these slices is typically acquired and reconstructed, forming a 3-D volume of anatomic information from the subject. The entire 3-D volume can then be "resliced" by computer software to display anatomy along any arbitrary orientation through the volume.

2.4.2 CT Acquisition

2.4.2.1 Basic Components of CT Scanners

A CT scanner typically requires four major components, an x-ray source (x-ray tube with high-voltage generator), an x-ray detection system (x-ray detector and data acquisition electronics), a patient table or sample-holding platform, and image reconstruction software and hardware. To acquire x-ray projection images from different angles around the object, two approaches to CT acquisition are used: (1) For clinical CT scanners, the x-ray tube and the detector are mounted on a rotating gantry while the patient table translates linearly through the open bore of the gantry. The rotating gantry is massive (typical weighs from 400 to 1000 kg) and requires precise engineering as it rotates at very high speed (normally faster than 2 rotations per second). For CT scanners designed for small animal imaging, often micro-CT scanners, the geometry is similar; however, the x-ray tube and detector typically rotate much more slowly, often pausing to collect each angular projection view in a so-called *step-and-shoot mode.* (2) For some experimental CT scanners, for example, many of those used in biospecimen imaging, the x-ray tube and the detector remain stationary, while a motorized platform rotates the object in the horizontal plane and, in some cases, can also translate the object vertically. With this design, the scanner can be quite compact. For both designs, rotation is driven by a motor, and an encoder accurately records the angle information associated with each acquired projection. Despite differences in the technical details, all these designs are conceptually identical and produce equivalent data. For brevity, the focus here will be on clinical CT systems. Micro-CT systems will be briefly discussed in Section 2.5.1.2.

The x-ray tube is a key component of a CT scanner. The x-ray spectrum and the effective x-ray photon energy determine the balance of the penetrability and absorption of photons, with the goal of achieving good image quality at reasonable dose levels. Typical clinical CT scanners utilize a kV range from 80 to 140 kV, with 120 kV being the most common setting.

The overall power of the x-ray tube characterizes its capability to produce the high x-ray flux required for rapid scanning. This reduces imaging time and decreases the chance of patient motion, which will cause blurring of the images. An extreme example is cardiac CT, in which very rapid scans are performed in order to capture stationary images of the beating heart. The x-ray focal spot size is another important parameter that impacts CT image quality. Similar to projection imaging, the spot size is a balance between x-ray tube power (requires larger spot size to generate higher x-ray flux) and image resolution (requires smaller focal spot size). The typical focal spot size for CT is between 0.6 and 1.6 mm.

As with all other system components, the x-ray detectors used in CT have experienced tremendous technological improvements over the past 40 years, and new designs are still evolving. Major improvements include the increased *z*-axis coverage (from a single detector array to large-area, multidetector arrays allowing multiple slices to be acquired simultaneously); the improved detection efficiency (e.g., solid-state scintillators coupled to photodiodes); and fast readout electronics. Currently the majority of CT scanners utilize the indirect detection concept as previously described. High-density ceramic solid-state scintillators, such as Gd_2O_2S and $CaWO_4$, are coupled with individually addressable photodiodes to meet the unique requirements of modern CT scanners, especially the high efficiency (>98% QDE) and the high readout speed (~100 to 200 μs for each detector element).

Although the time required for CT image reconstruction is independent of the time it takes to acquire the angular projection data, in the past, it was a bottleneck for patient throughput. With the development of modern computers, the reconstruction engine of a CT scanner has experienced tremendous improvement: The first CT images required 2.5 h to reconstruct one image slice. Currently, much-higher-resolution CT images can be reconstructed within a few milliseconds. Thus, reconstruction time no longer limits the throughput of a CT scanner.

Other important CT system components include the *slip ring* and the *bow-tie filter*. A slip ring enables electrical coupling (power and data) between the rotating gantry and the CT components (electronics, computer, and power supplies) that are mounted outside the gantry. With a slip ring, there are no physical cables connecting components, and the CT gantry can continuously rotate while the patient table is translating. This allows *helical* (also called *spiral*) CT acquisition for fast acquisition of large 3-D objects (patients, etc.) (Figure 2.16b). The bow-tie filter is a specially designed x-ray filter that has nonuniform thickness across the x-ray field in the fan-beam (across the width of the object) direction. A typical bow-tie filter has minimal thickness at the center of the x-ray beam and gradually increasing thickness toward the edge of the x-ray beam. The purpose of using a bow-tie filter is to reduce the radiation dose to the patient's periphery. X-rays passing through the periphery of the patient travel through less tissue thickness to reach the detector; thus, the total absorption is lower, and fewer x-rays are needed to achieve a given S/N level at the detector.

Important auxiliary equipment for CT scanning includes power injectors for the accurate control of the amount and timing of contrast agent injection, electrocardiogram (ECG) tracking devices for the cardiac-gated CT acquisitions, and laser alignment systems for patient positioning during setup.

2.4.2.2 Acquisition Schemes

Most modern whole-body CT scanners are so-called third-generation scanners, which utilize a rotating gantry in which the x-ray tube and the detector are rigidly mounted across from each other, as shown in Figure 2.15.

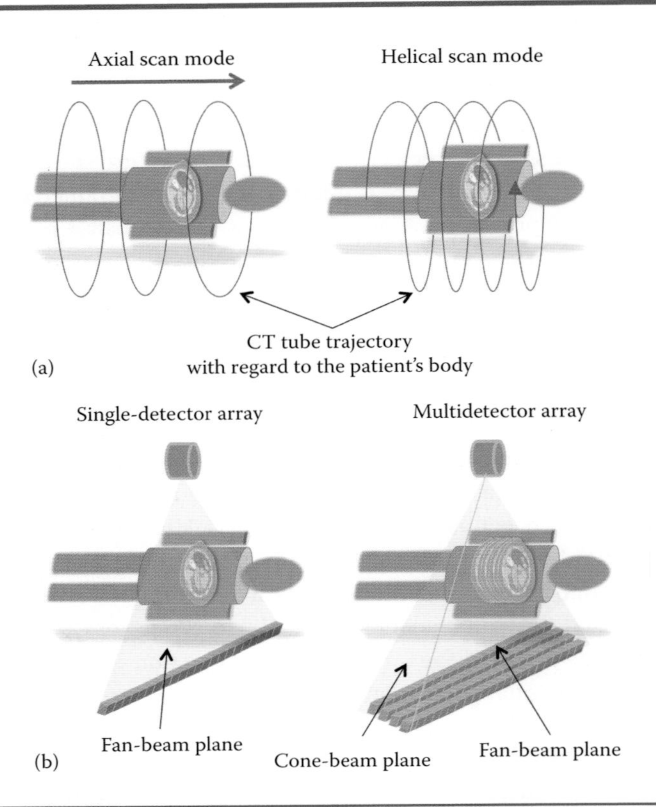

FIGURE 2.16 X-ray CT acquisition concepts. (a) The difference between axial and helical scanning. (b) The difference between a single-detector array and multidetector arrays (four rows of detectors are shown).

CT acquisition is fundamentally a combination of a series of x-ray projections, and thus, it inherits all of the geometric concepts of projection imaging, such as the divergent x-ray beam from a focused source and the magnification of the object onto the detector. For this acquisition geometry, the straight line that connects the focal spot to an individual detector element is called a "ray." The projection images during a CT scan are acquired with the gantry rotating around the *isocenter*. Each projection image (a collection of rays at a particular angle) is often called a "view" (as shown in Figure 2.15).

There are two CT scan modes that differ in the relative motion between the patient table and the rotating gantry: axial and helical (Figure 2.16a). For the axial (or sequential) scan mode, the patient table is stationary during a full rotation of the gantry. Traditionally, each slice within the object was scanned individually with the axial mode, and the patient table translated between each axial scan with the x-ray tube turned off before acquiring the next slice. The axial mode is less efficient in covering a relatively large volume in terms of acquisition time, and patient motion can become an issue, especially in the chest due to respiratory motion.

The introduction of slip ring technology enabled helical mode scanning, where the table is translated continuously while the gantry rotates. Comparing the trajectory of the x-ray tube, from the perspective of the patient, in the axial mode, the tube moves in discrete circles parallel to each other, while in the helical mode, the x-ray source moves along a spiral or helical trajectory. Since both the patient table and the gantry are moving during a helical scan, a very important parameter called *pitch* characterizes the relative speed between table translation and gantry rotation. Pitch is defined as

$$\text{Pitch} = \frac{\text{table translation (mm) per } 360^\circ \text{ gantry rotation}}{\text{collimation width (mm) at isocenter}} \quad (2.15)$$

Pitch is normally in the range from 0.75 to 1.5. A higher pitch indicates faster table translation and larger volume coverage per unit time. A lower pitch results in higher radiation dose level to the patient, better sampled data, and better image quality.

Another important concept related to CT acquisition is the detector-array configuration. A single linear array of detector elements was used in early-generation scanners. In this configuration, the x-ray beam is narrowly collimated onto a detector arc, resulting in a *fan beam* of data, which represents a 1-D projection of the object. With this geometry, only a single thin slice of the object can be imaged and reconstructed with each rotation. In a single-detector-array CT, the slice thickness was determined solely by the collimation of the x-ray beam onto the detector. Using a narrower collimator slit confines the x-rays to passing through a finer slice of tissue. With the rapid development of x-ray detector technology and increasing clinical demand for large volume coverage with very short scan times, multidetector-array CT (MDCT) has become the prevalent design concept for clinical CT scanners. For MDCT, a 2-D detector array accompanied with wider x-ray collimation enables larger volume coverage per rotation (Figure 2.16b). Since MDCT systems can resolve multiple slices along the table translation direction (z-axis), thinner CT slices can be reconstructed with relatively wide x-ray collimation. CT manufacturers have been competing with regard to the total number of detector arrays, initially producing 4- or 16-detector-array CT systems, with 64, 256, or even 320 detector-array CT systems available today.

A disadvantage of the increased x-ray collimator width of MDCT systems is degraded image quality because more scattered x-ray photons are detected due to solid angle considerations. Despite this limitation, several clinical and experimental systems utilize large-area flat panel detectors to perform an extreme case of multislice CT, in which the x-ray beam becomes a cone, rather than a fan. This is called *cone-beam CT* and is an active area of research.

2.4.3 CT Image Reconstruction

CT image reconstruction refers to the mathematical process that converts the CT projection data acquired at multiple angles into usable cross-sectional images of the subject. With all the preprocessing steps (described in Section 2.4.3.1) completed, the corrected projection data can be reconstructed by one of two different approaches [8], either using analytical algorithms such as filtered backprojection (FBP) or using optimization-based iterative approaches such as the algebraic reconstruction technique (ART). The entire process of image reconstruction has been extensively studied and continues to evolve, and the focus here is on basic concepts illustrated by two representative approaches for CT reconstruction, FBP and ART.

2.4.3.1 Preprocessing Steps

Several preprocessing steps have to be performed on the acquired projection data before reconstruction. Two major types of processing or correction will be discussed: detector response correction and the inverse logarithmic process.

The discrete elements within a digital x-ray detector do not provide identical output signals for a given input signal. This is due to the inevitable differences in the manufacture of individual electronic components and nonuniformities within the scintillator material. A straightforward approach to correct for the nonuniform response of x-ray detector elements is to acquire a series of images with incremental x-ray intensities and model the response of each detector element. With this information, the output of all the detector elements can be normalized such that they have the same linear proportionality to the incident x-ray intensity on each detector element. Thus, this step corrects for the nonuniformity between detector elements. This step also corrects for the inhomogeneous x-ray intensity across the detector arrays due to the bow-tie filter and the heel effect, as the measurement implicitly assumes that equal x-ray intensity is incident on all detector elements.

The inverse logarithmic process is a key step to process projection images for CT reconstruction. This process is basically the inverse of Beer's law (Equation 2.4) and converts the measured x-ray intensity at a detector element to a value that is proportional to the linear attenuation coefficient, μ, which is the desired quantity in the final reconstructed image. The inverse logarithmic process is performed using

$$P_n = \int \mu(t)\,dt = \ln\left(\frac{I_0}{I_n}\right), \tag{2.16}$$

where I_n is the x-ray intensity measured at the nth detector element and I_0 is the x-ray intensity measured without attenuation. Using the modified projections, P_n, as the input for reconstruction creates an image where the voxel intensity values directly reflect the attenuation coefficients at the corresponding location in the object.

2.4.3.2 Filtered Backprojection

FBP is the most straightforward and widely utilized approach for CT reconstruction. Intuitively, backprojection is the reverse geometric process of the CT acquisition. When x-ray projection data are acquired, the distributed attenuation coefficients within the object along one ray path are integrated into the signal detected by one detector element, which is determined as $P_n = \int \mu(t)\,dt$ (Equation 2.16). For backprojection, the process is reversed; thus, the measured signal P_n is "smeared" or distributed across the CT image matrix along a straight line in the direction it was measured. This process is carried out for every projection ray and view and eventually generates a reconstructed cross-sectional CT image representing the distribution of μ, and the t component gets cancelled out (as shown in Figure 2.17). A more rigorous mathematical derivation actually involves the so-called Fourier slice theorem and demonstrates that the exact reconstruction requires an additional processing step in which the projection data are mathematically filtered with a *ramp* or *Lak* filter before backprojection. For computational efficiency, this filtering step is usually performed in the spatial frequency domain. The filtering and backprojection processes are interchangeable. Thus, the filtering step can be performed before backprojection on the projection data or after backprojection on the image data.

FBP was first derived for parallel-beam geometry, which assumes that all the rays within one projection view are parallel to each other and equally spaced. This was the acquisition geometry of early CT systems. However, modern CT systems, as described in Section 2.4.2, acquire fan-beam projections where the measured projection lines at a particular angular

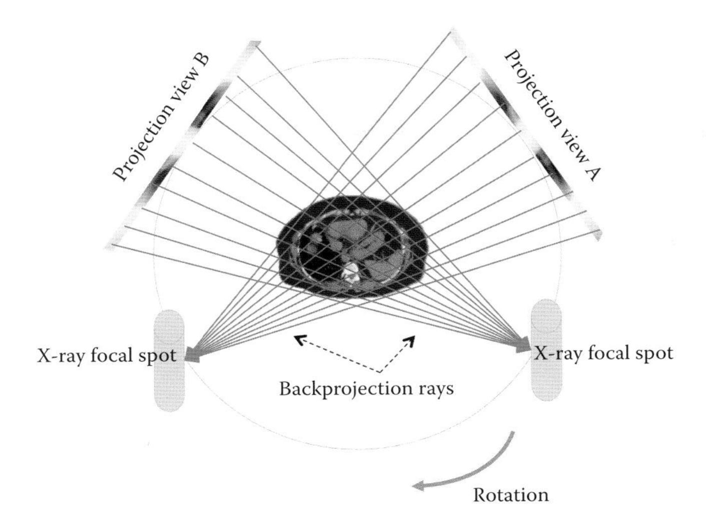

FIGURE 2.17 Concept of x-ray CT reconstruction. CT reconstruction is the mathematical process that reverses the x-ray projection image acquisition. It is normally called "backprojection" since the signal in each detector element from the projection image is "smeared" or "projected" back toward the x-ray focal spot along the specific line (ray) joining the detector element and focal spot. This backprojection is repeated for all the different projection angles.

view are not parallel to each other. The reconstruction of the fan-beam data can be converted back to the 2-D parallel-beam situation by grouping projection rays that are from different views but parallel to each other. For multislice and cone-beam CT scanners, cone-beam reconstruction extends the 2-D fan-beam reconstruction into FBP in 3-D. However, it is no longer an exact reconstruction for the entire object volume, except for the central image plane, and some artifacts can occur toward both ends of the reconstruction volume.

Even though in theory, a ramp filter (as shown in Figure 2.18) is sufficient for an exact reconstruction using FBP, it is rarely used as the reconstruction filter for clinical CT scanners. An apodization filter with decreasing magnitude at higher spatial frequencies is typically necessary to achieve satisfactory CT images. Because x-ray quantum noise is present at all frequencies, whereas the magnitude of the signal decreases with increasing spatial frequency (Figure 2.11) due to the limited detector resolution and focal spot size, S/N is worse at higher spatial frequencies. Therefore, to suppress image noise, the apodization filter reduces the contributions of higher-spatial-frequency noise. However, this also reduces the spatial resolution in the reconstructed image. Depending on the clinical application, different apodization filters are used to emphasize different levels of detail in the reconstructed images. For example, a *bone filter* has a relatively high magnitude in the higher frequency range in order to emphasize the sharp edges and high contrast of bones within the patient's body. *Soft tissue filters* suppress more of the high frequency range to reduce the high-frequency noise in the images. A comparison of several typical reconstruction filters is shown in Figure 2.18.

FBP has many advantages, especially its easy implementation and fast reconstruction times that can be achieved through parallel processing algorithms. The key disadvantage of FBP is that it does not fully model the scanner geometry or x-ray spectrum, and the reconstructed images therefore may not achieve the best possible S/N. FBP reconstruction also is susceptible to beam-hardening artifacts, as described in Section 2.4.4.

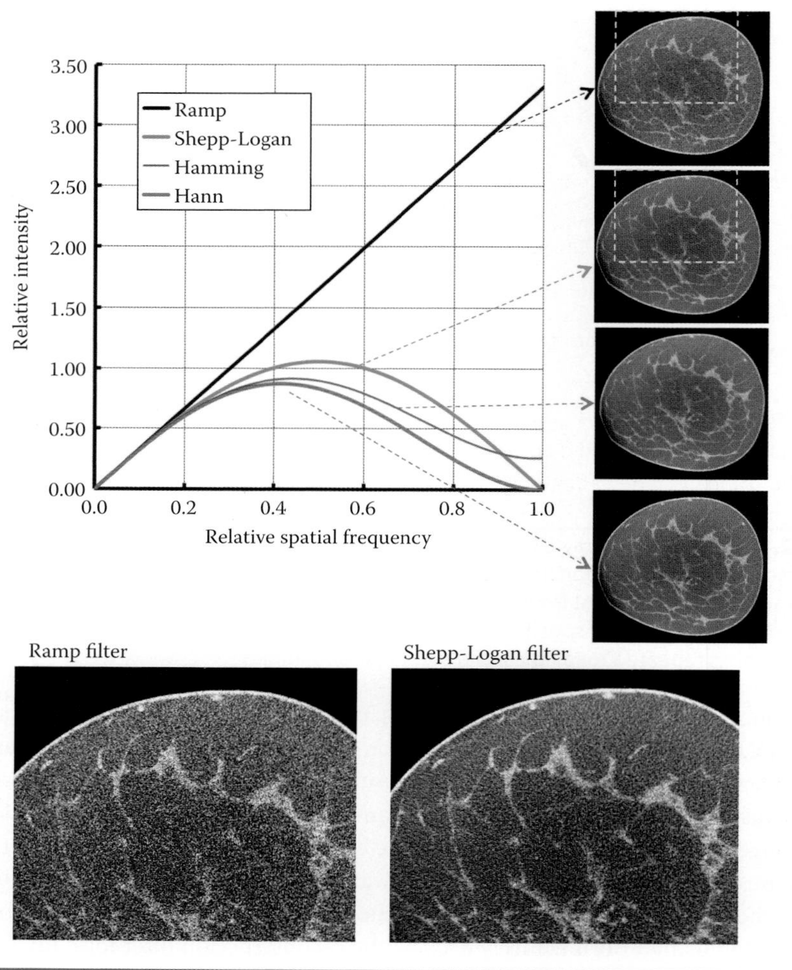

FIGURE 2.18 CT reconstruction filters. The "ramp" filter and three different apodization filters are shown together with example CT images of the breast reconstructed with these filters. A magnified view is shown (bottom) for the comparison between the ramp and Shepp-Logan filter. Notice the trade-off in image resolution and noise.

2.4.3.3 Algebraic Reconstruction Technique

In contrast to FBP, which is a transform-based approach, ART is a reconstruction algorithm that utilizes an iterative approach. Its goal is to incorporate prior known information about the CT scanner and the correlations between projection images acquired at different angles.

ART treats the CT reconstruction process as the solution to a series of linear equations. The effective linear attenuation coefficients for all of the pixels in the CT image are the unknowns to be solved for. The rays measured during CT acquisition are used to set up a system of equations. For example, to reconstruct a 2-D CT image matrix consisting of 512 × 512 pixels, a total of over 200,000 unknowns (the linear attenuation coefficient in each pixel within a circular area bounded by the square matrix) need to be solved. If 512 1-D angular projection views are acquired around the object using a detector with 512 elements, a total of 512 × 512 measurements are made. To solve this large series of equations is a formidable mathematical challenge. Thus, iterative algorithms use numerical methods to obtain reconstructed CT images by converging toward an image estimate that is consistent with the measured projection data.

Iterative reconstruction is typically composed of three major steps for each iteration: forward projection, comparison and adjustment of projection data, and backprojection. Each iteration starts with an estimate of the CT image to be reconstructed. The very first iteration might start with an image of uniform values in all pixels. A set of estimated angular projection images is then generated by forward projection of the estimated CT image. The forward projection models the acquisition of the projection data by integrating along the different ray paths. CT system and projection acquisition–related parameters, such as the x-ray spectrum, x-ray focal spot size and distribution, photon scattering, and so forth, can also be modeled and implemented in this step. The computed angular projections are then compared with the actual measured projections. If the image estimate matches the actual distribution of attenuation coefficients, then the computed forward projections will closely approximate (within noise) the measured projections. However, in the early iterations, this will not be the case. The differences between the computed and measured projection data are used to determine an error matrix, which is used to modify the next estimate of the CT image through a backprojection step. Ideally, the differences between the generated and acquired projections are reduced with each iteration. When these differences are below a predetermined threshold, an acceptable reconstruction of the CT image is achieved, and the algorithm stops (*converges*). A simple block diagram illustrating the concept of iterative reconstruction is given in Figure 2.19.

Iterative reconstruction is computationally intensive since it involves repeated forward and backward projection of the large CT data sets, while FBP only requires a single backprojection. ART therefore requires advanced computational hardware and software to achieve clinically acceptable speed for the commercial CT environment. However, iterative methods such as ART have the potential to produce improved reconstructed images

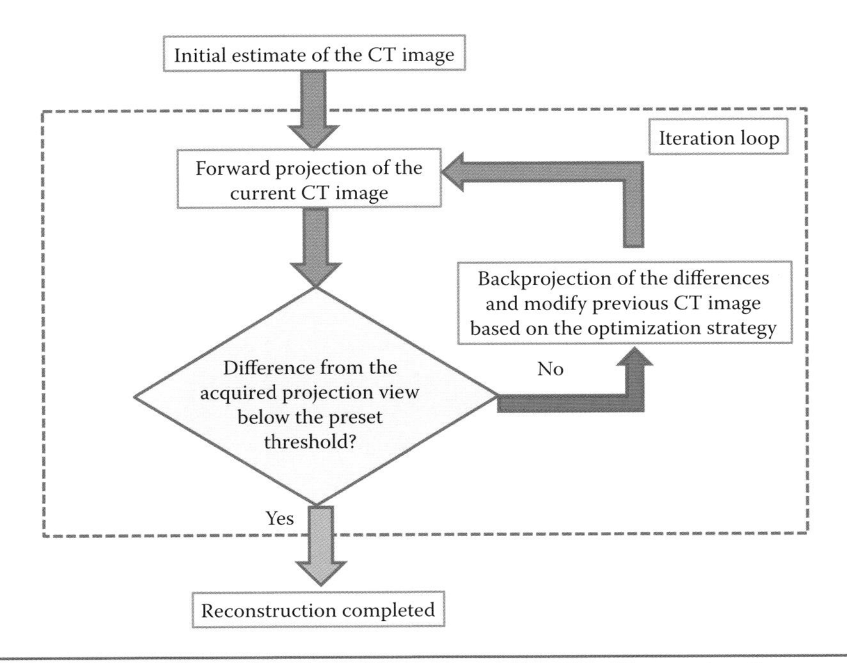

FIGURE 2.19 A block diagram illustrating the concept of iterative CT reconstruction.

because they can incorporate more realistic models of the CT acquisition geometry, as well as account for variations in the statistical quality of the data from one projection element to the next.

2.4.4 Image Artifacts in CT

An image artifact is an anomalous feature on an image that does not exist in reality. In CT images, there are a few characteristic artifacts that occur occasionally, including rings, streaks, and cupping. Figure 2.20 demonstrates examples of these artifacts.

The ring artifact, as its name indicates, appears as one or more concentric rings on a CT image, as shown in Figure 2.20a. It is a very common image artifact for third-generation CT scanners. The cause of a ring artifact is normally related to a defective detector element. Since the signal from one detector element falls along a straight line during backprojection at a given angle, and this line has a fixed distance to the isocenter (center of rotation of the scanner), for a full 360° reconstruction, the superposition of these straight lines back-projected from the same detector element forms an artificial ring in the CT image. Due to its unique appearance and highly localized effect, such a ring artifact can be effectively corrected by data interpolation in the original projection data prior to reconstruction.

Streak artifacts are commonly seen as straight lines across a CT image, as shown in Figure 2.20b. There are different possible causes for this artifact, but a common problem is *beam hardening*. Beam hardening is an effect that is caused by the polyenergetic nature of the x-ray spectrum that is used in CT. As the x-ray beam passes through an object, lower-energy x-rays are preferentially removed from the beam as they are more likely to interact (they have a higher linear attenuation coefficient) than higher-energy x-rays. Thus, as the x-ray beam progresses through the object, the spectrum changes to a higher effective energy (more penetrating), and this is not accounted for in the FBP reconstruction process. The effect is particularly noticeable when the x-ray beam passes through a dense material such as bone or a metal implant. Beam hardening is a fundamental issue with a continuous x-ray spectrum, and partial solutions include *prehardening* of the beam with added filtration or correction approaches during iterative reconstruction, where the effect of changes in the x-ray beam spectrum can be modeled.

The cupping artifact is a phenomenon that refers to reduced CT image intensity in the central region of the object, as shown in Figure 2.20c. Cupping is a gradual and subtle

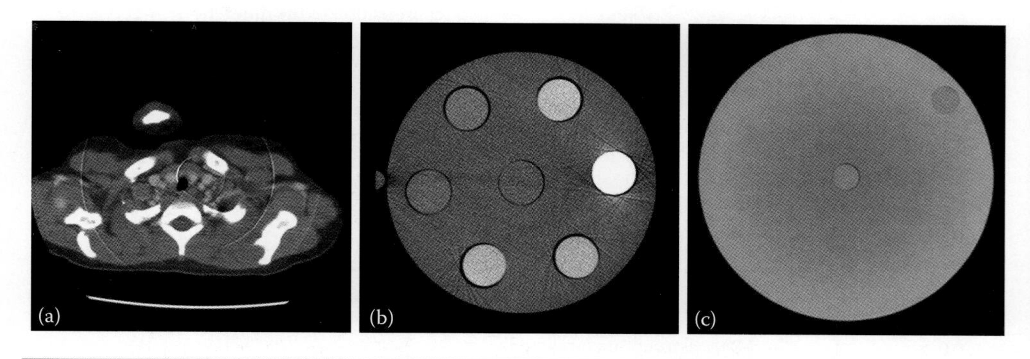

FIGURE 2.20 CT image artifacts. Example images are shown for (a) ring artifact, (b) streak artifact, and (c) cupping artifact.

artifact caused by various factors, including beam hardening and x-ray scattering. The signal from scattered x-rays is more likely to be detected toward the center of an object due to solid angle effects. The inverse Beer's law is no longer accurate in the presence of high scatter levels, because the measured x-ray intensity represents a combination of the primary x-ray signal plus scattered x-rays, thus leading to apparently lower linear attenuation values. This is still an active research topic in the x-ray CT field, and a variety of approaches have been proposed to correct/compensate for the cupping artifact. These include iterative reconstruction approaches in combination with system modeling, image processing approaches based on 2-D/3-D detrending (simply subtract the cupping trend from the images), and physical approaches such as scatter rejection and prehardening of the x-ray spectrum.

2.4.5 Trade-Off between Radiation Dose and Image Quality

Due to the statistical nature of x-ray photon detection (the Poisson distribution described in Section 2.2.3), image noise is improved by the detection of more x-rays; however, this also typically implies a higher radiation dose to the patient. In all practical situations, radiation dose to the patient is the essential limiting factor with regard to x-ray image quality, particularly for the critical metric of contrast-to-noise ratio (CNR). This trade-off is optimized by the selection and optimization of the x-ray spectrum (to generate the highest contrast for the object of interest at a given radiation dose) and the detector efficiency (to make maximum use of the x-ray signal striking the detector). Both projection imaging and CT share the same intrinsic limitation. However, because CT measures the distribution of effective linear attenuation coefficients within a patient, the mean reconstructed HU values in a CT image are independent of the incident x-ray intensity. However, the image noise (i.e., the standard deviation in HU) is strongly affected by the radiation dose level, as shown in Figure 2.21.

Each CT image is reconstructed from hundreds or even thousands of different projection images, and the final noise level is strongly influenced by those projection images with the higher noise levels. Because of this, it is possible to reduce radiation dose while maintaining the image noise level by modulating the x-ray intensity with *automatic exposure control* (*AEC*) throughout the acquisition. The idea is to use fewer

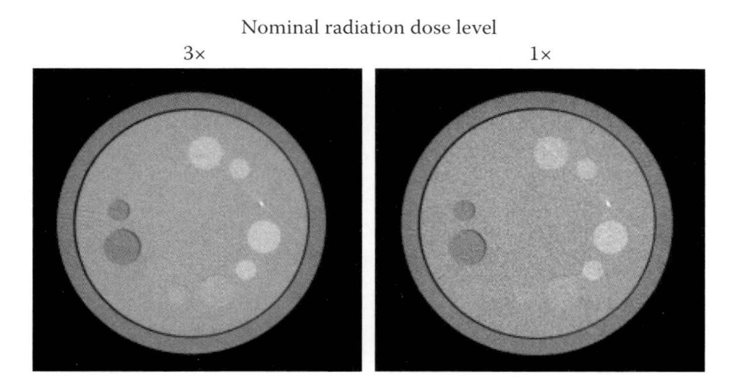

FIGURE 2.21 CT image noise versus radiation dose. Example CT images demonstrate that the image noise level is directly determined by the radiation dose produced by the acquisition. Using three-times-higher radiation dose, the image on the left shows much lower noise.

x-ray photons to acquire projection images from angles where there is less attenuation through the patient. For example, since the axial cross-section of a typical human body is an oval shape, fewer photons can be used to acquire projections parallel to the thinner anterior/posterior axis of the body, while more photons should be used for the projections parallel to the thicker lateral projection. Radiation dose modulation can also be implemented along the inferior/superior axis of a CT scan since the human body is not uniform in attenuation from head to toe; for example, the thoracic anatomy has a significant air component and therefore produces much less attenuation compared to other parts of the body.

This dose modulation approach can be extended to the concept of optimizing the interpatient dose level. Basically, for patients with smaller body sizes, a lower radiation dose is necessary to achieve acceptable image quality compared to larger patients. Approaches to reduce radiation dose with little to no effect on CT image quality have been a major focus of recent innovations in CT and are especially important for dose reduction in pediatric patients.

2.5 APPLICATIONS OF CT AND FUTURE DIRECTIONS

2.5.1 CT Applications

2.5.1.1 Clinical CT Applications

CT not only provides diagnostic information but also can provide guidance for many clinical procedures, including surgery, angiography, and radiation therapy. It also is used widely in clinical research, especially to monitor response to new therapies and interventions. A selection of CT images is shown in Figure 2.22 to demonstrate the wide application of the CT modality.

CT provides anatomical information from virtually all parts of the human body. Head and neck CT is typically ordered to detect tumors, calcifications, hemorrhage, and trauma. Thoracic CT can be used for detecting both acute and chronic changes in the lung parenchyma and for evaluating the function of the heart and great vessels. Cardiac CT can even provide dynamic 3-D (or 4-D) images of a beating heart. CT scans of the abdomen and pelvis are very useful to evaluate acute abdominal pain and to depict the internal anatomy (and function, when contrast agents are used) of important organs, such as the liver, kidneys, and bladder. Musculoskeletal CT provides a very accurate tool to evaluate complex fractures and congenital deformities.

The administration of agents that provide contrast enhancement has been widely utilized in CT for important clinical tasks, especially for cancer detection, diagnosis, and staging. Intravenously administered iodine-based contrast agents typically show strong signals (high absorption) in tumors, which usually develop additional blood vessels (through a process called *angiogenesis*) to support their excessive nutritional requirements. These vessels typically do not have the same circulation performance compared to normal vessels and often are leaky; therefore, the iodine-based contrast agent tends to accumulate in and around these vessels for a longer period. Thus, a properly designed scan protocol can maximize the image signal from these vessels and provide an indirect but very sensitive indication of certain tumors. Another common usage of contrast agents is to introduce x-ray-opaque gases (such as xenon) into the lung for cancer screening, diagnosis, and evaluation. This greatly increases the visibility of the lung parenchyma. Orally administered

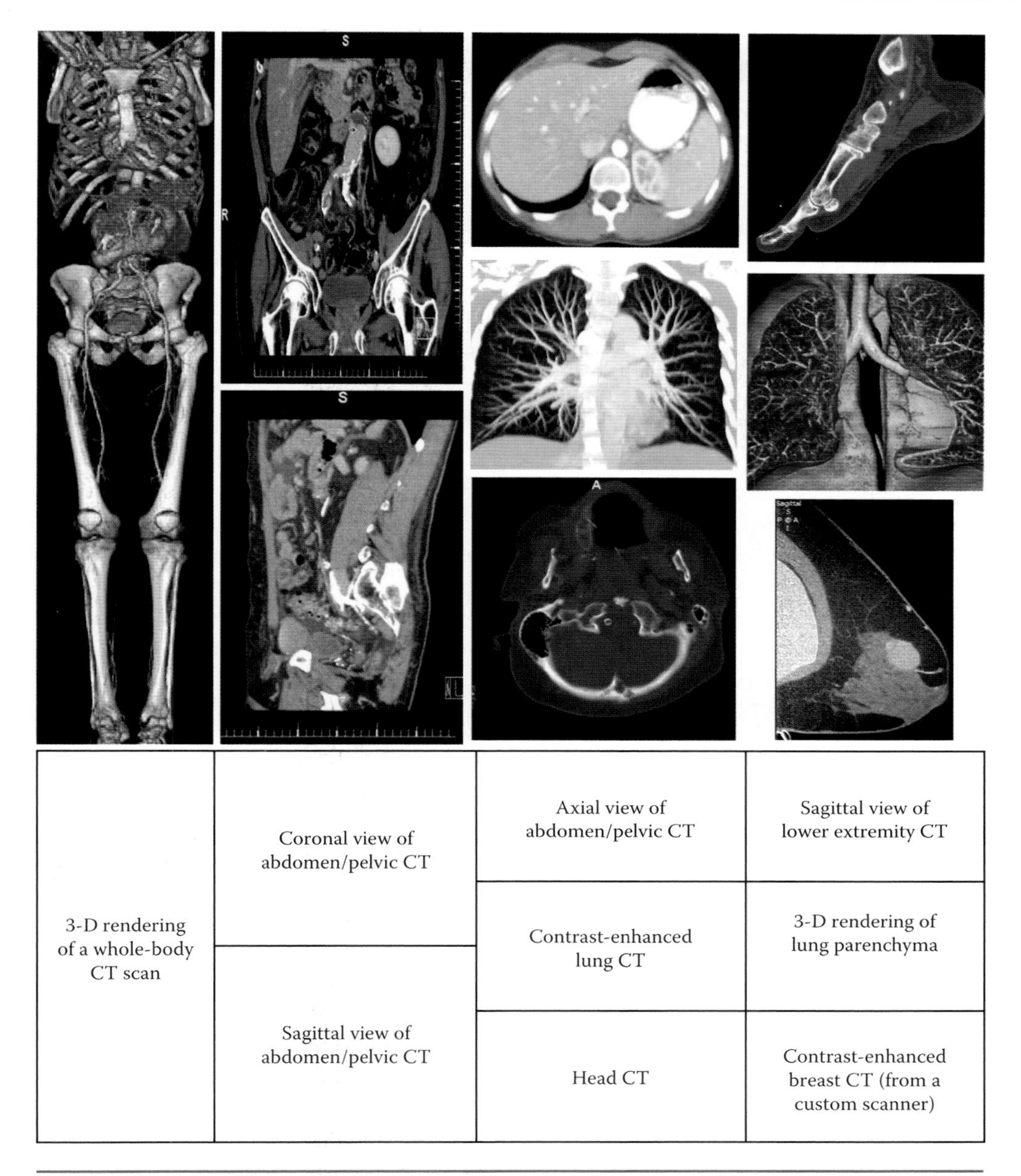

3-D rendering of a whole-body CT scan	Coronal view of abdomen/pelvic CT	Axial view of abdomen/pelvic CT	Sagittal view of lower extremity CT
		Contrast-enhanced lung CT	3-D rendering of lung parenchyma
	Sagittal view of abdomen/pelvic CT	Head CT	Contrast-enhanced breast CT (from a custom scanner)

FIGURE 2.22 Examples of CT images. A collection of CT images is shown to demonstrate the versatility of CT and its applications. The table describes what each image shows.

contrast agents based on another high-density material, barium, can be used to examine the digestive tract.

Besides being a primary diagnostic modality, CT is also utilized to guide other clinical procedures. CT angiography (CTA) utilizes contrast enhancement (mainly from iodine) to provide 3-D images of the vascular network. C-arm-based CT scanners provide intraoperative guidance for surgery. In the radiation therapy field, CT imaging of the patient provides an anatomical map of the tumor and surrounding tissues, which becomes the basis for treatment planning.

CT scanners also have been integrated with other imaging systems, notably with PET and SPECT. The CT component in these hybrid PET/CT and SPECT/CT scanners provides high-resolution anatomic context for areas of abnormal radiotracer accumulation observed by PET or SPECT. This can improve diagnostic accuracy and also be used for anatomically driven analysis of PET/SPECT data.

2.5.1.2 Micro-CT

Micro-CT is a CT technique primarily used for small animal preclinical research (mostly mice) and biospecimen imaging, designed to provide high-resolution 3-D anatomical information [4,9]. Micro-CT was first developed in the early 1980s, and its use was catalyzed by improved flat panel x-ray detectors and the so-called FDK (named after its three inventors, Feldkamp, Davis, and Kress) cone-beam reconstruction algorithm [10]. The spatial resolution of micro-CT typically ranges from ~1 to ~100 μm^3. The important role of micro-CT in life science research is evident from the increasing number of scientific journal articles reporting its use over a very broad range of applications. As illustrated by the examples in Figure 2.23, micro-CT can provide a robust tool for many biomedical tasks, including (1) disease detection and monitoring (especially for longitudinal studies) in animal models of human disease (Figure 2.23a); (2) phenotypic characterization of anatomy and material composition (Figure 2.23b,c,e); and (3) noninvasive evaluation of experimentally developed imaging and therapeutic agents before their clinical use in humans (Figure 2.23d). *In vivo* micro-CT also is often integrated in hybrid systems together with molecular imaging techniques such as PET, SPECT, and optical imaging.

Although micro-CT uses the same principles described in earlier sections for clinical CT scanners, the major components of a micro-CT system are typically distinctly different, reflecting the different requirements [5]. Micro-CT typically utilizes x-ray sources with much smaller focal spots, varying from 1 to ~100 μm, in order to achieve the required spatial resolution. Depending on the application, these sources may be very compact (e.g., micro-focus x-ray tubes) or also can be very large and sophisticated (such as the use of synchrotrons to provide monoenergetic and parallel-beam x-rays). Flat panel detectors (typically using either CCD or CMOS sensors) with detector element sizes ranging from ~$(5\ \mu m)^2$ to ~$(200\ \mu m)^2$ are almost universally chosen for micro-CT systems due to the combined advantages of high spatial resolution and relatively low cost. Many micro-CT systems for *in vivo* imaging of live animals utilize designs similar to a clinical CT scanner with a stationary animal bed and rotating scanner gantry. For *ex vivo* imaging of biospecimens, the design of a stationary gantry with a rotating sample platform is more common.

One important issue related to live animal imaging is the much faster breathing and cardiac rates compared to humans. For mice, the typical breathing rate is around 100 breaths per minute, and the heart rate can be up to 600 beats per minute. Thus, many micro-CT scanners for live animal imaging (which is performed under anesthesia) are equipped with respiratory or cardiac gating systems in order to synchronize the projection data acquisition with the animal's physiologic motion and to reduce the associated image blurring. With regard to the reconstruction algorithm, the FDK-based cone-beam reconstruction (a 3-D extension of the FBP algorithm) is most commonly used for micro-CT, while iterative reconstruction methods also are being actively studied. The FDK algorithm is theoretically only an approximate solution of the reconstruction problem and thus has a fundamental limitation of providing a nonexact reconstruction (also called cone-beam artifact). These artifacts start to become more noticeable with increasing cone angles beyond 10°.

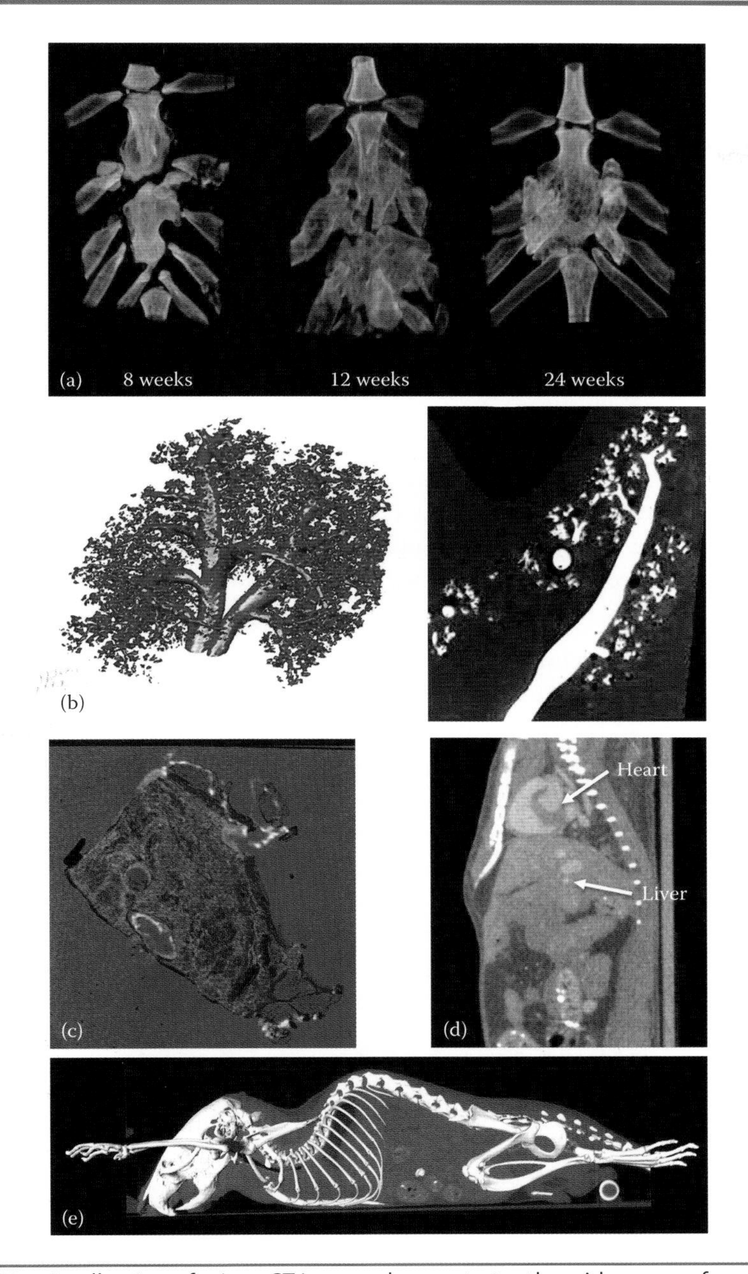

FIGURE 2.23 A collection of micro-CT images demonstrates the wide range of applications for high-resolution CT scanning. (a) Longitudinal micro-CT of rabbit sternum showing the healing process after fracture. (b) 3-D rendering and cross-sectional images of micro-CT data of a rat lung cast. By measuring the airway diameters, the effect of environmental factors (such as air pollution) can be studied in animal models. (c) Micro-CT image of a breast cancer tissue specimen in registration with pathology slide. (d) Contrast-enhanced micro-CT image of mouse abdomen using a preclinical vascular contrast agent that improves delineation of organs and soft tissues. (e) 3-D rendered *in vivo* micro-CT showing high-resolution whole-body imaging of the mouse. This can be useful in phenotyping genetically engineered mouse models. ([d, e] Courtesy of Jennifer Fung, University of California, Davis.)

2.5.2 New Directions in CT

2.5.2.1 Dual-Source CT

When the maximum power of a single x-ray tube becomes the primary limitation in further reductions of the total scan time, the concept of multisource CT provides a potential solution by simultaneously acquiring projection images with multiple x-ray sources. *Dual-source CT* (DSCT) scanners were first introduced in 2005 and rapidly gained market penetration as the primary choice for cardiac CT. For these systems, two sets of x-ray sources and detectors are installed approximately 90° apart with synchronized rotation and acquisition. Besides the doubled x-ray power and reduced gantry rotation angle and scan time (by a factor of 2) for general imaging, the key benefit of DSCT for cardiac CT and coronary CTA is the improved temporal resolution. With the scan time down to 75 ms per image and total acquisition completed within one single heartbeat, DSCT provides unprecedented image quality and greatly reduced radiation dose for cardiac CT. Since the two systems can be operated at different x-ray energy (kV), current (mA), and filtration, *dual-energy CT* (DECT) is another built-in capability of DSCT. DECT utilizes similar concepts to dual-energy radiography (see Section 2.3.4.2) with the aim of improving the contrast between different types of tissues and enhancing the visibility of structures of interest.

2.5.2.2 Photon Counting CT

The spectral distribution of the x-ray spectrum changes while passing through the body (i.e., beam hardening). Even though DECT improves the differentiation between different tissue types, it still is limited by the polyenergetic nature of the x-ray spectra. The development of energy-resolving photon counting detectors opens up new and exciting possibilities for CT applications. Multiple energy windows (currently around five) can be set up to resolve x-ray photon energies and measure attenuation values in each energy window. This information can be used to separate more than two types of tissues, with potentially reduced radiation dose. Compared to the energy-integrating detectors discussed previously, the challenge of photon counting systems is that each detected x-ray photon must be individually recorded and read out; thus, the x-ray flux that can be handled is currently limited. While this challenge has to be resolved before photon counting CT can be implemented for routine clinical use, impressive technological progress toward such detectors is being made, and they offer the prospect for significant dose reduction.

REFERENCES

1. Bushberg, J. T., J. A. Seibert, E. M. Leidholdt, and J. M. Boone. 2012. *The Essential Physics of Medical Imaging*. Wolters Kluwer Health/Lippincott Williams & Wilkins, Philadelphia, PA.
2. Poranski, C. F., E. C. Greenawald, and Y. S. Ham. 1996. X-ray backscatter tomography: NDT potential and limitations. *Mater Sci Forum* 210–212:211–218.
3. Thrall, D. E. 1994. *Textbook of Veterinary Diagnostic Radiology*. W.B. Saunders, Philadelphia, PA.
4. Badea, C. T., M. Drangova, D. W. Holdsworth, and G. A. Johnson. 2008. In vivo small-animal imaging using micro-CT and digital subtraction angiography. *Phys Med Biol* 53:R319–R350.
5. Holdsworth, D. W., M. M. Thornton, D. Drost, P. H. Watson, L. J. Fraher, and A. B. Hodsman. 2000. Rapid small-animal dual-energy x-ray absorptiometry using digital radiography. *Journal of Bone and Mineral Research* 15(12):2451–2457.
6. Peters, T. M., J. C. Williams, and J. H. T. Bates. 1998. *The Fourier Transform in Biomedical Engineering*. Birkhauser, Boston.
7. Beutel, J. 2000. *Handbook of Medical Imaging*. SPIE Press, Bellingham, WA.

8. Kak, A. C., and M. Slaney. 2001. *Principles of Computerized Tomographic Imaging*. Society for Industrial and Applied Mathematics, Philadelphia, PA.
9. Ritman, E. L. 2011. Current status of developments and applications of micro-CT. *Annual Review of Biomedical Engineering* 13:531–552.
10. Feldkamp, L. A., L. C. Davis, and J. W. Kress. 1984. Practical cone-beam algorithm. *J Opt Soc Am A* 1:612–619.

FURTHER READINGS

The physics of x-ray and CT imaging is discussed extensively in the following:

Bushberg, J. T., J. A. Seibert, A. M. Leidholdt, and J. M. Boone. 2012. *The Essential Physics of Medical Imaging*, 3rd ed. Wolters Kluwer Health/Lippincott Williams & Wilkins, Philadelphia, PA.

Topics related to CT and its applications are given extensive treatment in the following:

Kalender, W. A. 2011. *Computed Tomography: Fundamentals, System Technology, Image Quality, Applications*, 3rd ed. Publicis, Erlangen, Germany.

3

Magnetic Resonance Imaging

Jeff R. Anderson and Joel R. Garbow

3.1 INTRODUCTION

Magnetic resonance imaging, or MRI, is a technique that combines the use of magnetic fields and nonionizing, radio-frequency (RF) radiation to create high-resolution images. It does this by detecting and spatially encoding the nuclear magnetic resonance (NMR) signals from water within organs and tissues. MRI can be used to create images of a multitude of biological organisms, including humans (clinical studies) and a wide variety of laboratory animals (preclinical studies). The detected NMR signals arise from the hydrogen atoms of water molecules in the body following stimulation by RF electromagnetic fields. This is in contrast to all other major imaging modalities in which the detected signal is either generated outside of and subsequently transmitted through the tissues of the organism (e.g., x-ray, CT, and ultrasound) or is emitted by a molecule that is injected into the organism (e.g., PET and SPECT).

One of the most powerful aspects of MRI is that the observed contrast in a magnetic resonance (MR) image can be readily manipulated through the choice of experimental parameters. Selection of the type, order, strength, and duration of applied magnetic fields and RF irradiation will alter the contrast observed in the final image. The detailed pattern of RF and magnetic fields employed is called the "pulse sequence" and forms the experimental basis for every MRI scan. A large number of pulse sequences exist, and new sequences continue to be developed. Thus, versatility, in terms of pulse sequences, parameters, and image contrast, is one defining hallmark of MRI.

As noted, MRI is the only major imaging modality in which both signal and contrast are intrinsic (i.e., generated *by* the tissue itself). A consequence of the intrinsic nature of MR signals is that the availability of signal is independent of physical position or depth, so that if a homogeneous volume coil is used for excitation and detection (see Section 3.6), homogeneous tissue generates uniform signal intensity. Another defining characteristic of MRI relates to the experimental frame of reference of the acquired images. Unlike some imaging modalities, the experimental frame of reference for MR images need not correspond with the x, y, and z coordinates of the imaging hardware. The frame of reference, defined by the applied magnetic field gradients (see below), can be aligned with the patient (or his/her physiology), rather than requiring the subject to adapt to the instrument.

These advantages, coupled with the fact that MRI does not use ionizing radiation, are major strengths of the technique. MRI is, of course, not without its challenges, which include its relatively low sensitivity compared with optical or nuclear medicine imaging methods and its high costs. Further, the requirement of large magnetic fields means that the technology is not accessible to patients with ferromagnetic-metal implants or medical devices that do not function properly in magnetic fields. Nonetheless, MRI remains an exceedingly powerful and versatile imaging modality that is the method of choice in many research and clinical settings.

This chapter provides an overview of the concepts necessary to understand the theory and application of MRI. Section 3.2 conceptually introduces the source of signal for MRI, namely, NMR. Section 3.3 describes how NMR signals can be spatially encoded and detected to generate an image, leading into a detailed discussion of the physical phenomena (i.e., tissue characteristics) that can be encoded with MRI. Section 3.4 discusses basic MR imaging techniques, while Section 3.5 briefly introduces more advanced methodology. The chapter concludes with a practical section on MRI hardware (Section 3.6) and a brief discussion of future directions in MRI (Section 3.7).

3.2 NUCLEAR MAGNETIC RESONANCE

MRI is built upon the principles of NMR. First described in 1945 by Felix Bloch and Edward Purcell (winners of the 1952 Nobel Prize in Physics), NMR is an analytical technique used extensively by synthetic, medicinal, and protein chemists for a wide variety of applications, including identification of small molecules, assessment of product purity, and determination of protein structures. Both NMR and MRI detect stimulated, "resonant" signals from the nuclei of "NMR-active" atoms (e.g., ^{1}H, ^{2}H, ^{13}C, ^{19}F, and ^{31}P). Nuclei with either (1) odd nuclear mass or (2) even nuclear mass and an odd number of neutrons have the inherent property of nonzero spin and are, thus, detectable by NMR/MRI. Like the energy levels of the electron, nuclear-spin energy levels can only assume certain well-defined, discrete values (i.e., they are quantized). Transitions between levels require an exchange of energy equal to the exact energy difference between the two energy levels, the so-called resonant energy for that transition. The frequency of electromagnetic radiation corresponding to that energy, given by the Plank–Einstein equation,

$$E = h\nu \tag{3.1}$$

(where E is energy, h is Plank's constant, and ν is frequency), is called the *resonant frequency*.

While nuclei do not physically spin, the mathematical relationships that describe nuclear spin are similar to those of a spinning top. To better understand nuclear spin, it is, therefore, worth discussing the model of a spinning top or gyroscope.

3.2.1 Classical Angular Momentum—The Top and the Gyroscope

The ability of a spinning top to stand on a point is a manifestation of angular momentum. The angular momentum of a rigid body, a vector having both magnitude and direction, is equal to the product of its moment of inertia and its rotational velocity $\vec{L} = \overleftrightarrow{I}\vec{\omega}$ (here, the single arrow denotes a vector and the double arrow a tensor). The greater the angular momentum of a body, the longer that body will spin.

An interesting effect is observed when a spinning top begins to tip over. In response to the torque applied to the system by gravity (the twisting force causing the top to tip over),

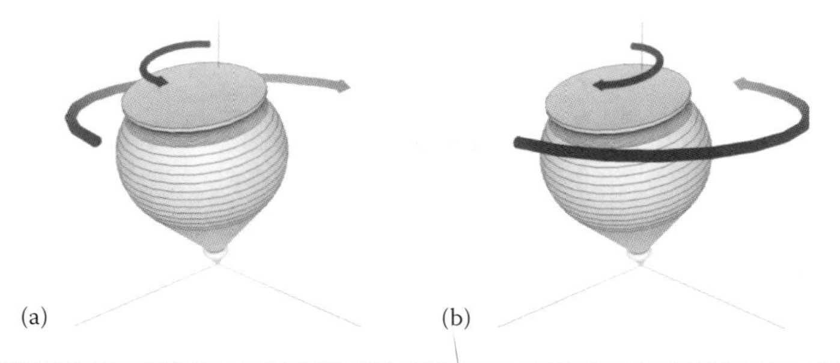

FIGURE 3.1 Precession of a toy top. The directions of rotation (small arrows) and precession (large arrows) are shown. (a) Top rotating counterclockwise. (b) Top rotating clockwise.

the spinning top slowly begins to rotate around the vertical axis (Figure 3.1). This rotation is called precession, and it occurs because angular momentum, like linear momentum, must be conserved. The direction of precession is determined by the directional relationship between the external force (in this case, gravity), the resultant torque, and the angular momentum of the system. One interesting, related point is that the direction of the precession is always opposite that of the spinning of the top. In other words, if the top is spinning counterclockwise, with its vector angular momentum pointing upward, the top will precess clockwise (Figure 3.1a), and vice versa (Figure 3.1b).

Precession can be even more clearly observed with a gyroscope in a gimbal. The gimbal allows one to manually apply torque to the system without slowing the spinning of the gyroscope. Consider the gyroscope shown in Figure 3.2a that is on its side (the extreme case of a top that is tipped over). The direction of the angular momentum is shown as a red arrow. A force exerted downward on the side of the gimbal (yellow arrow, Figure 3.2b) applies a torque to the system as shown (blue arrow, Figure 3.2b). In response, the gyroscope precesses to bring the angular momentum into alignment with the applied torque. Here, the applied force and resulting torque on the gyroscope are coincident with the force and torque applied on the spinning top by gravity; thus, the precession is also similar. A second example of precession is observed if a force is applied perpendicular to the downward force, that is, a force that pushes sideways on the gimbal housing (yellow arrow, Figure 3.2c). This force results in a torque perpendicular to that in the first example, and as a result, the gyroscope begins to tip upward.

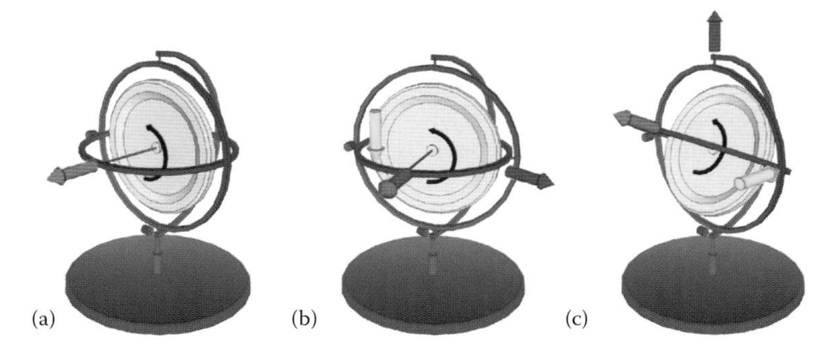

FIGURE 3.2 A gyroscope in a gimbal. The red arrows show the direction of the angular momentum, the yellow arrows the direction of the external force, and the blue arrows the direction of the external torque. The panels illustrate the motion of the gyroscope and gimbal. (a) Initial position of the gyroscope and gimbal. (b) In response to downward force (torque pointing to the side). (c) In response to sideways force (torque pointing up).

3.2.2 Quantum Mechanical Angular Momentum—Nuclear Spin

In many instances, nuclear spins behave analogously to the top or gyroscope described in Section 3.2.1. However, since nuclear spin is truly a quantum mechanical phenomenon, several distinct differences exist. The first major difference between nuclear spin and a spinning top is that nuclear spin is an intrinsic property of the nucleus. No force is required to induce nuclear spin, and no force can destroy it.

A second major difference is the existence of quantized spin energy levels. Only certain, discrete energy levels are available for nuclear spin. In the absence of a magnetic field, these energy levels are degenerate, such that spins in different levels are indistinguishable. In this state, the nuclear spin is analogous to a spinning gyroscope or top without any external force(s). Thus, in the absence of a magnetic field, nuclear spins do not precess.

Let us consider the case of ^{1}H, a spin-1/2 nucleus that has two discrete energy levels. Placing such a nucleus in a magnetic field causes these two energy levels to split, with one being higher in energy than the degenerate levels in the absence of the magnetic field and one being lower. An ensemble of nuclear spins will populate the energy levels according to a Boltzmann distribution,

$$N_{\text{upper}}/N_{\text{lower}} = e^{-\Delta E/(k \cdot T)}, \tag{3.2}$$

with spins preferentially in the lower energy level, aligned parallel to the magnetic field. In this equation, N_{upper} and N_{lower} are the number of spins in the upper and lower energy levels, respectively, ΔE is the energy difference between the two levels, T is temperature, and k is the Boltzmann constant (1.381×10^{-23} J/K). The difference between the numbers of spins aligned parallel and antiparallel to the magnetic field describes the *net magnetization* of the sample. It is this net magnetization that is available as signal in the NMR experiment. (It is important to reiterate at this point that it is the nuclear spins that orient parallel or antiparallel with the magnetic field. The nuclei, atoms, and molecules are otherwise completely unaffected by being placed in the field.)

For a typical clinical MRI field strength of 3 tesla (T) (for reference, the earth's magnetic field strength is approximately 25 to 65 μT), the splitting of the nuclear-spin energy levels (parallel vs. antiparallel to the field) for ^{1}H is 8.46×10^{-26} J. In 10 zeptoliters of water (10×10^{-21} L), there are about 666,667 ^{1}H nuclei. Using the Boltzmann distribution equation, at 35°C and 3 T, we expect that 333,337 spins will be aligned parallel to the magnetic field and 333,330 spins will be aligned antiparallel to the magnetic field. The difference in spin population between the two levels, the net magnetization available as signal in the NMR experiment, represents ~0.001% of the total spins in the sample. Thus, one can easily appreciate why NMR (and MRI) are relatively insensitive compared to other imaging modalities (e.g., nuclear medicine [Chapter 4] and optical imaging [Chapter 5]).

The net magnetization can be represented graphically as a vector, $\vec{M}$, the ensemble of spins that comprise the net magnetization processes at the resonant frequency of the spin system, designated as the Larmor frequency (128 MHz for ^{1}H in a 3 T magnet). The resonant frequency (and ΔE) is a linear function of magnetic field strength: $\nu_0 = -(\gamma B_0/2\pi)$ or $\omega_0 = -\gamma B_0$ (ν_0 is frequency, cycles/s, and ω_0 is angular frequency, radians/s). The nucleus-specific proportionality constant, γ, is called the *gyromagnetic (or magnetogyric) ratio.* Table 3.1 lists the gyromagnetic ratio and natural isotopic abundance values for several common NMR-active nuclei. The table shows why ^{1}H, which is ubiquitous in biological systems, is the most commonly studied NMR/MRI nucleus. ^{1}H has both a high natural abundance and a large gyromagnetic ratio.

TABLE 3.1
Natural Isotopic Abundance and Gyromagnetic Ratio Values of Several Common NMR-Active Nuclei

Nucleus	Natural Abundance (%)	$\gamma/10^6$ (rad/s/T)
^1H	99.9885	267.522128
^2H	0.0115	41.0662791
^3He	0.000137	−203.801587
^{13}C	1.07	67.28284
^{19}F	100	251.18148
^{31}P	100	108.394

Source: Winter, M., WebElements, 2012.

At equilibrium, the net magnetization is aligned with the direction of B_0, which, by convention, is designated as the positive z-axis. The z-component of the magnetization is labeled M_z; the equilibrium value of M_z is often written $M_z(0)$ (i.e., at time 0), or simply, M_0. Note, at equilibrium, $\vec{M}$ has no net xy-component.

3.2.3 NMR Experiment

In many ways, $\vec{M}$ behaves like a classic angular momentum vector. The basic NMR experiment, shown schematically in Figure 3.3, can be illustrated with this simple vector

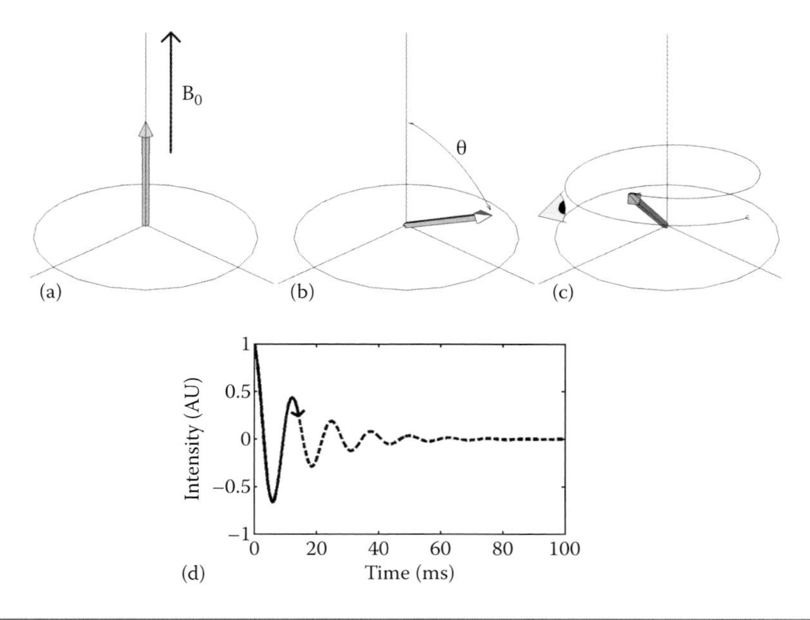

FIGURE 3.3 Magnetization vector description of the basic NMR experiment. (a) Magnetization vector, at equilibrium, pointed in the direction of the static magnetic field. (b) In response to an external force, applied by an RF pulse, the magnetization vector nutates to the position shown. (c) The magnetization vector precesses at the Larmor frequency and (d) is detected (from the perspective shown in panel [c]) as a sinusoidal signal that decays with time due to relaxation. The intensity of the signal corresponds to the projection of the magnetization vector on the xy-plane.

representation. For a sample placed in a large static magnetic field, the NMR experiment begins by applying a second, oscillatory magnetic field (electromagnetic radiation) to the sample. This oscillatory field is generated via a current running through a loop (or loops) of wire known as the *RF coil*. When the direction of this oscillatory field is perpendicular to B_0, it exerts a force on $\vec{M}$, which results in precession (or "nutation") of the bulk magnetization. From an energy-level perspective, the application of electromagnetic radiation induces transitions of spins from one energy level to another, thereby altering spin energy-level populations. In order to affect energy-level transitions, the energy of the applied electromagnetic radiation must match the resonant frequency of the spin system. For nuclear spins, energy-level differences typically correspond to radio frequencies. In practice, broadband RF radiation, distributed about the resonant frequency, is applied to the sample.

The nutation rate of $\vec{M}$ is directly proportional to the intensity of the applied RF radiation. Within certain broad limits, as long as the broadband energy is applied to the system, the magnetization continues to nutate. The extent of nutation of $\vec{M}$ is referred to as the flip angle (θ) and is measured in units of degrees, or radians, from the z-axis. It is apparent that some flip angles constitute special cases. For example, a flip angle of $\pi/2$ (90°) nutates the magnetization completely into the xy-plane, and a flip angle of π (180°) inverts the magnetization, while a flip angle of 2π (360°) returns $\vec{M}$ to its equilibrium state.

After nutation to an arbitrary angle, if $\vec{M}$ is not coincident with the z-axis, its projection onto the xy-plane precesses in that plane at the Larmor frequency (Figure 3.3c). This precession can be detected as an induced, sinusoidally oscillating current in an RF coil/antenna (either the same RF coil used for excitation or a different one, Figure 3.3d). With time, the magnetization, $\vec{M}$, will return to its equilibrium state—that is, net magnetization along the z-axis and no net magnetization in the xy-plane. Thus, the detected sinusoidal current decays with time. These damped, sinusoidal time-domain data are referred to as a *free induction decay* (FID). Often, the precessional frequencies are the quantities of interest in NMR and MRI experiments. These frequencies are obtained by performing a Fourier transform of the time-domain FID to generate a frequency spectrum. The Fourier transform can resolve multiple frequencies in time-domain data, as illustrated in the Section 3.3 on MRI.

3.2.4 MR Relaxation Mechanisms

The processes by which excited magnetization returns to equilibrium are referred to as relaxation. In magnetic resonance, two broad categories of relaxation exist, longitudinal and transverse. *Longitudinal relaxation* describes the return of the z-component of $\vec{M}$ to M_0 and is quantified by the time constant T_1 (or alternatively, by its reciprocal, the rate constant R_1). Transverse relaxation describes the loss of phase coherence of the xy-component of $\vec{M}$ in the xy-plane. *Transverse relaxation* can arise via two different mechanisms. First, it can result from spin–spin interactions within the sample. This relaxation is an inherent property of the sample and is quantified by the time (rate) constant T_2 (R_2). Second, it can result from magnetic susceptibility effects (distortions in the local magnetic field surrounding a collection of nuclear spins that lead to dephasing of magnetization and loss of signal) and imperfections in the static magnetic field. This relaxation is quantified by the time (rate) constant T_2' (R_2').

3.2.4.1 Longitudinal (T_1) Relaxation

Recall that nuclei, in terms of their physical size, are much smaller than atoms, which are, in turn, smaller than the molecules they form. On both the atomic and the molecular level, energy is constantly flowing to and from the surrounding world. Though a glass of water appears motionless on a macroscopic scale, on atomic and molecular scales, the situation is quite different. Molecules, diffusing randomly through space (Brownian motion), collide with one another and are attracted to/repelled from each other by intermolecular forces, such as hydrogen bonding. Molecules can change their conformations, and within molecules, atoms move as bond lengths "breathe" and bond angles scissor.

Conformational changes, translational and rotational motions, and interactions between molecules are all possible energy sources for nuclear-spin transitions (i.e., transitions from one spin state to another). In solution, these transitions are particularly facilitated by Brownian motion. As spin energy exchange occurs, the equilibrium Boltzmann distribution of spin populations is restored, a process referred to as longitudinal relaxation. We note that longitudinal relaxation is sometimes referred to as *spin-lattice relaxation*, a reference to solid materials in which, in lieu of Brownian motion, exchange of energy with the crystal lattice provides the relaxation mechanism.

T_1 relaxation of $\vec{M}$ after its inversion by a π pulse is shown schematically in Figure 3.4. The inverted vector initially shrinks in magnitude toward zero and then grows along the positive z-direction until it reaches M_0. The return to equilibrium is often a monoexponential process, describable as

$$M_z(t) = A - B \cdot e^{-t/T_1}, \tag{3.3}$$

where A and B are constants that depend on (1) the state of $\vec{M}$ before its initial excitation and (2) the flip angle of the RF excitation. For the special cases of relaxation following either a π or a $\pi/2$ pulse applied to a sample at equilibrium, the resulting equations are

$$M_z(t) = M_0(1 - 2 \cdot e^{-t/T_1}) \text{ and } M_z(t) = M_0(1 - e^{-t/T_1}), \tag{3.4}$$

respectively.

In solution, T_1 values for water are typically one to several seconds, while in biological samples, T_1 is typically on the order of hundreds of milliseconds. It should be noted that T_1 is dependent on both magnetic field strength (B_0) and temperature.

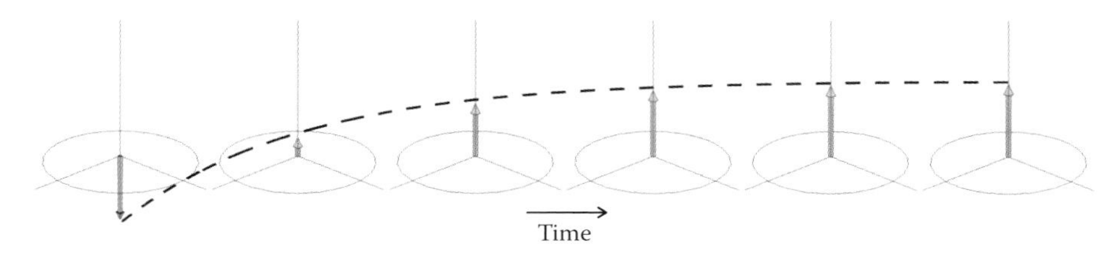

FIGURE 3.4 Schematic representation of the longitudinal relaxation of $\vec{M}$ following a π pulse that inverts the magnetization.

3.2.4.2 Transverse (T_2) Relaxation

The random motion of molecules in solution also facilitates transverse or spin–spin relaxation. This random molecular motion slightly alters the local magnetic fields within a sample. As a result, some spins precess at a frequency slightly above the Larmor frequency and others at a frequency slightly below the Larmor frequency. This distribution of precession frequencies causes the transverse magnetization to dephase or lose phase coherence, as it spreads out uniformly in the *xy*-plane.

T_2 relaxation and its mechanism can be visualized by considering a number of individual spin packets. Once nutated into the *xy*-plane, these spin packets precess at or near the Larmor frequency. Viewed in a rotating frame of reference, a Cartesian coordinate system that rotates about the *z*-axis at the Larmor frequency, on-resonance magnetization can be represented as a stationary vector in the *xy*-plane. (All subsequent magnetization vector figures in this chapter are described in this simplified frame of reference.)

Figure 3.5 shows magnetization vectors from five different spin packets. Immediately following a $\pi/2$ pulse, all the spin packets are in phase and precess about B_0 at the Larmor frequency. However, as the spin packets sample different local magnetic fields, some precess faster than the Larmor frequency and gain phase, and others precess slower than the Larmor frequency and lose phase. Eventually, the spin packets become uniformly distributed within the *xy*-plane. The process can be described mathematically as

$$M_{xy}(t) = A \cdot e^{-t/T_2} \tag{3.5}$$

and results in a damping of the FID signal. In this equation, the value A is dependent upon the initial condition of the experiment, that is, the state of $\vec{M}$ before its initial excitation, and the flip angle of the RF excitation. For the special case of a $\pi/2$ pulse applied to a sample at equilibrium, the equation reduces to the following equation.

$$M_{xy}(t) = M_0 \cdot e^{-t/T_2}. \tag{3.6}$$

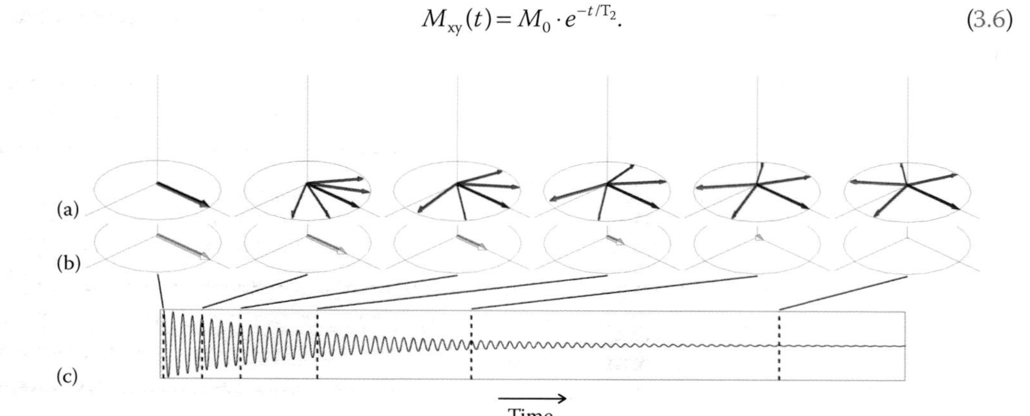

FIGURE 3.5 Schematic representation of the transverse relaxation of $\vec{M}$ following a $\pi/2$ pulse. (a) The transverse magnetization is assumed to be composed of five (arbitrary) spin packets. The behavior of the magnetization is displayed in a frame of reference rotating at the Larmor frequency, the so-called rotating frame. In this reference frame, the spin packet processing at precisely the Larmor frequency appears stationary (the center, black spin packet). The red and blue spin packets are precessing at frequencies slightly higher and lower than the Larmor frequency, respectively. (b) The observable transverse magnetization is the vector sum of the individual, component spin packets, corresponding to (c) the amplitude in the FID.

TABLE 3.2
T_1 and T_2 Values of Various Tissues at 1.5 and 3.0 T

	T_1 (ms)		T_2 (ms)	
Tissue	**1.5 T**	**3.0 T**	**1.5 T**	**3.0 T**
White matter (brain)	884 ± 50	1084 ± 45	72 ± 4	69 ± 3
Gray matter (brain)	1124 ± 50	1820 ± 114	95 ± 8	99 ± 7
Muscle	1008 ± 20	1412 ± 13	44 ± 6	50 ± 4
Fat[a]	288 ± 8	371 ± 8	165 ± 6	133 ± 4
Liver	576 ± 30	812 ± 64	46 ± 6	42 ± 3
Kidney	690 ± 30	1194 ± 27	55 ± 3	56 ± 4
Blood	1441 ± 120	1932 ± 85	290 ± 30	275 ± 50

Source: Stanisz, G.J. et al., *Magn. Reson. Med.*, 54, 2005.
[a] Gold, G.E. et al., *Am. J. Roentgenol*, 183, 2004.

T_2 is always less than or equal to T_1. In solution, T_2 values are often similar to T_1 values, being typically one to several seconds. In biological samples, T_2 is typically on the order of tens to hundreds of milliseconds. Table 3.2 shows T_1 and T_2 values for various tissues measured at 1.5 and 3.0 T. As illustrated in this table, T_1 has a significant magnetic field dependence, while T_2 is only weakly dependent on field. Both T_1 and T_2 are temperature dependent.

3.2.4.3 Apparent T_2 (T_2^*) Relaxation

In practice, the FID signal usually decays more rapidly than T_2 relaxation would predict. This decay is caused by magnetic susceptibilities and part-per-million-level inhomogeneities in the static magnetic field, B_0. Thus, on a macroscopic level, spin packets in a sample can experience magnetic field fluctuations due to an inhomogeneous B_0 that are comparable to those that generate true T_2 relaxation. The decay caused by these inhomogeneities is described by the time (rate) constant T_2' (R_2').

The experimentally measured transverse relaxation, which is a combination of true T_2 relaxation and this pseudo T_2 relaxation, is referred to as apparent T_2 or T_2^* (read "T_2 star") relaxation. R_2^* ($1/T_2^*$) is equal to the sum of R_2' and R_2 (i.e., $R_2^* = R_2 + R_2'$. Thus, the observed decay of the FID is actually described by the equation $M_{xy}(t) = A \cdot e^{-t/T_2^*}$. While T_2' is not directly dependent on magnetic field strength, the challenge of achieving a homogeneous magnetic field at higher B_0 creates an effective link between T_2' and B_0. In NMR, T_2^* decay is generally undesirable, as it broadens spectral lines, thereby lowering resolution. Spectroscopists generally "shim" the magnetic field to improve its homogeneity before starting an experiment, and samples in a high-resolution NMR spectrometer are usually spun to help eliminate the effects of inhomogeneities. As described in the next sections, in MRI, T_2^* is a potential source of contrast.

3.2.5 NMR Echoes and Pulse Sequences

Having reviewed the basic principles of transverse relaxation, it is now possible to understand a central precept of NMR (and MRI), the spin echo. NMR echoes are FIDs that occur even after T_2^* relaxation has caused signal decay. Figure 3.6 shows schematically, and with an NMR pulse sequence, the creation of an echo using a second RF pulse, a spin echo. In the figure, an initial excitation pulse creates an FID that decays due to T_2^* relaxation as

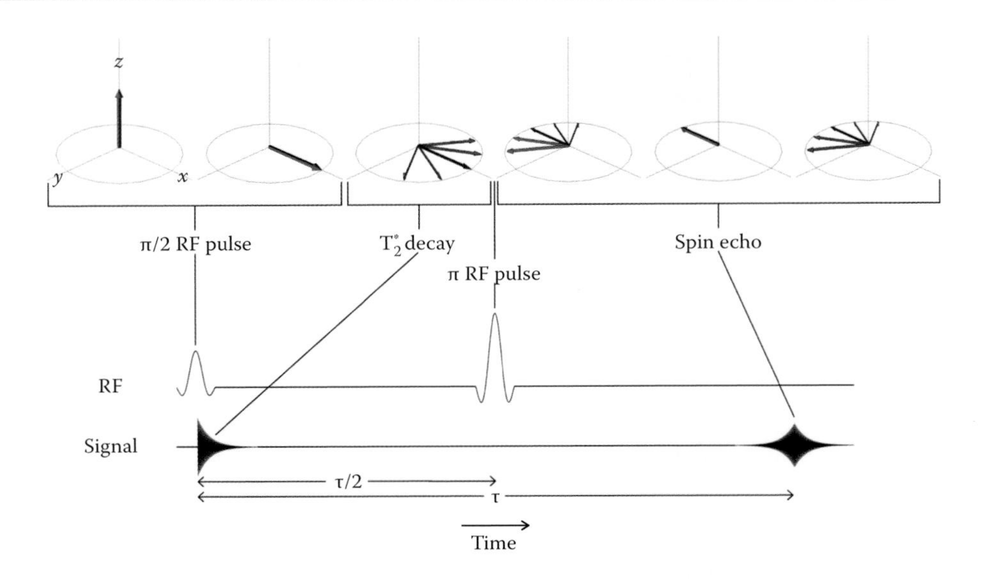

FIGURE 3.6 Creation of a spin echo with a π pulse after the initial excitation RF pulse. Application of a π pulse along the *x*-axis at time τ/2 following the initial RF excitation "flips" the magnetization vectors about the *y*-axis. Subsequent precession of the magnetization vectors generates a spin-echo at time τ.

described above. After a period of time, labeled τ/2, a π pulse is applied to the sample. As a result, an echo of the FID occurs, with a peak at time τ.

The echo occurs because the π pulse nutates all the spin packets across the *yz*-plane and back into the *xy*-plane. In their new positions, the same inhomogeneities in the magnetic field that caused the rapid T_2^* dephasing now cause the spin packets to rephase. As shown in the figure, the echo intensity will be smaller than the initial FID (by a factor of e^{-t/T_2} as the echo only rephases decay resulting from T_2' mechanisms [i.e., magnetic field inhomogeneities]). True transverse relaxation, due to the stochastic motion of the molecules, cannot be refocused.

3.3 MAGNETIC RESONANCE IMAGING

MRI is built upon the basic physical principles of NMR. Unlike NMR, however, the goal of the MRI experiment is to create images of spatially localized "packets" of water. On the surface, NMR would not seem to be a good candidate technique from which to build an imaging modality. In standard NMR experiments, one is able to measure frequency, amplitude, line width, and phase, none of which are directly related to spatial position. As described below, it is the application of magnetic field gradients that makes the "imaging" in MRI possible.

3.3.1 GRADIENTS

Gradients are magnetic fields having shaped-intensity profiles. They are designed to create well-defined, position-dependent variations in magnetic field, resulting in position-dependent variations in the NMR resonance frequency of water. MRI gradients are generally time varying (i.e., they are switched on and off) and produce a linearly varying magnetic field along a well-defined direction that adds to/subtracts from B_0. The amplitude of the gradient magnetic field is typically many orders of magnitude less than B_0. Nonetheless, by creating a

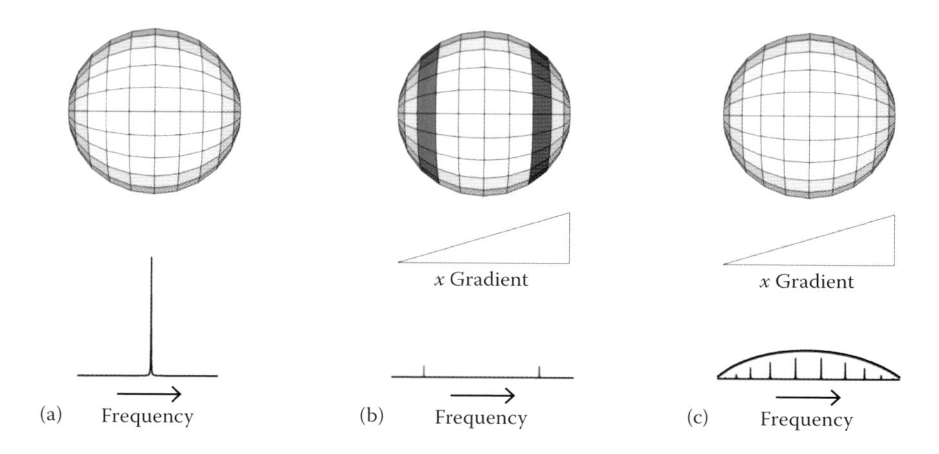

FIGURE 3.7 "NMR" spectra of a homogeneous sphere of water. (a) In the absence of any magnetic field gradients, a single peak (resonant frequency) is detected. (b) In the presence of an *x*-axis magnetic field gradient, water at different positions (red and blue bands) resonates at different frequencies and appears as different peaks in the spectrum. (c) The NMR signal across the entire frequency spectrum yields a 1-D projection image of the 3-D sphere.

well-defined dependence between spatial position and measured NMR frequency, the variations in field produced by gradients enable MRI. MRI instruments contain hardware to actuate gradients on the *x*-, *y*-, and *z*-axes. However, any arbitrary axis can be chosen as the effective gradient axis by activating two, or even all three, of the gradients simultaneously.

The effect of an applied magnetic field gradient on the detected MR signal can be understood by considering a homogeneous sample of water. By NMR, this sample would produce a single resonant frequency, that is, a single narrow peak in an NMR spectrum (Figure 3.7a). When the gradient is applied, however, multiple resonant frequencies result—one for each effective magnetic field value (recall that $\nu = -\gamma B/2\pi$). Thus, given an *x*-axis magnetic field gradient, signal from the homogeneous sample of water on the left-hand side would be detected at a frequency lower than signal from water on the right-hand side of the sample (Figure 3.7b). Of course, the actual result is continuous from left to right, resulting in what is, effectively, a projection image of the water sample (Figure 3.7c). In this projection image, the signal intensity shown on the *y*-axis of the plot corresponds to the amount of water in that "slice" of the sample (i.e., along the *x*-axis).

If multiple projection images are collected at unique "directions," it is possible, using appropriate mathematical methods, to reconstruct the entire shape of the sample. In other words, collecting multiple "NMR spectra" in the presence of well-defined, linear magnetic field gradients transforms NMR to MRI.

3.3.2 Phase Encoding

Section 3.3.1 demonstrated the use of magnetic field gradients to encode position into MRI-measurable frequency. It is also possible to encode positional information into the phase of detected MR signals. In order to understand this approach, a brief review of phase, as it relates to sinusoidal waves, is required. Phase is actually a familiar concept, describing the "difference" between sine and cosine waves. Phase provides a way to describe the progression of a sinusoidal wave through its cycle and is measured as an angle with respect to some reference point, for instance, the peak (or the zero-crossing) of the wave. Phase is illustrated

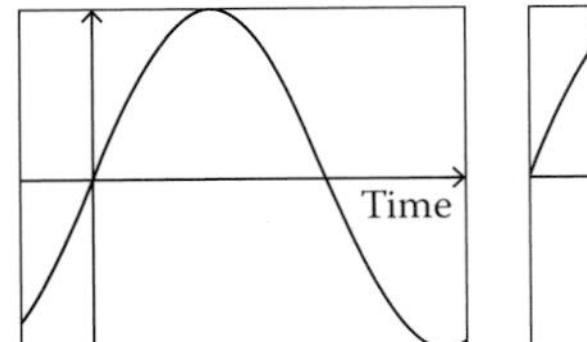

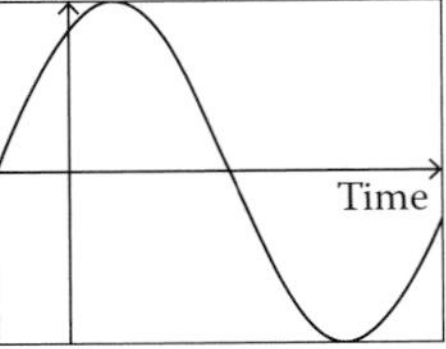

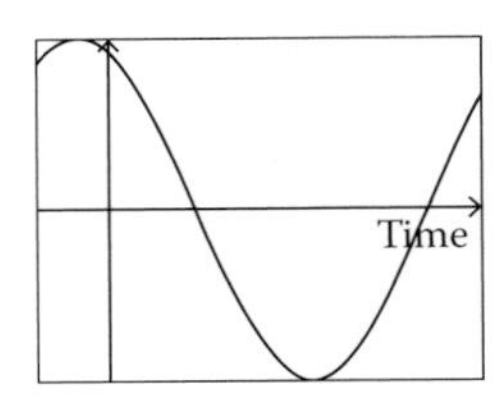

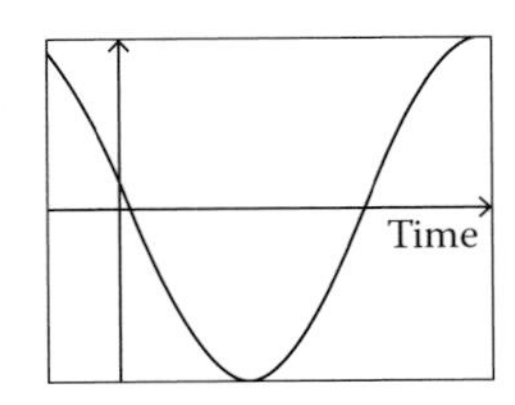

FIGURE 3.8 A sinusoidal wave with varying amounts of phase accumulation.

in Figure 3.8, which shows sinusoidal waves with the same frequency but with different phases (or with different amounts of phase accumulation).

There are two ways to trace, or record, the shape of the sine wave shown in Figure 3.8. The first is to continuously monitor the wave as a function of time, noting intensity versus time. This is like walking along the path of the sine wave in one of the panels in Figure 3.8 while recording intensities. A second way to sample the shape of the sine wave is to iteratively measure the intensity at time 0 and alter the phase in a controlled manner (e.g., through application of a magnetic field gradient). This approach is like taking a single measurement (at time = 0, as indicated by the vertical arrow) in each of the panels of Figure 3.8.

The first approach to sampling the shape of the sine wave is called frequency encoding; it is the projection image method described in Section 3.3.1. Frequency encoding is achieved by collecting the spin echo in the presence of an applied field gradient. The second sampling method, in which individual FIDs are collected at many unique phases, is called phase encoding. As we shall see in the following discussion of *k*-space, frequency and phase encoding can be combined to collect multidimensional MRI data.

3.3.3 *K*-Space

The concept of collecting time-domain data and then performing a Fourier transform to generate a frequency-domain spectrum is a familiar one in magnetic resonance spectroscopy (see Section 3.2.3). By analogy, MRI data are collected in a multidimensional spatial-frequency space, known as *k*-space, with a multidimensional mathematical transform then required to generate an image. Most often, magnetic field gradients and special frequency-selective RF pulses are used to select individual 2-D slices (imaging planes) of a 3-D volume for MR image acquisition. For each slice, MRI data are typically sampled on a Cartesian (x–y) grid in *k*-space, along "horizontal" trajectories defined by gradients applied in the readout (frequency-encode) and phase-encode directions.

Each line of data within the *k*-space matrix corresponds to the digitized signal at a particular phase-encoding value. That is, each line in *k*-space corresponds to a single MRI echo, and the "vertical" location of the line of data within *k*-space is determined by the phase of the data, which is altered in a stepwise fashion by the MRI instrument. Stepping the phase to fill *k*-space takes time, as a time delay is typically inserted after the acquisition of each line of *k*-space to allow the sample to partially return to equilibrium before the next RF pulse. For example, if a 128 × 128 image is desired, 128 individual phase-encoding steps must be performed. If it takes 10 ms to collect a single line of *k*-space, and a delay of 90 ms is required between the collection of consecutive lines of *k*-space, then the image will take about 13 s to acquire (100 ms × 128).

A major advantage of a square-grid sampling of *k*-space is that image reconstruction is easily realized via the Fourier transform. (Imaging data collected radially—along the spokes of a wheel—require either regridding of the data onto a Cartesian grid or the use of mathematical

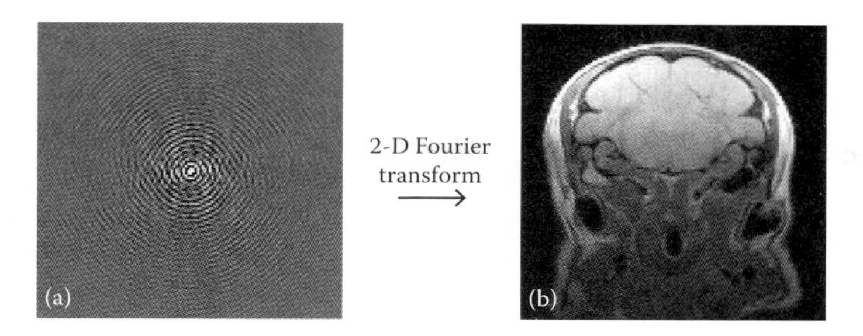

FIGURE 3.9 MRI *k*-space and its corresponding image (a transaxially oriented image of a mouse head). (a) MRI data are recorded in *k*-space (spatial-frequency space). (b) The image resulting from a 2-D Fourier transform of the *k*-space data shown in (a).

transforms, such as a Radon transform or filtered backprojection, to retrieve the image/location data within a 2-D imaging plane.) The resultant image avoids various artifacts (i.e., unwanted, systematic noise) that are troublesome for filtered backprojection images. An example of 2-D Cartesian *k*-space and its corresponding image is shown in Figure 3.9.

It can be noted from Figure 3.9 that even though a Cartesian sampling was used to collect *k*-space (in this case, a 128 × 128 grid), the pattern of the data is generally a circular ripple pattern, with the highest-intensity data located in the center of *k*-space. The relationship between raw MRI data and their corresponding images is not intuitive, but certain simple patterns can be observed. For instance, low spatial frequencies (central region of *k*-space) are required to reconstruct the bulk of an object, whereas high spatial frequencies (edges of *k*-space) are required to accurately reproduce rapid spatial variations (e.g., sharp edges). This is shown in Figure 3.10, where the *k*-space example shown in Figure 3.9 has been cropped before application of a Fourier transform.

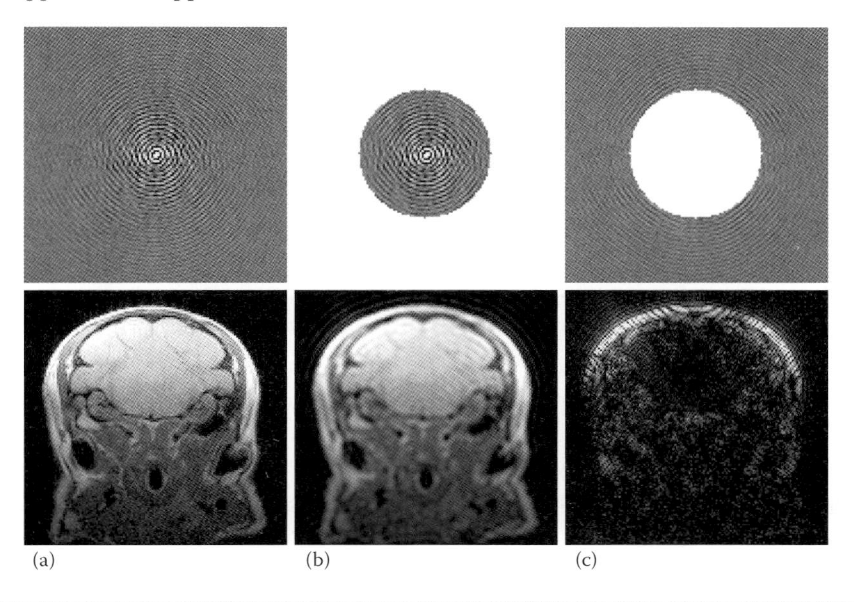

FIGURE 3.10 Different features of an object are encoded in the low- and high-frequency regions of *k*-space. (a) Full *k*-space sampling (top) and its corresponding MR image (bottom). (b) Sampling of only low frequencies in *k*-space generates a low-resolution image lacking sharp edges. (c) Sampling of only high frequencies in *k*-space generates an image with rapid spatial variations (sharp edges) but lacking the bulk of the object.

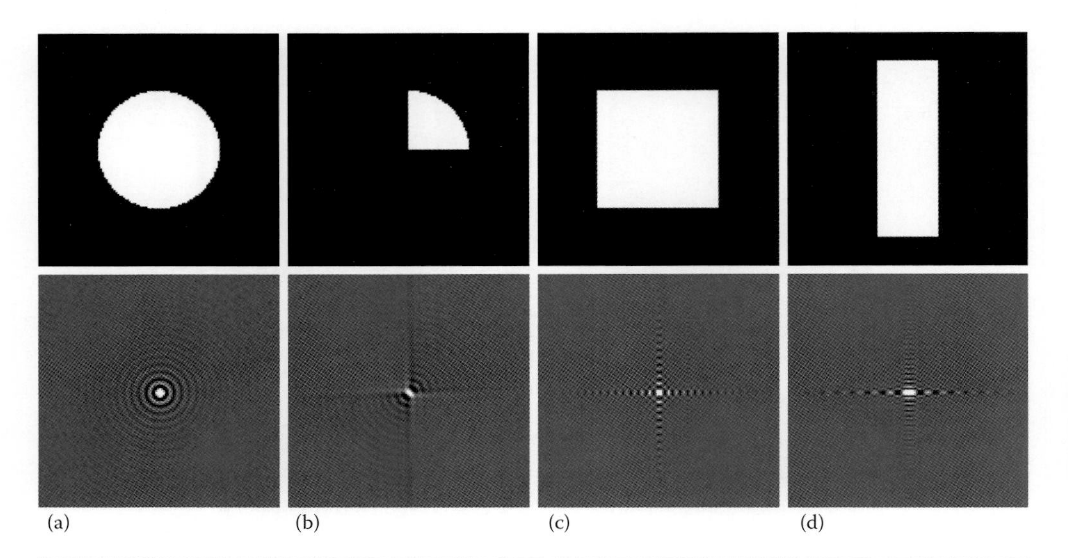

FIGURE 3.11 Images of basic geometric shapes and their corresponding *k*-space data. (a) Image/*k*-space of a circle. (b) Image/*k*-space of a single quadrant of a circle. (c) Image/*k*-space of a square. (d) Image/*k*-space of a rectangle.

While spatial relationships between *k*-space and image space exist, they are often not straightforward. This is demonstrated in Figure 3.11, which shows image/*k*-space pairs for several basic geometric images.

3.3.3.1 Creating 3-D Images

To create a 3-D image, *k*-space must be sampled in three dimensions. For this purpose, there are straightforward 3-D extensions of the two techniques for *k*-space sampling already described. Using gradients applied in a third, orthogonal direction, one can sample a sphere of projection images, or alternatively, a second phase-encoding direction can be added. However, as mentioned, the phase-encoding process takes time, and the time costs are multiplied in 3-D *k*-space. For example, to collect a 128 × 128 × 32 3-D image, 128 × 32 (4096) individual phase-encoding steps must be performed. If it takes 30 ms to collect each line of *k*-space and an additional delay of 970 ms is required between consecutive lines of *k*-space, this 3-D image will take approximately 68 min to acquire (1 s × 4096).

In view of this time requirement, it is common to collect multiple 2-D images (slices) and then "stitch" the slices together to create a 3-D image. Indeed, the primary reason multislice 2-D images, rather than true 3-D images, are frequently acquired is the significant time savings that can be achieved. In such experiments, frequency-selective RF pulses and magnetic field gradients are used, in combination, to select individual 2-D imaging planes. (As its name implies, a frequency-selective RF pulse is an RF pulse that only excites a narrow range of frequencies, in contrast to broadband RF pulses, which are often used in NMR to excite a broad range of frequencies.) A negligible time delay is required between the collection of individual lines of *k*-space from different 2-D slices. Sampling *k*-space via multiple 2-D slices, the same 128 × 128 × 32 image described here could be collected in just a few minutes. In a multislice MRI experiment, one line of *k*-space data is first collected for slice 1 and then, in turn, for each of the slices. The time required to collect a line of *k*-space for each of the slices ensures that necessary relaxation of the spins in slice 1 occurs before the acquisition of a second line of *k*-space in that slice. Extending the 3-D imaging example provided at the beginning of this section to this 2-D approach, the same image would take ~2 min (1 s × 128) to acquire. In practice, multislice 2-D images are often collected with fewer, thicker slices, resulting in lower image resolution in the slice-select direction compared with the in-plane (frequency- and phase-encode) directions.

3.3.3.2 Additional *k*-Space Sampling Schemes

Other schemes for sampling *k*-space are also possible. Many innovative methods exist, and new schemes continue to be developed and published. A complete description of *k*-space sampling methods is beyond the scope of this chapter. For more information, the interested reader is referred to *Handbook of MRI Pulse Sequences* [4].

3.4 BASIC COMPONENTS OF THE MRI SIGNAL

Section 3.3.3 introduced *k*-space and the central role of gradients in transforming the traditional NMR experiment into an imaging modality. As described earlier, a major strength of MRI is that the resultant images can be sensitized to any number of tissue-specific properties (i.e., contrast mechanisms) by manipulating the pulse sequence. This section describes the basic components of the MRI signal and discusses how MRI experiments are designed to produce images sensitized to one or more tissue-specific properties. Section 3.5 covers, in an analogous manner, more advanced contrast mechanisms and MRI techniques.

3.4.1 Pulse Sequences for MRI

Recall that, in NMR, the pulse sequence describes the pattern of RF pulses and delays used to excite the spin system and detect signals from the sample. The NMR pulse sequence is defined by precise timings and amplitudes; for example, how long is each RF pulse applied? What is the amplitude of each RF pulse? What are the time delays between pulses? When is the receiver turned on for signal acquisition? In addition to these components, MRI pulse sequences also include the pattern of magnetic gradients actuated on each of the three primary axes. Multiple gradients and RF pulses are switched on and off during a typical MRI experiment. Indeed, it is the rapid switching of the magnetic field gradients that produce the loud noise often associated with MRI scans.

Pulse sequences are most easily understood by example. Figure 3.12 shows the pulse sequence for the spin-echo MRI experiment, the imaging analogue of the RF-induced spin-echo NMR experiment that was described in Section 3.2.5. The pulse sequence consists of five lines, labeled RF, slice gradient, readout gradient, phase gradient, and receiver along a matched time axis. The RF line contains timing and amplitude information about all of the RF pulses. The slice, readout, and phase gradient traces show gradient timings and patterns on each of the three orthogonal axes. The receiver line shows when the receiver is turned on. (A sixth line is also included in the figure, depicting the MR signal itself.)

The RF line shows a $\pi/2$ pulse followed by a π pulse. Both RF pulses are frequency selective, that is, they are designed to excite only a narrow range of resonant frequencies. In concert with application of appropriate magnetic-field gradients, these frequency-selective pulses can be used to excite a single slice of spins. The time period between the center of the $\pi/2$ pulse (excitation) and the center of the echo (signal) is referred to as the *echo time*, TE, while the period between consecutive imaging experiments, that is, the collection of consecutive lines of *k*-space, is called the *repetition time* (TR). In a spin-echo pulse sequence, the π pulse is typically applied at TE/2.

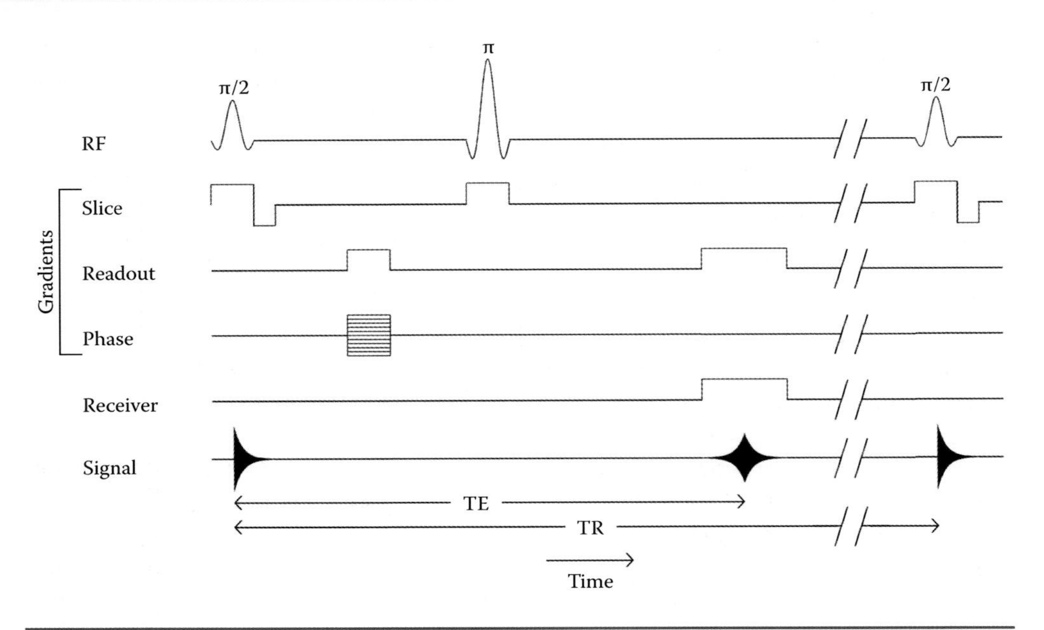

FIGURE 3.12 A spin-echo MRI pulse sequence.

In addition to enabling slice selection, the gradient associated with the π/2 pulse unavoidably acts as a phase-encoding gradient (see below). The small, negative lobe on the slice-select gradient pulse "rewinds" the unwanted phase accumulated during the π/2 pulse, reversing this phase-encoding and resetting the phase of all the excited spins within the slice to a common starting value. Because it must add phase opposite to the slice-selection gradient, it has negative amplitude. The gradient that accompanies the π pulse does not need a rewind gradient, as it is symmetric about the refocusing π pulse. Thus, any phase accumulated during the first half of the π pulse is refocused during the second half of the π pulse.

The readout and phase axes consist of gradients that produce a Cartesian sampling of *k*-space. In this example, the readout axis has two gradients, one that is applied before the refocusing π pulse and a second that is applied simultaneously with the collection of the echo (as the receiver is turned on). These gradients profile the image along the readout axis. Gradients applied along the phase axis are represented by a symbol that looks like many different gradients laid atop one another. This symbol represents a gradient that changes in amplitude (and polarity), but not duration, during the experiment. As explained in Section 3.3.3, the MRI experiment must be repeated many times to properly sample *k*-space in the phase-encode direction. The most negative and positive phase-encode gradients correspond to the bottommost and topmost lines of *k*-space, respectively. Sampling of the central line of *k*-space results from the application of zero phase-encode gradient.

3.4.2 Spin Echo

Spin echoes are central to many MRI experiments. The following simple analogy may help to explain conceptually the formation of a spin echo. Instead of thinking about spins and precession, imagine that you are watching a footrace on a circular track having the special property that it is easiest to run in the inside lane and hardest to run in the outside lane. (For simplicity, we ignore the difference in circumference between lanes.) When the official fires his/her gun, the race begins, and each runner starts running counterclockwise around the track at a constant pace determined by his/her lane position. With time, due to their different running speeds, the runners become randomly distributed (i.e., have random "phases") around the track. The official now fires his/her gun a second time, and we consider, in turn, two possible scenarios: (1) The runners all turn around and run, at their same fixed speeds, in the opposite direction. (2) The special property of the track is "reversed," so that it becomes easiest to run in the outside lane and hardest to run in the inside lane.

3.4.2.1 Runners Change Direction

After turning around, runners will be running clockwise on the track. At a time equal to the initial running period (between firings of the starter's gun), the runners will all cross the start/finish line in unison (i.e., "in phase"). The lanes of the track, corresponding to different running speeds, are analogous to the spectrum of precession frequencies in the readout direction when a magnetic field gradient is applied. Due to the gradient, runners can be distinguished by their speed; with time, they appear randomly dispersed around the track. The second firing of the starter's gun is analogous to the application of a π pulse, which causes the runners/precessing spins to "change direction," and the runners all crossing the line together corresponds to the formation of a spin echo.

3.4.2.2 Track Properties Are Reversed

In this second scenario, the runners' speeds will change instantaneously with the second firing of the starter's gun. At a time equal to the initial running period (between firings of the starter's gun), the runners will all be at the same position on the track ("in phase"), though not necessarily the start/finish line. This situation describes the formation of a gradient echo, whose pulse sequence is shown in Figure 3.13. As in the spin-echo experiment, the gradient echo begins with a slice-selective RF pulse, a refocus gradient, a phase-encode gradient, and an initial readout gradient. However, instead of applying an RF π pulse, the nuclear spins are rephased by reversing the polarity of the readout gradient, leading to the formation of a gradient echo.

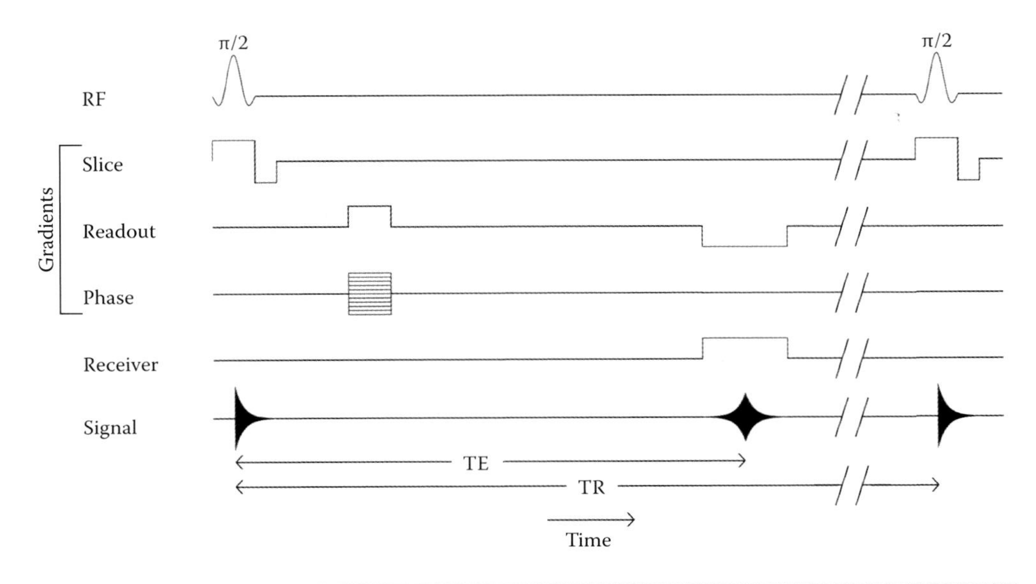

FIGURE 3.13 A gradient-echo MRI pulse sequence.

It is important to note that the effects of gradients are determined by their areas (amplitudes × lengths) rather than either their lengths or amplitudes alone. Thus, the readout gradients shown in the previous examples (Figures 3.12 and 3.13) need not have a length ratio of 1:2 but simply an *area ratio* of 1:2. For example, if one wants to image as quickly as possible (minimum TR), then gradients of maximum amplitudes (subject to hardware limitations) should be used. This can result in considerable time savings, especially for the gradient-echo pulse sequence.

3.4.3 Signal Equations

The differences that exist between the spin-echo and the gradient-echo pulse sequences produce images whose intensities have somewhat different dependencies on both spin-system relaxation parameters and NMR pulse-sequence parameters. These parameter dependencies can be described by the signal equation.

For a spin-echo experiment, the signal equation is

$$S = M_0\left(1 - e^{-\mathrm{TR}/\mathrm{T}_1}\right)e^{-\mathrm{TE}/\mathrm{T}_2}, \tag{3.7}$$

where S is the signal intensity in the image. The corresponding gradient-echo signal equation is

$$S = M_0\left(1 - e^{-\mathrm{TR}/\mathrm{T}_1}\right)e^{-\mathrm{TE}/\mathrm{T}_2^*}. \tag{3.8}$$

These equations can be derived from basic principles (see *Handbook of MRI Pulse Sequences* [4]) but are presented here, in their final forms, under the simplest experimental conditions. Under different conditions, the signal equations may be more complex. For instance, if a flip angle other than $\pi/2$ is used in the gradient-echo experiment, an additional parameter describing the flip angle, θ, is required:

$$S = \left[M_0 \sin\theta\left(1 - e^{-\mathrm{TR}/\mathrm{T}_1}\right)/\left(1 - \cos\theta \cdot e^{-\mathrm{TR}/\mathrm{T}_1}\right)\right]e^{-\mathrm{TE}/\mathrm{T}_2^*}. \tag{3.9}$$

Note that this equation assumes that steady state is reached prior to the acquisition of data (see Section 3.4.4.4).

The parameters in these signal equations have all been presented previously in this chapter but are summarized here for clarity. They can be separated into two general groups: parameters that are under control of the experimenter (pulse-sequence parameters) and those that are properties of a particular tissue or organ, at a given temperature and magnetic field strength (spin-system parameters).

3.4.3.1 Pulse-Sequence Parameters

TR—The repetition time between acquiring consecutive lines of *k*-space. TR is typically on the order of tens or hundreds of milliseconds to seconds.

TE—The echo time is the time from the initial excitation pulse to the center of the acquired echo. TE, which is always shorter than TR, is typically on the order of milliseconds or tens of milliseconds.

θ—The excitation flip angle, which depends on both the amplitude and the duration of the RF pulse.

3.4.3.2 Spin-System Parameters

M_0—The equilibrium magnetization of a sample; it is proportional to the number of NMR-active nuclei in the sample.

T_1—The longitudinal relaxation time constant ($1/R_1$) measures the return of the bulk magnetization, M_z, to equilibrium, M_0, after an excitation pulse.

T_2—The transverse relaxation time constant ($1/R_2$) measures the true loss in phase coherence (decay of signal in the transverse plane) following an excitation pulse. Spin echoes are sensitive to T_2 relaxation, which is not affected by magnetic field inhomogeneities.

T_2^*—The apparent transverse relaxation time constant ($1/R_2^*$) measures the apparent (observed) loss in phase coherence following an excitation pulse. Gradient echoes are sensitive to T_2^* relaxation, which includes the effects of T_2 relaxation, plus the loss in phase coherence due to magnetic field inhomogeneities. T_2^* is always equal to or less than T_2.

3.4.4 Signal Weighting

From the signal equations, it is clear that signal intensity is dependent upon multiple tissue-dependent parameters. Often, however, experimental parameters are chosen such that image intensity is dictated *primarily* by a *single* tissue-dependent parameter, referred to as the signal weighting. Examples of proton density, T_1, T_2, and T_2^* weighting are shown in Figure 3.14 and discussed in Sections 3.4.4.2 and 3.4.4.3.

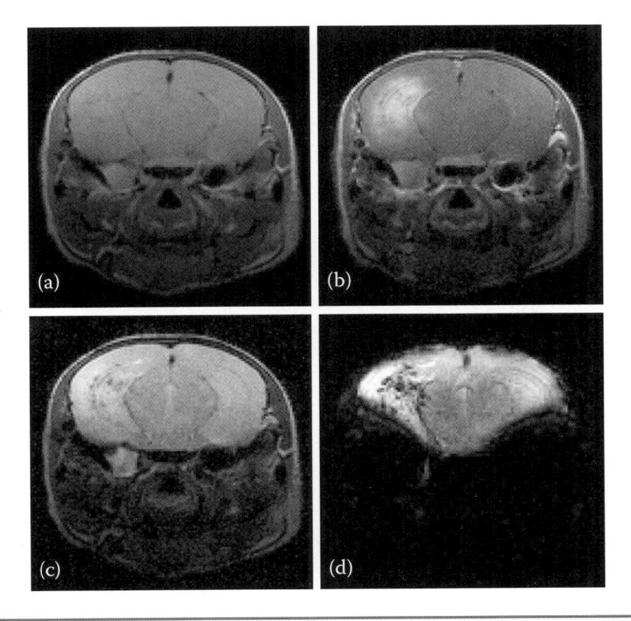

FIGURE 3.14 Single slice from (a) proton density–, (b) T_1-, (c) T_2-, and (d) T_2^*-weighted images of an *in vivo* mouse brain, collected at a field strength of 4.7 T. The left hemisphere of this mouse brain was irradiated with a single fraction of 60 Gy of radiation approximately 6 weeks prior to MR imaging. Radiation necrosis, a late time-to-progression pathology, can be seen a hyperintense region in the left hemisphere.

3.4.4.1 Proton Density Weighted

To create an image whose intensities accurately reflect M_0, the proton density, the effects of T_1 and T_2/T_2^* on the image must be minimized. From the signal equation, it is clear that this can be achieved by minimizing the value of TE and selecting the value of TR to be long enough so that the term $(1-e^{-TR/T_1})$ is effectively equal to 1. In the resulting proton density–weighted image, high intensity corresponds to high proton density.

3.4.4.2 T_1 Weighted

To create an image whose intensities are sensitive to T_1, TR must be chosen to be significantly less than the T_1 value(s) of the relevant tissue or organs. Also, the effects of T_2/T_2^* on the image must be minimized by minimizing TE, in Section 3.4.4.1. In a T_1-weighted image, high intensity corresponds to a short T_1 (i.e., a fast R_1; rapid recovery). In the case of a gradient-echo pulse sequence with arbitrary flip angle, θ, a larger flip angle results in greater T_1 weighting of the image.

3.4.4.3 T_2/T_2^* Weighted

To create an image that is purely T_2/T_2^* weighted, the effects of T_1 must be minimized by increasing TR to a value that exceeds roughly $2 \times T_1$ of the relevant tissue/organs. As noted in Section 3.4.4.2, for gradient-echo experiments, using a small flip angle further minimizes T_1 weighting. The remaining parameter to adjust is TE, with increasing TE producing greater T_2/T_2^* weighting in an image. While T_2/T_2^* contrast increases with TE, overall signal levels drop. Thus, the optimum TE for a T_2/T_2^*-weighted image is a balance between contrast and signal-to-noise ratio (S/N). In T_2/T_2^*-weighted images, tissue or organs having longer T_2/T_2^* (i.e., slower R_2/R_2^*; less rapid dephasing) correspond to high image intensity. Typically, a spin-echo sequence is used for collecting T_2-weighted images, while a gradient-echo sequence is used for T_2^*-weighted images.

3.4.4.4 Steady State and the Ernst Angle

As noted, there exists an obvious trade-off between the amount of T_2/T_2^* weighting and the total available signal. A similar, though less visible, relationship exists for T_1-weighted images. Recall that before RF excitation is applied to a sample, its bulk magnetization, M_z, is equal to the equilibrium magnetization, M_0. After a $\pi/2$ excitation pulse, a time delay, determined by T_1, is required before the equilibrium magnetization is fully recovered. (After a delay of $5 \times T_1$, 99% of the magnetization will be recovered.) If the value of TR is set shorter than this time, the available magnetization for all subsequent experiments will be less than the full equilibrium magnetization.

The situation becomes a little more complex when shorter excitation pulses (less than $\pi/2$) are used, though it remains true that longer values of TR mean that more signal is available for subsequent experiments. The added complexity arises because the available magnetization does not reach a fixed, steady-state value immediately. Depending on the flip angle and TR, many pulses may be required before this "steady state" is achieved. While this approach to steady state can be described mathematically, the signal equations in Section 3.4.3 assume a steady-state magnetization. To assure that steady state is reached, a series of dummy scans (i.e., the pulse sequence is run, but data are not acquired) is often inserted immediately prior to the start of an experiment.

Given the T_1 of a particular organ or tissue, there exists for every TR value a flip angle, known as the Ernst angle, which maximizes the steady-state magnetization. The Ernst angle thus serves to maximize S/N for any particular ratio T_1/TR. For long TR values, the Ernst angle is always $\pi/2$, but it takes on smaller values as TR is shortened. It is worth noting that many spin-echo pulse sequences begin with a $\pi/2$ excitation pulse regardless of T_1 or TR. In these situations, the optimal TR for maximizing S/N may be shorter than $5 \times T_1$ (fully relaxed), as time saved by shortening TR can be used to acquire additional signal averages.

3.4.5 Signal-to-Noise Ratio, Resolution, and Scan Time

MR image quality always represents a trade-off between S/N, spatial resolution, and scan time. In practice, scan time is often fixed by the imaging protocol, leaving the experimenter to balance spatial resolution and S/N. S/N can be defined as the ratio of the mean intensity in a region of interest (ROI) and the standard deviation of the measurement. Thus, a common way to estimate the S/N is to take the mean intensity of an ROI and divide it by the standard deviation calculated from a separate ROI collected over an area in the image that contains only noise (i.e., the background of the image—outside of the body).

Because of the nature of the MRI signal and noise, S/N is halved by decreasing the volume of the voxel by a factor of two. In an image, though, resolution is typically halved or doubled along both in-plane axes. Therefore, an image with twice the in-plane spatial resolution (e.g., 200 × 200 = 40,000 voxels vs. 100 × 100 = 10,000 voxels) will have a quarter the S/N. If the third axis, the slice thickness, is also altered, S/N can further suffer.

Signal averaging increases S/N. Averaging two consecutive images results in an increase in S/N by a factor of $\sqrt{2}$. This is because the signal increases linearly, as a factor of the number of scans, N, but the noise, being random, increases as a function of the square root of the number of scans, $\sqrt{N}$ (in this case, $\sqrt{2}$). Thus, doubling S/N requires quadrupling the number of images (a situation of diminishing returns). With these different factors in mind, decisions must be made about how many averages to acquire versus spatial resolution versus S/N requirements for the images. A final, unrelated way to increase S/N is to increase the magnetic field strength—recall that M_0 scales linearly with B_0, hence the drive for ever higher-field MRI systems.

3.4.6 Quantitative Measurement of Tissue-Dependent Parameters

A weighted image (e.g., proton density, T_1, T_2) is qualitative, in that the resultant image intensity is a function of several different tissue and/or experimental parameters. Nonetheless, weighted images can be used in a quantitative manner, that is, signal intensity from images acquired under identical experimental conditions can be compared and contrasted. Sometimes, however, the goal is to make absolute, quantitative measurement of a specific time constant or parameter. A variety of different methods exist for generating such parametric maps, as described in Sections 3.4.6.1 and 3.4.6.2.

3.4.6.1 T_1

The most traditional methods for measuring the T_1 at every voxel in an image are inversion-recovery-type experiments that incorporate a spin-echo pulse sequence with a preparation pulse preceding the $\pi/2$ excitation pulse. Multiple images are acquired, each with a unique weighting created by the delay time between the preparation pulse and the excitation pulse. Once a set of images is collected, the intensities at each voxel are used to construct a T_1 map. The intensities are typically fit to an exponential recovery curve as a function of delay time; the time constant of the recovery is T_1. There are many variations on this basic experiment.

If the initial preparation pulse is a π pulse, the technique is called inversion recovery; if it is a $\pi/2$ pulse, then the technique is called saturation recovery. Other variations include modified inversion recovery, fast inversion recovery, and "Look-locker"-style experiments.

A second major method of T_1 mapping consists of collecting multiple, steady-state gradient-echo images at different excitation flip angles. Image intensities are again fitted to an equation, this time, the gradient-echo signal equation that includes θ. The technique is commonly referred to as a variable flip angle or variable nutation angle experiment. The major advantage of the approach is that it is generally much faster than inversion-recovery-style experiments. Its major disadvantage is that it is more susceptible to errors, especially those due to spatially inhomogeneous excitation pulses, resulting from imperfect RF coils.

3.4.6.2 T_2/T_2^*

To quantitatively measure T_2 or T_2^*, multiple images are collected as a function of increasing echo time, TE. As TE increases, the signal intensity decays exponentially, and the resulting signal intensities at each voxel are modeled with the appropriate pulse-sequence signal equation. The time constant of the decay is equal to T_2 for a spin-echo pulse sequence and T_2^* for a gradient-echo experiment.

Proton density can be measured from either gradient-echo or spin-echo experiments. T_1 or T_2/T_2^* parametric maps are acquired, and the resulting values are used to adjust image intensities according to the appropriate signal equation to extract a parametric map of M_0, the proton density.

3.4.7 Image Contrast and MRI Contrast Agents

Water relaxation characteristics in different tissues and organs can serve as an important source of image contrast. Consider, for example, a system consisting of equal populations of water in two different environments, A and B, having T_1 relaxation time constants of 400 and 2000 ms, respectively. In a proton density image, collected with a long TR (TR > 10 s), the water protons in environments A and B will generate signals of equal intensity because TR is long enough for full T_1 relaxation to occur in each. However, in T_1-weighted experiments performed with shorter TR values, this difference in T_1 value will generate a difference in signal intensity (i.e., contrast) between environments A and B. This is illustrated in Figure 3.15, which shows the normalized signal intensity for A and B spins (left) and the difference between these signal intensities—contrast—as a function of TR. For these particular T_1 values, optimal contrast is achieved for a TR of ~800 ms. Similar arguments lead to generating contrast as a function of TE for species having different T_2 values.

As described in this section, differences in water T_1 (or T_2) relaxation time constants are a source of image contrast. Thus, agents introduced into biological systems that change these relaxation rates can serve as MRI contrast agents.

3.4.7.1 T_1 Contrast Agents

To better understand MRI contrast agents, we briefly describe the factors that affect the relaxation of the hydrogen spins in water. Recall that each hydrogen spin can be thought of as a nuclear magnetic dipole. The dominant relaxation mechanism for hydrogen nuclides in water is the through-space, dipole–dipole interaction, with the water ^1H magnetic dipole sensing the magnetic field from nearby magnetic dipoles. The interaction energy between neighboring dipoles is proportional to (1) the size of the interacting dipoles; (2) the inverse cube of the distance between the dipoles; and (3) $[1 - 3(\cos^2\theta)]$, where θ is the angle a

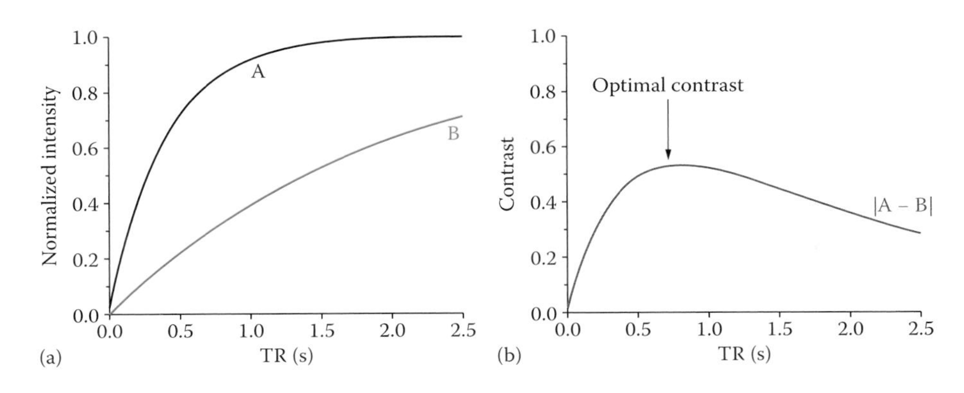

FIGURE 3.15 (a) Recovery of *z*-magnetization (M_z) as a function of pulse sequence repetition time (TR) for two different ensembles of spins, "A" and "B," having longitudinal relaxation time constants (T_1) of 400 ms and 2000 ms, respectively. (b) Contrast, $|(\text{signal intensity})_A - (\text{signal intensity})_B|$, as a function of TR, assuming equal numbers of A and B spins. The optimal contrast, at TR = 800 ms, is indicated by the black arrow.

vector connecting the two dipoles would make relative to the static magnetic field, B_0. (Mobile water molecules are tumbling rapidly, so the third term is typically averaged over all possible angles, θ.) With the exception of radioactive tritium, hydrogen has the largest nuclear magnetic dipole, so ^1H—^1H relaxation between neighboring water molecules is generally dominant in water. However, the magnetic dipole of an electron is 660 times stronger than that of the hydrogen nuclide. Thus, the presence of materials having one or more unpaired electrons (e.g., lanthanide metals) can significantly reduce the T_1 relaxation times of neighboring water molecules, leading to image contrast. In particular, Gd^{3+}, with its seven unpaired electrons, is the building block for most T_1 relaxation agents.

While lanthanides such as Gd have many unpaired electrons that make them effective MR relaxation agents, they are also reactive with, and toxic to, living systems. Thus, MR contrast agents built around Gd centers must be chelated with stable, inert organic cages. At the same time, because the dipolar interaction is dependent upon the inverse cube of the distance, effective relaxation requires that water be close to the caged metal. The solution is to use a six-coordinate ligand to chelate the Gd, leaving an open site for the close approach of water molecules.

A variety of different chelated, Gd-centered molecules have been approved for clinical use as MRI contrast agents. Initially, it would perhaps seem unlikely that MR contrast agents could ever be effective, since water is abundant in tissue and each contrast-agent molecule can relax only one water molecule at a time. The success of MR contrast agents is due, in large part, to the short residence times of individual water molecules near the Gd^{3+} center of the agent and fast chemical exchange amongst water molecules. Together, these lead to a significant "amplification effect," allowing each contrast-agent molecule to relax many water molecules.

Approximately 15 years ago, a condition known as nephrogenic systemic fibrosis (NSF) was first described. NSF is rare and serious syndrome that involves fibrosis of skin, joints, eyes, and internal organs that occurs in patients having compromised kidney function. While the exact cause of NSF remains unclear, a high association has been observed between the syndrome and Gd-based MR contrast agents, the hypothesis being that free Gd ions released from chelated agents trigger the condition. This observation has led to an increased use of contrast agents having cyclic, rather than linear, chelates because of their increased stability. Kidney function is now routinely measured in patients prior to the use of MRI contrast agents, and such agents are contraindicated in patients with severely comprised function.

3.4.7.2 T_2/T_2^* Contrast Agents

Agents can also generate contrast by shortening the T_2/T_2^* relaxation time constant of water. Typically such T_2/T_2^* contrast agents are built around superparamagnetic iron oxide (SPIO) cores. The mechanism of action of these agents is magnetic susceptibility—The agents produce distortions of the local magnetic fields surrounding them, leading to a dephasing of magnetization, shortening of T_2^*, and concomitant loss of signal in T_2^*-weighted experiments. SPIOs are used in a wide variety of cell labeling and tracking experiments and as contrast agents in dynamic susceptibility contrast (DSC) experiments designed to measure tissue perfusion, as described in Section 3.5.6. Gd-based contrast agents can also be used, though less effectively, as T_2/T_2^* contrast agents.

3.5 ADVANCED COMPONENTS OF THE MRI SIGNAL

Gradient-echo and spin-echo sequences with linear sampling of *k*-space are the basic building blocks of most MRI experiments. However, many advanced MRI methods have been introduced to augment these basic sequences. The goals of these advanced methods are multifold and include (1) decreasing acquisition time and reducing artifacts (i.e., unwanted signal aberrations) in T_1-, T_2-, and proton density–weighted images; (2) encoding information beyond primary (T_1, T_2, and proton density) weighting of signal intensity; and (3) combining multiple images (e.g., images collected as a function of time) to further extend existing contrast mechanisms. Examples of (1) that are addressed include multiecho approaches in MRI, motion suppression, and fat signal suppression. Examples of (2) to be discussed herein include localized spectroscopy, chemical shift imaging (CSI), diffusion-weighted imaging (DWI), magnetic resonance angiography (MRA), and blood oxygen level dependence (BOLD). Examples of (3) include diffusion tensor imaging (DTI), arterial spin labeling (ASL), dynamic contrast-enhanced MRI (DCE-MRI), dynamic susceptibility contrast MRI (DSC-MRI), and functional MRI (fMRI).

A detailed discussion of these techniques is beyond the scope of this text. Therefore, more frequent references, focusing on dedicated texts, are included in this section. A useful textbook that provides a more in-depth overview of many of these advanced techniques is *Quantitative MRI of the Brain: Measuring Changes Caused by Disease* [5].

3.5.1 Fast Acquisition

While conceptually simple, experiments that acquire only a single line of *k*-space per excitation pulse are inherently slow. A significant savings in acquisition time can be realized if multiple lines in *k*-space are acquired following a single excitation. This can be achieved by refocusing the FID multiple times, that is, by creating a train of sampled echoes. Recall the footrace analogies for spin and gradient echoes. In both paradigms, there is a point in time where all the runners are in synchrony, corresponding to the formation of an echo that can be sampled (a single line of *k*-space). If the race continues beyond this point, the individual runners will again distribute themselves along the track (dephasing of signal). If the track/runners respond to another firing of the starter's gun, a second realignment of runners (i.e., a second echo that can be sampled, a second line of *k*-space) will subsequently occur. Continuing this pattern leads to the formation of multiple gradient or spin echoes. Sampling of multiple echoes following a single excitation pulse, which corresponds to collecting data from multiple lines of *k*-space in a single experiment, can result in significant time savings.

There are limitations to this multiecho approach, imposed by T_2/T_2^* relaxation, which lead to attenuation of echoes. In addition, the sampling of long echo trains can increase sensitivity to image artifacts, including those arising from sample motion. Multiecho spin-echo experiments are commonly referred to as fast spin-echo or turbo spin-echo sequences, while multiecho gradient-echo MRI is termed echo planar imaging (EPI). With modern equipment, EPI experiments are sufficiently fast to permit acquisition of an entire image following a single excitation pulse (single-shot EPI).

A second, time-saving technique is to acquire data from only a fraction of *k*-space. Most commonly, the edges of *k*-space are either not sampled at all or undersampled (i.e., sampled in a less than optimal fashion). The edges of *k*-space, corresponding to high spatial frequencies, are specifically targeted because they contain information about the fine details (sharp edges) of an image. Thus, images derived from an undersampling or omission of lines near the edge of *k*-space retain all of the coarse features observed in an optimized image.

3.5.2 Artifact Suppression

3.5.2.1 Motion

A major challenge associated with MRI is minimizing image errors or artifacts. This is especially true for artifacts that result from patient motion. Examples of motion that can occur during an MRI scan include those due to the involuntary movements of anxious or restless patients, respiratory (lung) motion, and circulatory (heart or blood) motion. Strategies for reducing motion artifacts include physical approaches, such as physical restraint and breath holding or shallow breathing, and hardware/software methods, such as gating and postprocessing registration techniques. In addition, general anesthesia is frequently used in preclinical (animal) imaging experiments, and sedation is sometimes used for human patients experiencing a high level of anxiety, including those that suffer from claustrophobia.

Gating describes the process of triggering the start of the MRI scan, or the acquisition of data, with some physiological marker. Respiratory gating triggers off the tidal volume of the breathing cycle. The result is that all the lines of *k*-space are acquired at approximately the same "location" within the breathing cycle. Analogously, cardiac gating triggers off the signal from an electrocardiogram (ECG), and each line of *k*-space is acquired with the heart in the same "location" within the heartbeat.

An advanced version of cardiac gating is called CINE imaging, in which multiple images are acquired sequentially following a single trigger from the cardiac cycle. The process is repeated many times, and images occurring at the same point in the cardiac cycle are averaged together. If the heart rate is regular, each averaged image captures a unique time point in the heartbeat, resulting, collectively, in a high-resolution "movie" of the beating heart.

Postprocessing techniques that attempt to register different images to one another are not unique to MRI. However, in MRI the registration process can be aided through the use of a "navigator echo," typically a single projection image of a subject. Navigator echoes acquired periodically throughout an imaging protocol can be used to track and subsequently correct for various types of translational motion.

3.5.2.2 Fat Suppression

A second, common type of artifact in MRI results from detecting signal from hydrogen spins in species other than water, such as those found in fat/lipid. The hydrogen spins of the fat resonate at a different frequency than those of water. In the MRI experiment, this difference in frequency is interpreted as a difference in spatial position, leading to a spatial misregistration of the signal. Because of this, it is often desirable to suppress all fat signals. Many fat-suppression methods employ a series of RF "prepulses" applied before the main RF excitation pulse. One such approach uses a frequency-selective RF prepulse to selectively null or saturate the fat signal immediately before the RF excitation pulse so that only the water magnetization is observed following the excitation pulse. A second approach uses a broadband RF pulse to invert both the fat and water magnetization. A delay is then inserted into the pulse sequence corresponding to the exact time needed to null the fat signal (recall that in an inversion-recovery experiment, the magnetization passes through zero on its way back to equilibrium), after which the RF excitation is applied and data collected. Both of these approaches can also be used to selectively suppress the water signal in order to acquire a "fat-only" image. Lastly, the "Dixon method" leverages phase differences between fat and water signals to separate these signals from one another [6].

3.5.3 Localized Spectroscopy and CSI

Localized spectroscopy is the collection of NMR spectra from a specified, localized region within a sample or patient. Though an image is not acquired, per se, these NMR spectra contain valuable information about different chemical analytes (apart from water) within the chosen ROI. Importantly, ratios of signal intensities within a spectrum are quantifiable. With sufficient resolution and signal-to-noise, one can monitor changes in the ratios of these analytes in response to biological processes.

In a localized spectroscopy experiment, a voxel or ROI is selected by using multiple spatially selective pulses. The first RF pulse, the excitation pulse, excites an entire slice. Additional RF pulses are used to sequentially select a single column from within this slice and then a single voxel from within the column. This spatial localization is generally done in conjunction with water suppression to reduce the very large signal associated with water.

CSI is an extension of localized spectroscopy wherein a separate NMR spectrum is acquired from each voxel in an image. CSI yields data-rich, multidimensional images. However, the image resolution must often be significantly reduced in order to maintain sufficient S/N in each voxel. Like localized spectroscopy, the spectra can give detailed information about specific chemical species *in vivo*. Figure 3.16 illustrates CSI data from a prostate cancer patient, showing a spectroscopy grid superimposed on a T_2-weighted axial MR image of the prostate [7]. The region outlined in yellow, displaying an abnormal ratio of choline and creatine to citrate, identifies the cancerous region of the prostate.

CSI data can be collected by simply repeating a localized spectroscopy protocol many times. However, time-saving alternative data acquisition schemes exist. A valuable reference for both of these modes of *in vivo* spectroscopy is *In Vivo NMR Spectroscopy: Principles and Techniques* [8].

3.5.4 DWI and DTI

Water molecules in a container (e.g., a glass of water) diffuse due to random Brownian motion, a process often described as unrestricted, or free, diffusion. While water also

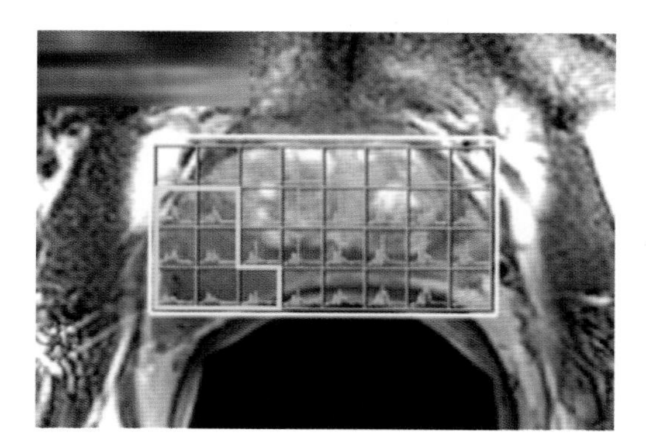

FIGURE 3.16 CSI data from a prostate cancer patient, showing a spectroscopy grid superimposed on a T_2-weighted image of the prostate. The cancerous region of the prostate is outlined in yellow. (Reproduced from Verma, S. et al., *Am. J. Roentgenol.*, 194, 2010. With permission.)

moves incoherently in tissue, this motion is often highly restricted by frequent "collisions" with cellular structures, including organelles and cell membranes. While the resulting water motion is not diffusive, in a formal sense, it is, nonetheless, useful to describe it using the language (and mathematical formalism) of diffusion. Recognizing that water motion in tissue is restricted and is thus only "pseudo" diffusive, this motion is often described by an *apparent* diffusion coefficient (ADC). As a consequence of tissue structure, water motion/diffusion *in vivo* can be highly anisotropic, with average motion in some directions being much faster than in others. For example, apparent diffusion parallel to white-matter axons in the brain and spinal cord is much faster than that perpendicular to the axons. This anisotropy can serve as a valuable source of contrast, for example, helping to distinguish white and gray matter (high versus low anisotropy, respectively) within the brain.

Traditionally, DWI is enabled by the addition of a pair of equal-amplitude, equal-length diffusion-sensitizing gradients on either side of the refocusing pulse in a spin-echo experiment. Signal from water molecules that are static during the time period between these gradients refocus fully, while signal from water that moves between gradient pulses is attenuated. Thus, regions of low apparent diffusion appear bright on a diffusion-weighted image, while those corresponding to relatively mobile spins appear dark. Altering either the intensity (i.e., magnitude or duration) of the diffusion-sensitizing gradients or their spacing alters the effective time/length scale over which apparent diffusion is effective. The combined effect of gradient length, strength, and spacing is captured in a quantity referred to as the *b-value*.

DTI experiments are designed to probe *both* the magnitude and directionality of water diffusion within tissue. This combination can serve as an important source of contrast and can provide important insights into tissue microstructure. An interesting use of the preferred directionality (anisotropy) of diffusion in brain white matter is MR tractography, in which one attempts to map/trace the paths of white-matter tracts/bundles. DTI experiments are carried out by acquiring many (at least six, and up to hundreds) diffusion-weighted images with different *b*-values and/or directions of the applied gradients. Postprocessing methods are then used to appropriately combine the images and extract parameters that quantitatively describe the rates and directionalities of water motion. Figure 3.17 shows MR tractography data collected in the mouse brain [9].

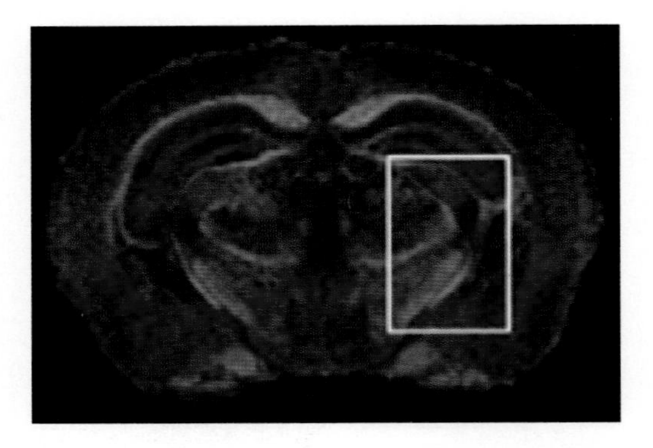

FIGURE 3.17 DTI tractography map of a mouse brain. Colors describe the orientation of the primary eigenvector (direction of maximum water diffusion): green, left to right; red, rostral to caudal; blue, dorsal to ventral. (Reproduced from Jiang, Y. and G.A. Johnson, *NeuroImage*, 50, 2010. With permission.)

As detailed in the figure caption, the colors in this figure specify the direction of the primary eigenvector (direction of maximum water diffusion) in the brain. The reader is referred to the edited textbook *Diffusion MRI: Theory, Methods, and Applications* for an in-depth description of diffusion MRI [10].

3.5.5 MRA AND ASL

A number of MRI methods exist for mapping the vasculature, techniques known collectively as MRA. MRA experiments in which signal from the vasculature is enhanced are referred to as "bright blood"; when signal from the vessels is suppressed, the experiments are referred to as "black blood." MRA experiments are used to evaluate the integrity of the vasculature and can reveal occlusions, aneurysms, and hemorrhages.

Blood is unique in that it is constantly flowing through vessels within the body. As a consequence, protons excited in one slice during an imaging experiment may be in a completely different slice when the echo is formed and *k*-space data are acquired. Most MRA methods rely upon this flow to distinguish blood from other tissue. An exception is contrast-enhanced MRA (CE-MRA), an experiment that involves injecting a contrast agent that remains primarily intravascular during the subsequent collection of T_1-weighted images. The CE-MRA image of a mouse head collected using a surface-conjugated gadolinium liposomal contrast agent, shown in Figure 3.18, allows the vascular network in the brain to be visualized [11].

ASL is an advanced MRA method designed to quantify how well perfused a tissue is by blood. ASL experiments, which are applied predominantly in the brain or kidney, are more complex and time consuming than simple MRA techniques. The ASL experiment begins by applying an RF pulse to the blood of the feeding arteries (e.g., in the neck, if the brain is to be imaged), thereby labeling or tagging that blood before it enters the tissue of interest. Shortly thereafter, a T_1-weighted image is acquired of the tissue of interest as it is perfused by the tagged blood. This image is compared to a control image collected without blood tagging. Appropriate mathematical equations are used to derive voxel-by-voxel maps of three related parameters: cerebral blood flow, cerebral blood volume, and mean transit time (i.e., the time it takes for the tagged blood to reach the imaged slice). A suggested text on the topic of MRA is *Magnetic Resonance Angiography: Principles and Applications* [12]. The interested reader is also referred to a review article that provides an excellent overview of ASL [13].

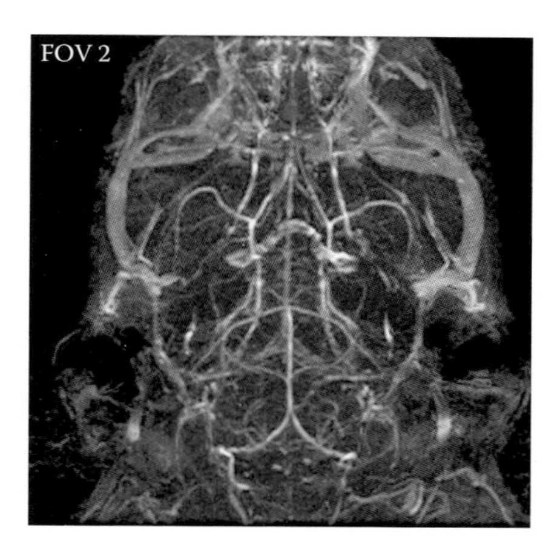

FIGURE 3.18 CE-MRA image of a mouse head collected using a liposomal nanoparticle with surface-conjugated gadolinium, a long-circulating blood-pool contrast agent. (Reproduced from Howles, G.P. et al., *Magn. Reson. Med.*, 62, 2009. With permission.)

3.5.6 DCE-MRI and DSC-MRI

Both DCE and DSC are MRI techniques in which a contrast agent is injected intravenously and multiple, sequential (time-resolved) images are acquired. Both techniques produce quantitative maps of multiple tissue parameters. DCE-MRI yields quantitative maps of three related parameters that measure how the contrast agent leaks from the vasculature and distributes into the extravascular extracellular space (EES). These three tissue-specific constants are the volume transfer constant (K^{trans}), the rate constant (k_{ep}), and the fractional volume of the EES (v_e). The standard unit for K^{trans} and k_{ep} is min^{-1}, whereas, v_e, being a fractional volume, is unitless. Ad hoc parameters, such as area under the curve (AUC), initial area under the curve (IAUC), and time to peak, are also often reported.

DCE-MRI experiments involve injecting a bolus of contrast agent and then collecting a series of time-resolved, T_1-weighted images. Images are typically collected for 5 to 30 min (or longer) following contrast-agent injection. Prior to contrast-agent injection, a T_1 map and a set of T_1-weighted control images must be acquired. The resulting DCE-MRI time-course data can then be analyzed to yield maps of K^{trans}, k_{ep}, and v_e. Figure 3.19 shows the DCE-derived parametric map of K^{trans}, superimposed on the anatomic image of a 4T1 mammary carcinoma cell tumor in the flank of a mouse [14].

A variety of different models of varying complexity—reflecting contrast-agent dynamics and various tissue properties, including the rate of intracellular–extracellular water

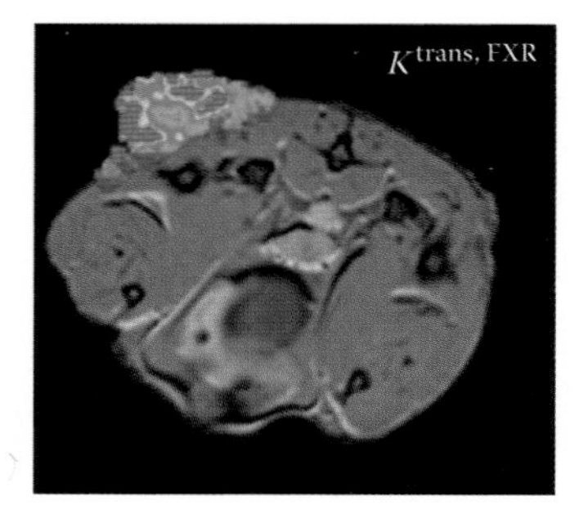

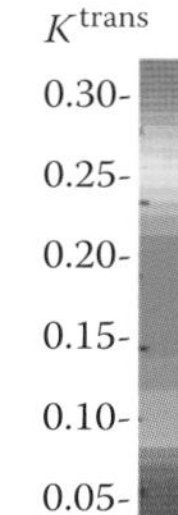

FIGURE 3.19 Parametric map of K^{trans}, superimposed on the anatomic image of a 4T1 mammary carcinoma cell tumor in the flank of a mouse. K^{trans} values were derived from DCE-MRI data using a four-parameter model, FXR-RR. (Reproduced from Yankeelov, T.E. et al., *Magn. Reson. Med.*, 59, 2008. With permission.)

exchange—have been developed to analyze DCE-MRI data. The development and validation of these models in different tissues/organs continues to be an active area of research. An additional challenge of DCE-MRI is the requirement that the contrast agent concentration in the blood plasma, referred to as the arterial input function (AIF), be known throughout the entirety of the experiment. The AIF is usually assumed or measured directly from blood-only voxels in the image.

The tissue vascular parameters modeled by DSC-MRI are different from those of DCE-MRI. Like ASL, DSC-derived parameters describe blood perfusion into tissue and bear similar names: blood flow, blood volume, and mean transit time. Like DCE, DSC-MRI experiments begin with injection of a bolus of contrast agent. However, DSC uses rapid, T_2^*-weighted imaging to observe just the first pass (and, sometimes, the second pass) of the bolus through a tissue of interest. The observed effect is a transient decrease in signal intensity, caused by magnetic susceptibility changes induced by the bolus of contrast agent. As in DCE-MRI, an AIF must be assumed or measured, and parametric maps are derived by modeling the data with appropriate mathematical equations. The book *Dynamic Contrast-Enhanced Magnetic Resonance Imaging in Oncology* is recommended to the interested reader [15].

3.5.7 BOLD and fMRI

The term *blood oxygen level dependence* refers to the facts that the oxygen content of the blood: (1) varies with time and (2) can alter MRI signal intensity. This is true because deoxyhemoglobin is paramagnetic (i.e., it has unpaired electrons, due to the high spin state of the heme iron) and, therefore, creates local magnetic field inhomogeneities. Oxyhemoglobin is diamagnetic (i.e., it has low spin and no unpaired electrons) and has no such effect on the local magnetic field. Thus, in T_2^*-weighted images, relative signal changes are observed that correlate with the ratio of oxyhemoglobin to deoxyhemoglobin.

fMRI, which has had a major impact in the social sciences, as well as radiology, involves monitoring the response of the brain to an external stimulus, thereby providing a map of brain "function." Most often, a rapid series of T_2^*-weighted images are collected continuously during alternate periods of stimulation (e.g., tap fingers, observe blinking lights) and rest. During periods of stimulation, blood flow increases in activated regions of the brain, thereby raising tissue oxygenation. This change in oxygenation can be observed with MRI via the BOLD effect. Brain activation maps in mild traumatic brain injury (MTBI) patients (Figure 3.20, bottom) and control subjects (Figure 3.20, top) during performance of a working memory task demonstrate increased activation in bilateral frontal and parietal lobes, consistent with activation of working memory circuitry [16]. Less activation of activation is observed in the MTBI patients than in the control group.

3.5.7.1 Resting State MRI

A recently observed phenomenon, known as resting state or functional connectivity MRI (fcMRI), has generated considerable excitement and activity in the MRI community. The fcMRI experiment is identical to fMRI, except that there is no external stimulus. A series of T_2^*-weighted images of the brain are collected over a period of several minutes, with a typical time resolution of several seconds. For any given voxel in the brain, a plot of signal intensity versus time appears to be simply random noise. However, this pattern of "noise" is highly correlated amongst voxels that are functionally "connected" within the brain. fcMRI

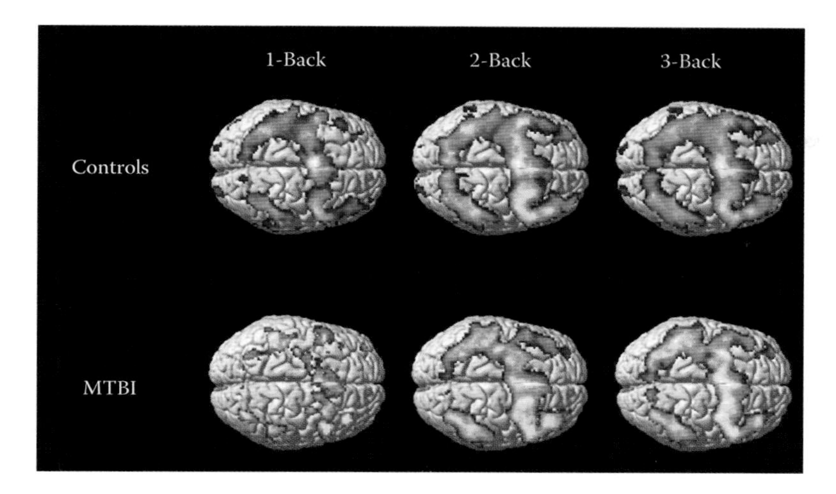

FIGURE 3.20 Brain activation maps for MTBI patients (bottom) and control subjects (top) during performance of a working memory task. Increased activation, observed in the bilateral and parietal lobes, is greater in the control group than in the MTBI patients. (Reproduced from Chen, C.J. et al., *Radiology*, 264, 2012. With permission.)

experiments are being used to create functional connectivity maps of the brain, connectivities that may be disrupted or altered by various brain pathologies. Readers interested in fMRI are referred to the textbooks *Functional Magnetic Resonance Imaging* [17] and *Introduction to Functional Magnetic Resonance Imaging* [18].

3.6 MRI HARDWARE

The part of the MRI scanner actually seen by a patient is a hollow cylinder consisting of (1) the magnet that produces B_0; (2) the gradient coils; and (3) a large, whole-body RF coil. Additional RF coils are also available for local excitation and/or detection. The electronics console, high-power amplifiers, and computers used to control the scanner are typically found in a control room adjacent to the magnet assembly.

3.6.1 B_0 Magnet

Modern MRI scanners are built around superconducting magnets, in which the magnetic field is generated by electrical current flowing through superconducting wire. The most common magnet geometry is a solenoid that produces a static magnetic field, B_0, parallel to its length. The solenoid of superconducting wire is submerged in a dewar containing liquid helium (~4 K), which is surrounded by a vacuum jacket and a second dewar containing liquid nitrogen (~77 K) to minimize boil-off of the helium. (Liquid nitrogen is much less expensive than liquid helium.) Finally, the nitrogen dewar is surrounded by another vacuum jacket and additional insulation. Ports for the periodic filling of cryogens are located on the top of the magnet, along with vents to release gaseous helium and nitrogen. Many newer magnets use a cryocooler, a device that can refrigerate cryogens to very low temperatures, to minimize helium boil-off, thus decreasing operating costs.

Typical field strengths for preclinical superconducting magnets include 4.7, 7.0, 9.4, 11.7, and 15.0 T, with 4.7, 7.0, and 9.4 T being the most common field strengths. Due to the larger required magnet bore sizes, clinical scanners operate at lower field strengths, typically 1.5, 3.0, or 7.0 T. Currently, most clinical MRI scanners operate at either 1.5 or 3.0 T.

3.6.1.1 Magnet Safety

Since current runs continuously through its main coil, a superconducting magnet is always "on." In addition, the magnetic field created by the superconducting magnet extends outside the bore of the magnet. This so-called fringe field rapidly loses intensity as a function of distance from the center of magnet, though it remains sufficiently strong (especially near the two open ends of the magnet bore) to pose a significant hazard, as this (invisible) field attracts ferromagnetic material. Devices such as floor buffers, ladders, pressurized gas tanks, hand tools, and chairs can be drawn rapidly toward the bore of the magnet, creating a danger for patients or researchers caught in their path. Furthermore, the objects often cannot be removed without a costly demagnetization and subsequent reenergizing of the magnet. To help make for a safer working environment and to reduce the size of the room required to house the magnet, a variety of schemes have been developed to "shield" the magnet and, thereby, reduce the fringe fields. Nonetheless, keeping all ferromagnetic objects, including keys, pocket knives, and personal electronic devices, away from the magnet is an absolute safety requirement. In addition, individuals with surgical implants, such as pins and screws, must avoid MRI instruments.

High magnetic fields can also cause electronic and magnetic devices to malfunction. Patients with pacemakers should not be near an MRI magnet due to the risk of device failure. Magnetically coded media, such as credit cards, memory sticks, and ID badges, may be erased if brought close to the magnet.

3.6.2 Gradients

The gradients are coils of wire typically wound around a polymer or plastic form and positioned just inside the superconducting B_0 solenoid. In many preclinical scanners, gradients are modular and removable, allowing gradient sets of different sizes and performance characteristics to be used with the same magnet. Like the B_0 solenoid, magnetic fields are generated in the gradient coils via electric currents in the wires. Unlike the B_0 solenoid, the gradient coils are made of resistive wires, since these magnetic fields must be switched on and off very rapidly. Because of this resistance, gradients generate considerable heat and must be cooled with cooling lines filled with circulating chilled water.

As described earlier, three gradients coils, creating fields in three orthogonal directions, are required to generate MR images. By convention, the direction of the static magnetic field, B_0, is defined as the z-axis (thus, the z-axis runs down the bore of the magnet). The performance characteristics of the gradient magnetic fields (e.g., strength, linearity, rise time) greatly affect the quality and spatial resolution of the resulting MR images. The maximum amplitude of the gradient magnetic fields, known as the gradient strength, determines the maximum spatial resolution. Furthermore, the gradient amplitude and switching time combine to help determine the minimum time required to execute a particular pulse sequence. It is the mechanical motion of gradient coils in response to the rapid switching of electrical currents that generates the loud buzzing sounds that accompany MRI scans. The sounds come from the gradients vibrating within their plastic or epoxy forms in response to being rapidly switched on and off.

In addition to the *x*-, *y*-, and *z*-axis gradients, MRI instruments also contain a number of additional coils (typically 2 to 10), referred to as shim coils. These generate magnetic fields designed to shim (level) the static B_0 field, thereby improving its homogeneity. Individual vendors have established standards for field homogeneities that are measured by calculating the width of the NMR peak of a sample (an MRI "phantom") containing a couple of liters of contrast agent–containing solution. Typically, magnet homogeneity of 5 ppm or less is required for routine imaging on clinical scanners, though multiecho sequences (e.g., EPI) or spectroscopy is more demanding in terms of homogeneity. Both the shims and gradients receive current from power amplifiers located in the control room. The shim coils and cooling lines are positioned within the same form around which the gradients are wound.

3.6.3 RF Coils

RF coils, the antennas used to excite a sample and detect the MRI signal, are another major component of an MRI instrument. The excitation portion of the experiment is driven by high-power RF amplifiers located in the control room. The detection portion of the experiment is supported by an electronics console that contains amplification, filtering, and digitization hardware. RF coils, or resonators, are coils of resistive wire or strips of resistive metal and must satisfy three basic requirements: (1) They must physically fit the patient or sample. (2) They must produce a magnetic field perpendicular to B_0 (B_1 must be perpendicular to B_0 to satisfy the geometry rules for torque, that is, if B_1 is not perpendicular to B_0, then no nutation occurs). (3) They must be tuned to the resonant frequency of water at the field strength B_0.

The need to simultaneously surround the patient and produce a field perpendicular to B_0 drives the use of nontraditional RF coil geometries. For example, in order to generate a field perpendicular to B_0 with a solenoid-shaped RF coil (Figure 3.21b), the coil must be positioned either sideways or up-and-down within the magnet bore. However, in neither of these orientations is it possible to slide a large sample into the coil. A "birdcage"-style coil (Figure 3.21d) is a hollow cylinder consisting of two rings at the ends of the cylinder connected by rungs (typically 6 to 32) that run the length of the cylinder, thus making it appear like a circular birdcage. Because a birdcage coil aligned with the static field produces a field perpendicular to B_0, the birdcage is a common design for head coils and whole-body coils. Schematic drawings of four common RF coil geometries are shown in Figure 3.21.

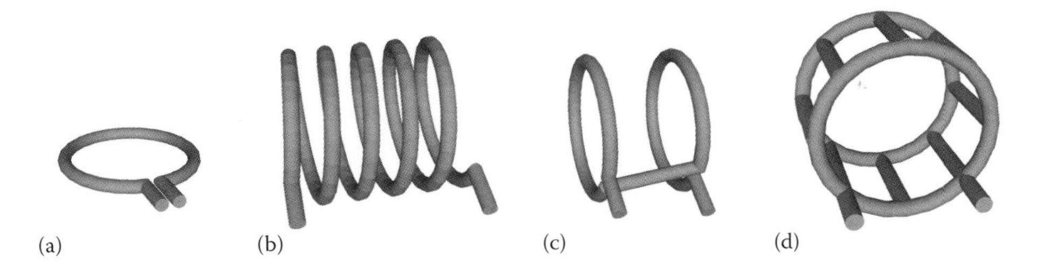

FIGURE 3.21 Common RF coil geometries. (a) Surface coil, (b) solenoid coil, (c) Helmholtz coil, and (d) birdcage coil.

RF coils that completely surround a sample, like birdcage coils, are referred to as volume coils. A major advantage of volume coils is that they generate very homogeneous magnetic fields over, potentially, very large sample volumes. This makes them ideal for the excitation portion of an MRI experiment. However, because the sensitivity of an RF coil decreases as its size increases, volume coils are not always the preferred coil for signal detection. Instead, surface coils (Figure 3.21a) are often used.

The simplest surface coil is a single loop of wire that can be physically placed on the surface of a sample or patient. Depending upon the shape of the sample, more complex, custom-manufactured surface coils might be used. Surface coils have very high local sensitivity, though that sensitivity drops off rapidly as a function of distance from the coil.

The use of a pair of RF coils, one for excitation and a second for signal detection, permits one to achieve both high excitation homogeneity and high detection sensitivity. However, to avoid unwanted signal noise, a pair of coils used in this manner must be decoupled from one another, so that their RF fields do not interact. Passive decoupling is achieved by physically orienting the RF coils such that the magnetic fields produced by the coils are perpendicular to each other (while simultaneously maintaining perpendicularity to B_0). Active decoupling is achieved using circuitry to rapidly tune and detune the two coils during their respective periods of transmit and receive activity.

An effective compromise between the homogeneity and size of the volume coil and the sensitivity of the surface coil is realized by a phased-array coil, which combines a bed of multiple surface coils into a single, large coil (Figure 3.22). Unlike the volume/surface coil pair, phased-array coils require the simultaneous activation of all of the individual component RF coils. The resulting technique, known as parallel imaging, requires additional hardware and postprocessing software but returns greater sensitivity and significant savings in terms of scan time.

A simple way to think about parallel imaging is to imagine that each coil within the phased array simultaneously takes a small, high-resolution image and that these images are stitched together after data acquisition. In reality, the technique begins by serially acquiring reference images from each individual coil, since the coils cannot be completely isolated from each other and interact in a complex manner. Furthermore, the images from different coils can overlap significantly with one another; thus, a simple stitching together of the images is not possible. Instead, the reference images are used, after parallel image acquisition, to correctly combine the multiple, overlapping images into a single image. The hardware requirements for parallel imaging include multiple RF generators, amplifiers, and digitizers (one per component coil). Multiple postprocessing algorithms exist for parallel imaging; two common

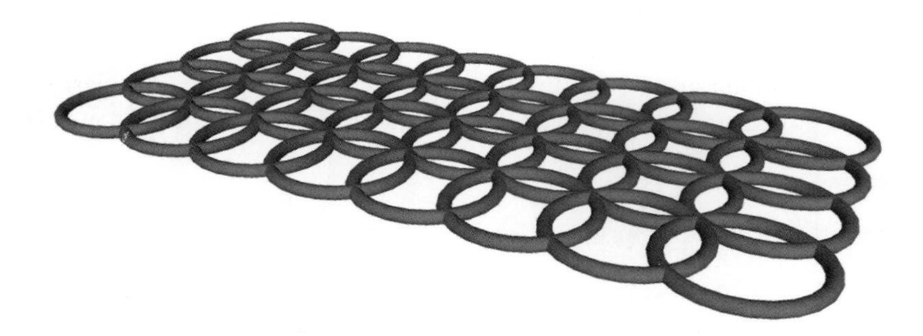

FIGURE 3.22 A phased-array RF coil made up of 32 individual surface coils.

postprocessing techniques are SENSE and GRAPPA. Phased-array coils of 32 or more channels are becoming increasingly common clinically, especially for brain imaging.

As mentioned, a final requirement for an RF coil (or coil array) is that it be tuned to the resonant frequency of water at the field strength of the MRI magnet. The process is similar to tuning an old radio to a particular station by adjusting a variable capacitor within the radio. A related operation, matching, is required to "match" the impedance (complex resistance, typically 50 Ω) of the coil to that of the MR electronics console. Effective matching is required to ensure that RF energy (i.e., transmitted pulses and received signals) is efficiently exchanged between the scanner console and the coil.

While the majority of the transmitted RF energy excites nuclear spins, some is, unavoidably, absorbed by other molecular mechanisms, which results in tissue heating. As a consequence, restrictions have been imposed to limit the amount of heating to completely benign levels. On clinical MRI systems, these restrictions are implemented by calculating the expected specific absorption rate (SAR; the rate of energy absorption) of the patient and using these results to limit RF levels for the experiment [19]. These calculations, and the resulting SAR limits, are built into the console software of all clinical scanners.

Many different coil designs and technologies exist. Relative to the MRI consoles and magnet, RF coils are inexpensive (tens of thousands of dollars, as opposed to millions) and more accessible to a small-business model. Thus, there are many different commercial RF coil companies. The book *NMR Probeheads for Biophysical and Biomedical Experiments* is an invaluable reference for RF coil design and construction [20].

3.6.4 MRI Systems

There are several major manufacturers of clinical MRI scanners with varied field strengths and price points. For human imaging, the most common field strengths are 1.5 and 3 T; however, many major medical research centers now have 7 T scanners that are being used for a range of research applications. In the more specialized field of preclinical imaging, where bore sizes are typically in the range of 20–30 cm, 7 and 9.4 T are the most common field strengths. These high-field, small-animal, and biospecimen MRI systems have been traditionally built around cryogen-cooled superconducting magnets. Recently, a new generation of MRI instruments, with field strengths up to 7 T, have been introduced that use cryogen-free, cryocooled magnets. Scanners built around 1 T permanent magnets are available for extremity imaging (arms and legs) and preclinical applications.

MRI systems are also being integrated with other imaging modalities and technologies. Hybrid scanners that combine MRI with positron emission tomography (PET) are commercially available, for both clinical and preclinical uses, and there have been successful combinations of MRI with magnetoencephalography (MEG) and optical imaging. MRI has also been combined and integrated with other nonimaging technologies, including radiation therapy, high-intensity focused ultrasound (HIFU), and dynamic nuclear polarization (DNP).

3.7 FUTURE DIRECTIONS IN MRI

The broad field of MRI remains filled with opportunities for further scientific development and advancement. In this final section, five general areas of MR research, with particular promise (and unique challenges) for the next generation, are briefly discussed. The areas are clinical field strengths greater that 7 T, new contrast mechanisms and pulse sequences, new contrast agents, hyperpolarization, and intraoperative MRI. In view of the vast array of quality research in each of these areas, the following research highlights are meant purely as examples.

3.7.1 Clinical Field Strengths Greater Than 7 T

Currently, clinical MRI instruments with 7 T magnetic fields are in the final stages of development before general release. Yet already, prototypes exist that push the B_0 field strength even higher. The opportunity is clear: the potential to produce an MRI scanner that can achieve higher-quality images than the current generation. The greatest challenge is not only to create higher-intensity magnetic fields, per se, but also to create them while maintaining homogeneity.

An interesting idea that may prove to be key for addressing this challenge is being pioneered in the area of ultra-high-field NMR. Here, field strengths of greater than 20 T have been achieved, and greater than 30 T are expected in the near future. One approach has been to build nested, or hybrid, magnets. These magnets are comprised of two individual magnets nested one inside the other. One of the magnets is a traditional superconductor, while the other is a highly specialized, resistive magnet. A second technology that may play a role is high-temperature superconductors. As the name implies, the critical temperature of these semiconductors is higher than traditional semiconductors. High-temperature superconductors could be used to make magnets that do not require liquid helium or even liquid nitrogen. Some hybrid magnet designs marry the two technologies by nesting low- and high-temperature superconducting magnets.

Increasing magnetic field strengths drive the need for RF coils that resonate at higher frequencies. Traditional RF coil methods do not translate well to very high fields. Thus, alternative coil designs, such as transverse electromagnetic (TEM) coils, and coil-enhancing technologies, such as cryogen bathing of coils, are very active areas of research. Furthermore, RF tissue penetration is hampered at higher field strengths due to the so-called standing wave effects [21,22]. Standing wave effects, which are also referred to as dielectric effects, can result in unwanted image artifacts. These artifacts are more pronounced in certain tissues and body regions, including the abdomen. Some view the development of higher-frequency RF coils as a greater challenge than the development of higher-field magnets. For a further discussion of imaging at high fields, please see Refs. [23–25].

3.7.2 New Contrast Mechanisms and Pulse Sequences

In addition to instrumentation, novel MRI contrast mechanisms and techniques continue to be invented and developed. A good example is magnetic resonance elastography (MRE), which uses MRI to image the elastic properties of tissue [26]. These properties are measured using specialized pulse sequences, in which standing acoustic waves are introduced in the tissue. Such innovative methods open MRI to yet unknown diagnostic and prognostic capabilities. The challenges, though, are often steep, as many new techniques require development of novel pulse sequences, hardware, *and* postprocessing algorithms.

Another example of a recent, novel contrast mechanism is called delta relaxation–enhanced MR (dreMR) [27]. This technique combines the use of common T_1 contrast agents and field cycling to amplify the effects of some contrast agents. Field cycling, or fast field cycling, is the practice of varying B_0 in a cyclical fashion during an MR experiment.

dreMR offers a new contrast mechanism but, like MRE, also requires advances in hardware. (On a related note, hybrid magnets would, in principle, be able to field cycle, thus facilitating dreMR and related approaches.)

A final example of a new contrast mechanism that pushes the technological limits of many aspects of MRI is sweep imaging with Fourier transform (SWIFT) [28]. The method employs an inventive pulse sequence that refocuses the MR signal and acquires *k*-space data simultaneously. Thus, the technique allows images to be collected with essentially zero echo time, which allows visualization of short-T_2 species, such as bone, tendons, and ligaments. The resultant MRI scan requires dedicated RF coils, animal/patient holders, and innovative image reconstruction methods for the pulse sequence to run properly.

3.7.3 New Contrast Agents

New contrast agents, like contrast mechanisms, are continually being discovered and developed. While the simplified goal is to improve upon currently approved, Gd-based contrast agents, the available strategies are wide and varied. One approach is to synthesize contrast agents that react to their surrounding environments, that is, agents that are functional in some way. A second common tack is the development of contrast agents that are more specific than the current generation, that is, so-called targeted contrast agents.

Currently approved contrast agents are nonreactive. They affect the relaxation properties of water but are intended to be totally inert in every other way. Functional contrast agents have reactivity built into their molecular structure. In general, the functionality is intended to allow the contrast agent to "report" on some biological mechanism. Examples include contrast agents intended to alter MR relaxation (and thus image intensity) in response to pH or the local concentration of some biologically relevant analyte, such as Zn^{2+} (a surrogate marker of insulin secretion from pancreatic β-cells).

Currently approved contrast agents are nonselective. That is, after being injected into the blood stream, they passively leak out of the vasculature (if the blood vessels are leaky) and/or are cleared from the body, primarily through the kidneys. By contrast, targeted contrast agents are chemically designed to be more selective. The most common approach has been to attempt to synthesize contrast agents that are taken up preferentially in one tissue by adding reactive functional groups to the chemical compound. Thus, the contrast agent concentrates in the tissue of interest, thereby increasing its effectiveness. A common, parallel goal of such contrast agents is to also locally deliver chemical therapeutics. For a more in-depth discussion of contrast-agent design and synthesis, the interested reader is directed to Refs. [29–31].

3.7.4 Hyperpolarization

Hyperpolarization is a method to dramatically, though transiently, increase the signal intensity of NMR-active nuclei. Recall that nuclear-spin energy states are normally populated according to a Boltzmann distribution (Section 3.2.2). The difference in population between the levels, corresponding to the net magnetization (signal) available for the MR experiment, is very small. Hyperpolarization techniques, using lasers and/or microwave fields, can create differences in spin energy-level populations that are hundreds- to thousands-fold greater than the Boltzmann distribution, permitting MR experiments that would otherwise be difficult or impossible. For instance, inert gases such as helium-3 and xenon-129 are NMR active. However, because gases are so diffuse, the signal of inhaled helium-3 or xenon-129 is undetectable. Via hyperpolarization, the signal detected from these inert gases can be amplified thousands of times, transforming these gases into very interesting and sensitive "contrast agents." Hyperpolarization occurs outside of the MRI scanner, in a purpose-built instrument. The gas is then quickly transferred to the MRI patient or animal for inhalation and subsequent imaging.

Many compounds containing carbon-13 can also be polarized through DNP. Due to the natural abundance of carbon-13 (~1%) and its low gyromagnetic ratio, carbon-13 imaging is very difficult without hyperpolarization. With DNP, solutions of biochemically interesting analytes, such as pyruvate, can be first hyperpolarized (at low temperature) and then quickly warmed and injected into the patient or animal. The injected analytes can then be monitored by localized MR spectroscopy experiments to track their metabolic fate. Refs. [32–37] provide more information about current, state-of-the-art hyperpolarization techniques and applications.

3.7.5 Intraoperative MRI

A final example of an area in MRI research with particular promise is intraoperative MRI. As already noted, MRI is more benign than techniques like CT, PET, and SPECT, as no ionizing radiation is involved. Also, the resolution of MR images is generally superior to that of PET, SPECT, and US. Thus, MRI is a prime candidate for integration into the surgical suite.

Of course, many complications exist. For instance, practically every piece of instrumentation used in a traditional surgery suite is incompatible with the strong magnetic fields of MRI. Thus, nearly every tool found in an operating room must be reengineered, from hand tools (e.g., scalpels and forceps) to anesthesia equipment to surgical lights. Furthermore, the narrow bores of many MRI magnets restrict patient access. Potential approaches to the latter challenge include large or open-bore magnets and/or the use of (robotic or nonrobotic) laparoscopic-like methods. Of course, laparoscopic tools or robots also need to be reengineered to be MR compatible. Also, intraoperative MRI requires images from extremely rapid MRI scans to be displayed to the surgeon in real time. Finally, methods must be developed to display surgical tools in these images, if desired. All of these issues are outweighed by the utility of intraoperative MRI, making its development a very active field of research. The interested reader is directed to Refs. [38–40].

REFERENCES

1. Winter, M. 2012. WebElements.
2. Stanisz, G. J., E. E. Odrobina, J. Pun, M. Escaravage, S. J. Graham, M. J. Bronskill, and R. M. Henkelman. 2005. T1, T2 relaxation and magnetization transfer in tissue at 3T. *Magn Reson Med* 54:507–512.
3. Gold, G. E., E. Han, J. Stainsby, G. Wright, J. Brittain, and C. Beaulieu. 2004. Musculoskeletal MRI at 3.0 T: Relaxation times and image contrast. *Am J Roentgenol* 183:343–351.
4. Bernstein, M. A., K. F. King, and Z. J. Zhou. 2004. *Handbook of MRI Pulse Sequences*. Academic Press, Amsterdam; Boston.
5. Tofts, P. 2003. *Quantitative MRI of the Brain: Measuring Changes Caused by Disease*. Wiley, Chichester, West Sussex; Hoboken, NJ.
6. Dixon, W. T. 1984. Simple proton spectroscopic imaging. Radiology 153:189–194.
7. Verma, S., A. Rajesh, J. J. Futterer, B. Turkbey, T. W. Scheenen, Y. Pang, P. L. Choyke, and J. Kurhanewicz. 2010. Prostate MRI and 3D MR spectroscopy: How we do it. *Am J Roentgenol* 194:1414–1426.

8. De Graaf, R. A. 2007. *In Vivo NMR Spectroscopy: Principles and Techniques*. John Wiley & Sons, Chichester, West Sussex, England; Hoboken, NJ.
9. Jiang, Y., and G. A. Johnson. 2010. Microscopic diffusion tensor imaging of the mouse brain. *NeuroImage* 50:465–471.
10. Jones, D. K. 2010. *Diffusion MRI: Theory, Methods, and Application*. Oxford University Press, Oxford; New York.
11. Howles, G. P., K. B. Ghaghada, Y. Qi, S. Mukundan, Jr., and G. A. Johnson. 2009. High-resolution magnetic resonance angiography in the mouse using a nanoparticle blood-pool contrast agent. *Magn Reson Med* 62:1447–1456.
12. Carr, J. C., and T. J. Carroll. 2012. *Magnetic Resonance Angiography: Principles and Applications*. Springer, New York.
13. Wheaton, A. J., and M. Miyazaki. 2012. Non-contrast enhanced MR angiography: Physical principles. *J Magn Reson Imaging* 36:286–304.
14. Yankeelov, T. E., J. J. Luci, L. M. DeBusk, P. C. Lin, and J. C. Gore. 2008. Incorporating the effects of transcytolemmal water exchange in a reference region model for DCE-MRI analysis: Theory, simulations, and experimental results. *Magn Reson Med* 59:326–335.
15. Jackson, A., D. Buckley, and G. J. M. Parker. 2003. *Dynamic Contrast-Enhanced Magnetic Resonance Imaging in Oncology*. Springer, Berlin; New York.
16. Chen, C. J., C. H. Wu, Y. P. Liao, H. L. Hsu, Y. C. Tseng, H. L. Liu, and W. T. Chiu. 2012. Working memory in patients with mild traumatic brain injury: Functional MR imaging analysis. *Radiology* 264:844–851.
17. Huettel, S. A., A. W. Song, and G. McCarthy. 2008. *Functional Magnetic Resonance Imaging*. Sinauer Associates, Sunderland, MA.
18. Buxton, R. B. 2009. *Introduction to Functional Magnetic Resonance Imaging: Principles and Techniques*. Cambridge University Press, Cambridge; New York.
19. Hartwig, V., G. Giovannetti, N. Vanello, M. Lombardi, L. Landini, and S. Simi. 2009. Biological effects and safety in magnetic resonance imaging: A review. *Int J Environ Res Public Health* 6:1778–1798.
20. Mispelter, J. L., M. Lupu, and A. Briguet. 2006. *NMR Probeheads for Biophysical and Biomedical Experiments: Theoretical Principles & Practical Guidelines*. Imperial College Press, London.
21. Merkle, E. M., and B. M. Dale. 2006. Abdominal MRI at 3.0 T: The basics revisited. *Am J Roentgenol* 186:1524–1532.
22. Schick, F. 2005. Whole-body MRI at high field: Technical limits and clinical potential. *Eur Radiol* 15:946–959.
23. Moser, E., F. Stahlberg, M. E. Ladd, and S. Trattnig. 2012. 7-T MR—from research to clinical applications? *NMR Biomed* 25:695–716.
24. Bandettini, P. A., R. Bowtell, P. Jezzard, and R. Turner. 2012. Ultrahigh field systems and applications at 7 T and beyond: Progress, pitfalls, and potential. *Magn Reson Med* 67:317–321.
25. Vaughan, T., L. DelaBarre, C. Snyder, J. Tian, C. Akgun, D. Shrivastava, W. Liu, C. Olson, G. Adriany, J. Strupp, P. Andersen, A. Gopinath, P. F. van de Moortele, M. Garwood, and K. Ugurbil. 2006. 9.4T human MRI: Preliminary results. *Magn Reson Med* 56:1274–1282.
26. Mariappan, Y. K., K. J. Glaser, and R. L. Ehman. 2010. Magnetic resonance elastography: A review. *Clin Anat* 23:497–511.
27. Hoelscher, U. C., S. Lother, F. Fidler, M. Blaimer, and P. Jakob. 2012. Quantification and localization of contrast agents using delta relaxation enhanced magnetic resonance at 1.5 T. *MAGMA* 25:223–231.
28. Idiyatullin, D., S. Suddarth, C. A. Corum, G. Adriany, and M. Garwood. 2012. Continuous SWIFT. *J Magn Reson* 220:26–31.
29. Viswanathan, S., Z. Kovacs, K. N. Green, S. J. Ratnakar, and A. D. Sherry. 2010. Alternatives to gadolinium-based metal chelates for magnetic resonance imaging. *Chem Rev* 110:2960–3018.
30. Que, E. L., and C. J. Chang. 2010. Responsive magnetic resonance imaging contrast agents as chemical sensors for metals in biology and medicine. *Chem Soc Rev* 39:51–60.
31. Merbach, A. E., and E. v. Tóth. 2001. *The Chemistry of Contrast Agents in Medical Magnetic Resonance Imaging*. Wiley, Chichester; New York.
32. Viale, A., F. Reineri, D. Santelia, E. Cerutti, S. Ellena, R. Gobetto, and S. Aime. 2009. Hyperpolarized agents for advanced MRI investigations. *Q J Nucl Med Mol Imaging* 53:604–617.

33. Fain, S., M. L. Schiebler, D. G. McCormack, and G. Parraga. 2010. Imaging of lung function using hyperpolarized helium-3 magnetic resonance imaging: Review of current and emerging translational methods and applications. *J Magn Reson Imaging* 32:1398–1408.
34. Yen, Y. F., K. Nagasawa, and T. Nakada. 2011. Promising application of dynamic nuclear polarization for in vivo (13)C MR imaging. *Magn Reson Med Sci* 10:211–217.
35. Nelson, S. J., J. Kurhanewicz, D. B. Vigneron, P. E. Z. Larson, A. L. Harzstark, M. Ferrone, M. van Criekinge, J. W. Chang, R. Bok, I. Park, G. Reed, L. Carvajal, E. J. Small, P. Munster, V. K. Weinberg, J. H. Ardenkjaer-Larsen, A. P. Chen, R. E. Hurd, L. Odegardstuen, F. J. Robb, J. Tropp, and J. A. Murray. 2014. Metabolic imaging of patients with prostate cancer using hyperpolarized [1-^{13}C]pyruvate. *Sci Transl Med* 5(198):198ra108.
36. Rodrigues, T. B., E. M. Serrao, B. W. C. Kennedy, D. Hu, M. I. Kettunen, and K. M. Brindle. 2014. Magnetic resonance imaging of tumor glycolysis using hyperpolarized ^{13}C-labeled glucose. *Nat Med* 20:93–97.
37. Merritt, M. E., C. Harrison, A. D. Sherry, C. R. Malloy, and S. C. Burgess. 2011. Flux through hepatic pyruvate carboxylase and phosphoenolpyruvate carboxykinase detected by hyperpolarized ^{13}C magnetic resonance. *Proc Natl Acad Sci* U S A 108 (47):19084–19089.
38. Duchin, Y., A. Abosch, E. Yacoub, G. Sapiro, and N. Harel. 2012. Feasibility of using ultra-high field (7 T) MRI for clinical surgical targeting. *PLoS One* 7:e37328.
39. Avula, S., C. L. Mallucci, B. Pizer, D. Garlick, D. Crooks, and L. J. Abernethy. 2012. Intraoperative 3-Tesla MRI in the management of paediatric cranial tumours—Initial experience. *Pediatr Radiol* 42:158–167.
40. Seifert, V., T. Gasser, and C. Senft. 2011. Low field intraoperative MRI in glioma surgery. *Acta Neurochir Suppl* 109:35–41.

FURTHER READINGS

For a more complete treating of the subject of NMR, readers are referred to any number of quality texts on the subject, for instance, the following:

Farrar, T. C. 1989. *Introduction to Pulse NMR Spectroscopy*. Farragut Press.
Levitt, M. H. 2008. *Spin Dynamics: Basics of Nuclear Magnetic Resonance*. Wiley, Chichester, UK.

For more information on MRI and pulse sequences, the interested student is directed toward the following:

Bernstein, M. A., K. F. King, and Z. J. Zhou. 2004. *Handbook of MRI Pulse Sequences*. Academic Press, Amsterdam.
Haacke, E. M. 1999. *Magnetic Resonance Imaging: Physical Principles and Sequence Design*. Wiley, New York.
Vlaadingerbroek, M. T., and J. A. den Boer. 2003. *Magnetic Resonance Imaging: Theory and Practice*. Springer, Berlin.

Other relevant textbooks on advanced MRI topics have been cited at the end of major sections within the main text.

4

Ultrasound

K. Kirk Shung

4.1 INTRODUCTION

Since the early pioneering work of Wild and Reid along with others on ultrasound in the late 1950s [1], ultrasound has become one of the most important diagnostic imaging tools. Ultrasound not only complements the most utilized imaging approach, x-ray, but also possesses unique properties that are advantageous in comparison to other competing modalities discussed in this book, namely, x-ray computed tomography (CT), radionuclide emission tomography, and magnetic resonance imaging (MRI). More specifically, (1) ultrasound is nonionizing and is considered safe to the best of present knowledge, (2) it is less expensive than imaging modalities of similar capabilities, (3) it produces images in real time, (4) it has a resolution in the millimeter range for the frequencies being clinically used today and may be improved at higher frequencies, (5) it is capable of yielding blood flow information by applying the Doppler principle, and (6) it is portable and thus can be easily brought to the bedside of a patient.

Ultrasound also has several drawbacks. Most notable among them are that (1) organs containing gases and bony structures cannot be adequately imaged, (2) only a limited window is available for ultrasonic examination of certain organs such as the heart and neonatal brain, (3) it is highly operator skill dependent, and (4) it is sometimes impossible to obtain good images from certain types of patients such as obese patients.

The many advantages that ultrasound can offer have allowed it to become a valuable diagnostic tool in such medical disciplines as cardiology, obstetrics, gynecology, surgery, pediatrics, radiology, emergency medicine, and neurology. Ultrasound is the tool of choice in obstetrics primarily because of its safe, noninvasive nature and real-time imaging capability. Ultrasound also enjoys similar success in cardiology. Ultrasound is also becoming an indispensible tool in emergency rooms. Today, ultrasound is the second-most-utilized diagnostic imaging modality in medicine, next only to conventional x-ray. High-frequency ultrasound also is widely used in clinical research and in preclinical imaging applications.

Although ultrasound has been in existence for more than 40 years and is now considered a mature technology, technical advances in ultrasound are still constantly being made. The introduction of contrast agents, harmonic imaging, flow imaging, multidimensional imaging, elasticity imaging, and high-frequency imaging are just a few examples [2–4]. In this chapter, these new developments along with fundamental physics, instrumentation, system architecture, biological effects of ultrasound, and biomedical applications will be briefly reviewed.

4.2 FUNDAMENTALS OF ACOUSTIC PROPAGATION

Ultrasound is a sound wave whose frequency is higher than 20 kHz, out of the human audible range, and is characterized by such acoustic parameters as pressure, particle (or medium) velocity, particle displacement, density, and temperature. Since ultrasound is a wave, it transmits energy just like an electromagnetic wave with the difference that ultrasound requires a medium in which to propagate and thus cannot propagate in a vacuum. When a medium is disturbed by sound, the distance traveled by an elemental volume or "particle," defined as a volume much smaller than a wavelength, is called the *particle* or *medium displacement* and usually is very small. The velocity at which this elemental volume is traveling is called the *particle* or *medium velocity* and again is very small—on the order of a few centimeters per second in water. This velocity is different from the rate at which the energy is propagating through the medium, defined as the *phase velocity* or the sound propagation velocity, typically denoted as c. In water, $c \sim 1500$ m/s. It follows, then, that

$$f \cdot \lambda = c, \tag{4.1}$$

where f is the sound frequency and λ the wavelength. For an ultrasonic wave at 5 MHz, the wavelength is about 300 μm. As will be discussed in Section 4.3, the spatial resolution of an ultrasonic imaging system (that is, its capability to spatially resolve an object) is primarily determined by the wavelength. The ultimate resolution that a 5 MHz ultrasonic imaging system can achieve is 300 μm. To improve the resolution, one option is to increase the frequency.

4.2.1 Compressional Wave

In a medium where there is only a force in the z-direction with no forces along the x- and y-directions, the 1-D wave equation along the z-direction, or the *longitudinal wave equation*, is given by

$$\frac{\partial^2 W}{\partial z^2} = \frac{\rho}{\nu + 2\mu} \frac{\partial^2 W}{\partial t^2}. \tag{4.2}$$

The equation describes a wave in which the medium displacement, W, is in the z-direction as well. In Equation 4.2, ρ denotes the mass density of the elemental volume,

t is time, and ν and μ are so-called Lamé's constants, which are commonly used elastic constants of a material. If there also are forces in the x- and y-directions, then the wave equation becomes 3-D. The solutions for this equation have the form of $f(z \pm ct)$, where the negative sign indicates a wave traveling in the $+z$-direction, whereas the positive sign indicates a wave traveling in the $-z$-direction. This type of wave is called a *compressional* or *longitudinal wave* because the displacement W is in the same direction as the wave propagation direction, z. The sinusoidal solution for this equation is

$$W^{\mp}(z, t) = W_0 e^{j(\omega t \pm kz)}, \tag{4.3}$$

where W^- and W^+ denote displacements for positive and negative going waves, respectively; ω = angular frequency = $2\pi f$; $k = \omega/c$ is the wave number; and the sound velocity c is given by

$$c = \sqrt{\frac{\nu + 2\mu}{\rho}}. \tag{4.4}$$

For a fluid that is generally assumed not capable of supporting a shear force, the shear modulus $\mu = 0$,

$$c = \sqrt{\frac{1}{G\rho}}, \tag{4.5}$$

where G represents the compressibility of a medium ($1/\nu$ in a fluid) and is defined as the negative of the change in volume per unit volume per unit change in pressure. This equation shows that the sound velocity in a medium is determined by the density and the compressibility of a medium.

4.2.2 Shear Wave

For the case where there is only a shear force in the z-direction producing a displacement W that is perpendicular to the direction of propagation x with no longitudinal forces, the wave equation is given by

$$\frac{\partial^2 W}{\partial x^2} = \frac{\rho}{\mu}\frac{\partial^2 W}{\partial t^2}, \tag{4.6}$$

where μ is one of Lamé's constants and is also termed the *shear modulus*. This equation describes a wave, called a *shear* or *transverse wave*, traveling in the x-direction with a displacement in the z-direction. The sinusoidal solution to Equation 4.6 is

$$W^{\mp}(x,t) = W_0 e^{j(\omega t \pm k_t x)}, \tag{4.7}$$

where $k_t = \omega/c_t$ and c_t is the shear wave propagation velocity given by

$$c_t = \sqrt{\frac{\mu}{\rho}}. \tag{4.8}$$

Equation 4.8 shows that a shear wave can only exist in a medium with nonzero shear modulus. In general, it is assumed that fluids do not support the propagation of a shear wave.

Both the longitudinal and shear velocities, apparent from Equations 4.4 and 4.8, are affected by the elastic properties of a tissue. Pathological processes that change these properties can cause the sound velocity to change. Therefore, if the velocity can be accurately measured, the result can be used to infer or diagnose the pathology. A few ultrasonic devices on the market today for diagnosing osteoporosis use this principle since osteoporosis causes a loss of bone mass leading to a reduction in bone density affecting the sound velocity, as shown by Equation 4.4.

4.2.3 Characteristic Impedance

The *specific acoustic impedance* of a medium, Z, is defined as

$$Z^{\pm} = \frac{p^{\pm}}{u^{\pm}}, \tag{4.9}$$

where p is pressure and u is the medium velocity. For a fluid, $Z = \pm\rho c$. This product is also called the *characteristic acoustic impedance* of the medium. The acoustic impedance has units of kg/m^{2}s or Rayl. The positive and negative signs are for the positive and negative going waves, respectively. The acoustic velocity and impedance for a few common materials and biological tissues are listed in Table 4.1 [4]. The acoustic velocity in a medium is a sensitive function of the temperature, but its dependence on frequency is minimal over the frequency range from 1 to 15 MHz. The acoustic impedance is a very important parameter in ultrasonic imaging since it determines the amplitude of the echoes that are reflected or scattered by tissue components. A majority of clinical ultrasonic imaging devices today are of the pulse-echo type, in which very short pulses of ultrasound consisting of a few sinusoidal cycles are transmitted and the returned echoes are received to form an image.

TABLE 4.1
Acoustic Properties of Biological Tissues and Relevant Materials

Material	Speed, at 20–25°C (m/s)	Acoustic Impedance (MRayl)	Attenuation Coefficient (Np/cm) at 1 MHz	Backscattering Coefficient (cm^{-1} sr^{-1}) at 5 MHz
Air	343	0.0004	1.38	———————
Water	1480	1.48	0.00025	———————
Fat	1450	1.38	0.06	———————
Myocardium (perpendicular to fibers)	1550	1.62	0.35	8×10^{-4}
Blood	1550	1.61	0.02	2×10^{-5}
Liver	1570	1.65	0.11	5×10^{-3}
Skull bone	3360	6.00	1.30	———————
Aluminum	6420	17.00	0.0021	———————

4.2.4 Intensity

The intensity of an ultrasonic wave is the average energy carried by a wave per unit area normal to the direction of propagation over time. It is well known that energy consumed by a force F, which has moved an object by a distance L, is equal to the product $F \times L$. The power is defined as energy per unit time. It is therefore fairly intuitive that the intensity $i(t)$ carried by an ultrasonic wave is given by

$$i(t) = p(t) \cdot u(t), \tag{4.10}$$

where p is pressure and u is the medium velocity. For the case of sinusoidal propagation, the average intensity I can be found by averaging $i(t)$ over one cycle

$$I = p_0 u_0 \cdot \frac{1}{T} \int_0^T \sin^2 \omega t = \frac{1}{2} p_0 u_0, \tag{4.11}$$

where p_0 and u_0 denote peak pressure and medium velocity, respectively, and T is the period = $1/f$.

The intensity within an ultrasound beam in general is not spatially uniform. As a result, several definitions of ultrasound intensity have been introduced as indicators of ultrasound exposure level. These are illustrated in Figure 4.1a and b. The *spatially averaged intensity* I_{SA} is defined as the average intensity over the ultrasound beam.

$$I_{SA} = \frac{1}{A} \int_0^a I(r) \mathrm{d}A(r), \tag{4.12}$$

where r is the radial distance from the beam center. A is the area of the beam that is often defined as the spatial extent in which the intensity is within −3 or −6 dB of the spatial peak (SP) value at the beam center. The unit of decibels, abbreviated as dB, is defined as 10 log $I(r)/I_{SP}$. For a beam of circular symmetry, $A = \pi a^2$. At $r = a$, the intensity value $I(a)$ is −3 or −6 dB below the spatial peak value, I_{SP}. Figure 4.1a shows that the spatial peak intensity

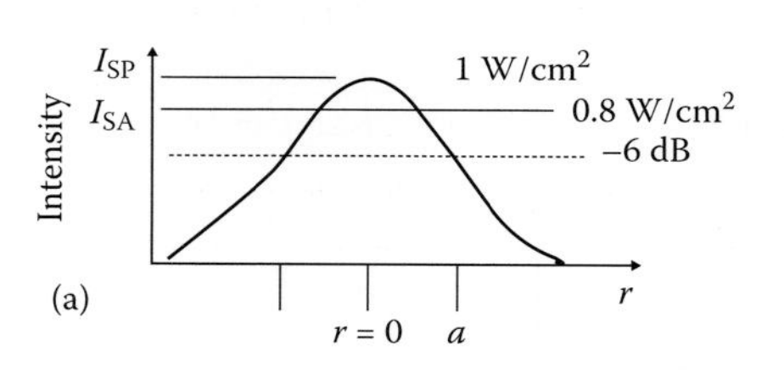

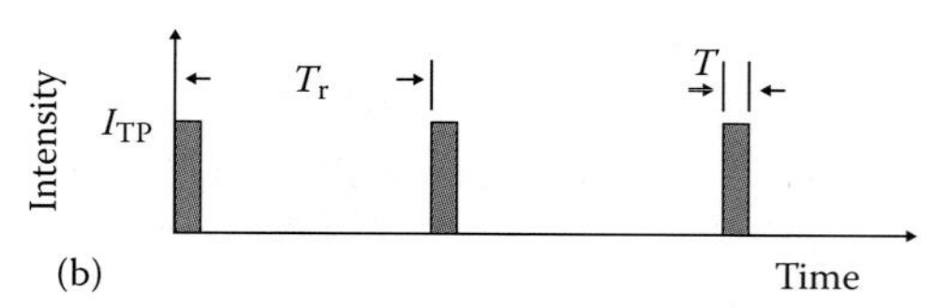

FIGURE 4.1 (a) Ultrasonic lateral beam profile in a direction perpendicular to the propagation direction. (b) An ultrasonic pulse train in time with temporal peak intensity I_{TP}, pulse duration T, and pulse repetition period T_r.

of this particular beam is I_{SP} = 1 watt/cm^2, while the spatial average intensity I_{SA} is only 0.8 watt/cm^2.

The *temporal average intensity* (I_{TA}) is defined as the average intensity over a pulse repetition period T_r and is given by the product of duty factor and temporal peak intensity, where the duty factor is defined as duty factor = pulse duration (T)/pulse repetition period (T_r). Figure 4.1b shows an example where the duty factor is 0.1. Thus, the temporal average intensity at a certain spatial location within the beam is then I_{TA} = 0.1 I_{TP}, where I_{TP} is the temporal peak intensity.

Whenever biological effects of ultrasound are discussed, it is necessary to state or understand which definition of intensity is being used [5]. In general, *spatial average temporal average intensity* I_{SATA} and *spatial peak temporal average intensity* I_{SPTA} are preferred. The US Food and Drug Administration (FDA) also requires reporting of a few other numbers when submitting an ultrasonic device for approval, including *spatial peak pulse averaged intensity* (I_{SPPA}) and the total power emitted by an ultrasonic probe. The spatial peak pulse averaged intensity is defined as the spatial peak intensity averaged over the pulse duration. As an example, a probe at 5 MHz emitting a total power of 1.1 mW at a pulse repetition frequency of 3 kHz may have an I_{SPTA} of 1.1 mW/cm^2, I_{SPPA} of 25 W/cm^2, and spatial peak and temporal peak pressure of 0.66 MPa at the focal point. The atmospheric pressure is approximately 0.1 MPa. These numbers indicate that the time-averaged intensity may be low but the instantaneous peak pressure and intensity, which is a concern for mechanical biological effects, can be very high. The potential biological effects of ultrasound will be discussed in Section 4.6.

4.2.5 Radiation Force

An acoustic wave exerts a body force to a medium (unit, newton/m^3) due to nonlinear interactions when there is a decrease in ultrasonic intensity in the direction of wave propagation due to attenuation [6]. At higher frequencies, tissues are not capable of following the motion of the ultrasound signal, resulting in the deposition of energy into tissues and causing a momentum in the direction of wave propagation and tissue heating. For a plane wave of intensity I propagating in a medium with a sound velocity c, the radiation force per unit area or *radiation pressure* p_r acting on a planar reflector is given by

$$p_r = \zeta(I/c), \tag{4.13}$$

where ζ is a constant depending upon the acoustic properties of the reflector [6]. If it is made of a material of large acoustic impedance such as steel, and almost completely reflects the ultrasound beam propagating in water, ζ = 2. If it is a perfect absorber, ζ = 1. The reason for this can be understood by simply considering the momentum transfer that occurs at the boundary. Since the momentum transfer occurring at the interface for a perfect reflector is twice as large as that for an absorber, the value of p_r acting on a perfect reflector should be twice as large.

Equation 4.13 has been used for determining the power of an ultrasonic beam in a device called a radiation force balance [4]. More recently, radiation force has been used as a means to perturb an object remotely so that from the displacement of the object, which can be estimated ultrasonically, an assessment of elastic properties of the medium surrounding the object may be made [6,7].

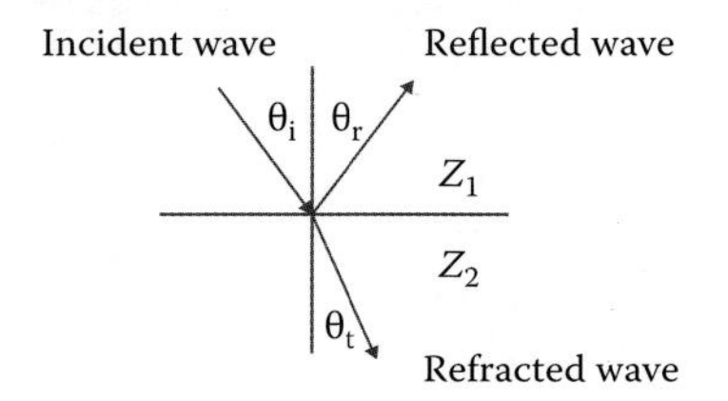

FIGURE 4.2 A plane ultrasonic wave reflected and refracted by a flat interface between two media.

4.2.6 Reflection and Refraction

As a plane ultrasonic wave encounters an interface between two media with different acoustic impedance, Z_1 and Z_2, it is reflected and refracted. Part of the energy carried by the incident wave is reflected and travels at the same velocity as the incident wave. The transmitted or refracted wave in the second medium travels at a different velocity. The directions of the reflected and refracted waves are governed, just as in optics, by Snell's law. This is illustrated in Figure 4.2, where the subscripts i, r, and t refer to incident, reflected, and transmitted or refracted waves, respectively. Analogously to optics,

$$\theta_i = \theta_r \quad \text{and} \quad \sin\theta_i/\sin\theta_t = c_1/c_2 \tag{4.14}$$

When $\theta_t = \pi/2$, $\sin\theta_t = 1$, and $\theta_{ic} = \sin^{-1} c_1/c_2$ if $c_2 > c_1$. For any incident angle greater than θ_{ic}, there is no transmission, that is, total reflection occurs. Therefore, θ_{ic} is called the critical angle.

The pressure (p) reflection and transmission coefficients, R and T, can be easily found by using the boundary conditions that the pressure and particle velocity must be continuous across the boundary.

$$R = \frac{p_r}{p_i} = \frac{Z_2\cos\theta_i - Z_1\cos\theta_t}{Z_2\cos\theta_i + Z_1\cos\theta_t} \tag{4.15}$$

$$T = \frac{p_t}{p_i} = \frac{2Z_2\cos\theta_i}{Z_2\cos\theta_i + Z_1\cos\theta_t} \tag{4.16}$$

For normal incidence, $\theta_i = \theta_t = 0$, and Equations 4.15 and 4.16 become

$$R = \frac{p_r}{p_i} = \frac{(Z_2 - Z_1)}{Z_2 + Z_1} \tag{4.17}$$

$$T = \frac{p_t}{p_i} = \frac{2Z_2}{(Z_2 + Z_1)} \tag{4.18}$$

These relationships show that as ultrasound encounters an interface, part of its energy will be lost due to reflection at the boundary. Since air and bone have acoustic impedances very different from soft tissues, more than 90% of the energy will be lost at bone or air–soft tissue boundaries. Organs containing air (e.g., lungs) or surrounded by bone (e.g., brain) for this reason are, in general, not readily interrogated by ultrasound.

4.2.7 Attenuation, Absorption, and Scattering

When an ultrasonic wave propagates in a heterogeneous medium, its energy is reduced or attenuated as a function of distance. The energy may be diverted by reflection or scattering or absorbed by the medium and converted to heat. The reflection and scattering of a wave by an object actually are referring to the same phenomenon, the redistribution of the energy from the primary incident direction into other directions. This redistribution of energy is termed *reflection* when the wavelength and wave front of the wave are much smaller than the object and is termed *scattering* if the wavelength and the wave front are greater than or similar to the dimension of the object.

4.2.7.1 Attenuation

The pressure of a plane monochromatic wave propagating in the z-direction decreases exponentially as a function of z:

$$p(z) = p(0)e^{-\alpha z}, \tag{4.19}$$

where $p(0)$ is the pressure at $z = 0$ and α is the pressure attenuation coefficient, which has units of neper (Np)/cm or dB/cm, where 1 Np/cm = 8.686 dB/cm. This equation is identical to that for the attenuation of x-rays, gamma rays, and optical radiation. Typical values of the attenuation coefficient in several materials are given in Table 4.1.

Investigations to date have shown that scattering (see Section 4.2.7.3) contributes little to attenuation, at most a few percent, in the majority of soft tissues. Therefore, it is safe to say that absorption is the dominant mechanism for ultrasonic attenuation in biological tissues in the clinical ultrasound frequency range from 1 to 50 MHz.

4.2.7.2 Absorption

Ultrasound propagating in an inhomogeneous medium loses energy in part because of absorption of the energy by the medium, which subsequently is converted to heat. The absorption mechanisms in biological tissues are quite complex and have been observed experimentally to be approximately linearly proportional to frequency. This behavior cannot be explained by classical absorption due to viscosity that occurs in a homogeneous medium like air or water, which has a dependence on the square of the frequency. Therefore, it has been theorized that ultrasonic absorption arises from two terms, classical absorption and a relaxation phenomenon, but is dominated by the latter [8].

The relaxation phenomenon may be described simplistically in the following manner. When an elemental volume or particle in a medium such as a molecule is pushed to a new position by a force and then released, a finite time is required for the particle to return to its neutral position. This time is called the relaxation time of the particle. For a medium that is composed of the same type of particles, this relaxation time also corresponds to the relaxation time of the medium. If the relaxation time is much shorter than the period of the wave, its effect on the wave should be small. However, if the relaxation time is comparable to the period of the wave, the particle may not be able to completely return to its neutral state before a second wave arrives. When this occurs, the wave is moving in one direction, whereas the particles are moving in the other direction. More energy is thus required to overcome the particle motion. At the other extreme, if the frequency is so high that the particles simply cannot follow the wave motion, the relaxation effect again becomes negligible. Maximum absorption occurs when the relaxation motion of the particles is completely out of synchronization with the wave motion. It is plausible that the relaxation process is characterized by a relaxation frequency where the absorption is maximal and is negligibly

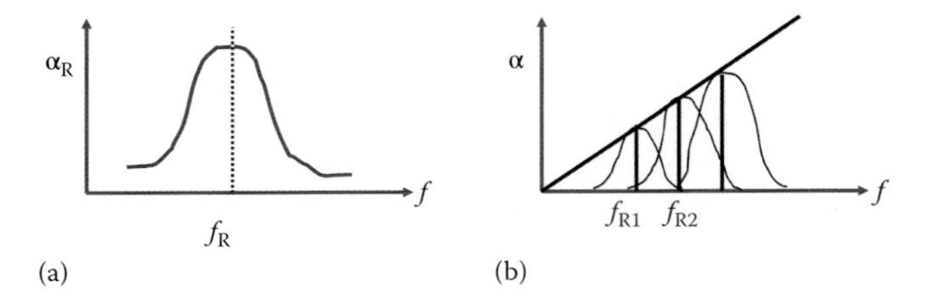

FIGURE 4.3 (a) Ultrasonic absorption caused by a relaxation process characterized by relaxation frequency f_R. (b) Ultrasonic absorption caused by multiple relaxation processes characterized by relaxation frequency f_{R1}, f_{R2} ....

small for extremely low-frequency and high-frequency regions, illustrated in Figure 4.3a. Mathematically, the relaxation process can be represented by the following equation:

$$\alpha_R = \frac{Bf^2}{1+(f/f_R)^2}, \tag{4.20}$$

where α_R is the component of the absorption coefficient due to the relaxation process, f_R is the relaxation frequency, t_R (= $1/f_R$) is the relaxation time, and B is a constant. In a biological tissue, there are many components, giving rise to many relaxation frequencies. The *absorption coefficient* in a tissue, α_a, can be expressed as

$$\alpha_a/f^2 = A + \sum_i \frac{B}{1+(f/f_{Ri})^2}, \tag{4.21}$$

where A is a constant associated with classical absorption and the B_i's and f_{Ri}'s are the relaxation constants and frequencies associated with different tissue components [8]. A possible scenario for this equation is illustrated in Figure 4.3b, where many relaxation processes may overlap, resulting in an approximately linear increase in attenuation in the diagnostic ultrasound frequency range, that is, the product of the attenuation coefficient and wavelength ($\alpha\lambda$) is a roughly a constant. Attenuation in general is not desirable, because it limits the depth of penetration of ultrasound into the body. However, it may yield useful information for diagnostic purposes because it carries information about the properties of the tissues if it can be accurately estimated.

4.2.7.3 Scattering

As a wave is incident on an object as shown in Figure 4.4, part of the wave will be scattered, and part of the wave will be absorbed by the object. Since an ultrasonic image is constructed from echoes backscattered by biological tissues, an understanding of the scattering characteristics of tissues is crucial, and the phenomenon of ultrasonic scattering in biological tissues has been studied extensively. In the frequency range from 1 to 15 MHz, the ultrasound wavelength in general is much larger than the size of biological components (e.g., red blood cells) with which the ultrasound interacts. It has been found that scattering from a dilute suspension of red blood cells is related to ultrasound frequency and the acoustic properties of the sample,

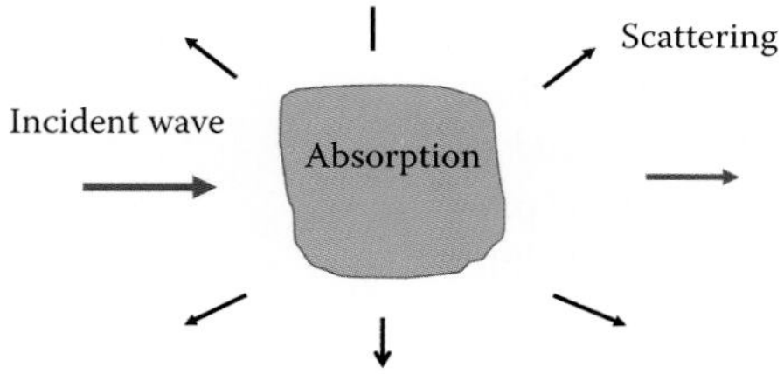

FIGURE 4.4 Scattering resulting from a plane wave incident upon a scatterer.

namely, sound speed and density of the cells. Typically, the scattering has a strong dependence on frequency [9]. In a dense distribution of scatterers like biological tissues, for instance, human blood (consisting of 5×10^9 red blood cells in 1 cm^3), scatterer-to-scatterer interaction cannot be ignored. In this case, the scattering phenomenon becomes very complicated.

Ideally, the structure and acoustic properties of the scatterers can be deduced from measuring their scattering properties. Although the potential of characterizing a tissue structure from its scattering properties was appreciated in the biomedical ultrasound community almost three decades ago, little progress has been made, primarily due to the complex nature of biological tissues.

Since pathological processes in tissues involve anatomical variations, it is likely that they will result in corresponding changes in ultrasonic backscatter and, thus, the characteristics of the detected echo. This is, in fact, the reason why ultrasonic imaging is capable of providing diagnostic information in the body.

4.2.8 Nonlinearity

In *linear acoustics perturbation*, the acoustic properties of a medium caused by the propagation of an acoustic wave are assumed to be small so that they are linearly related. One example would be that the density change produced by a pressure change is linear. Since the spatial peak temporal peak intensity I_{SPTP} of a modern ultrasonic diagnostic instrument may sometimes reach a level of more than 100 W/cm^2, nonlinear acoustic phenomena are bound to occur as ultrasound propagates in tissues [9,10]. Although *nonlinear* or *finite amplitude acoustics* has been studied for a long time, it has not been given much attention in diagnostic ultrasound until recently, when harmonic imaging (in which harmonics generated as a result of the nonlinear interaction between ultrasound and tissues are used to form an image) became popular [4,7,11]. In fact, it has been shown that in many instances, harmonic imaging that utilizes the harmonics generated due to the nonlinear propagation yields better images than conventional ultrasonic imaging.

The nonlinearity of ultrasonic wave propagation can cause the distortion of waveforms. A sinusoidal waveform of high amplitude becomes a sawtooth waveform as a result of the generation of higher harmonics after propagating in a medium because the sound velocity during the compressional phases (denser region of medium) is greater than in the rarefactional phases. As a result, the wave in the denser region will catch up with the wave in the less dense region. The harmonic amplitudes relative to that at the fundamental or first harmonic frequency are known to increase, whereas the amplitude of the fundamental frequency drops as a function of the propagation distance [12].

4.2.9 Doppler Effect

The Doppler effect describes a phenomenon in which a change in the frequency of sound emitted from a source is perceived by an observer when the source or the observer is moving or both are moving. This effect is used in ultrasonic Doppler devices for the measurement and imaging of blood flow transcutaneously, that is, without penetrating the skin in any manner. In these devices, ultrasonic waves are launched into a blood vessel, and the scattered radiation from the moving red cells undergoes a Doppler shift (since the blood is moving) and is detected. Appropriate instrumentation is incorporated to extract the Doppler frequency. The fractional change in frequency is directly proportional to the blood flow velocity. Doppler measurements are discussed in more detail in Section 4.3.2.

4.3 GRAYSCALE ULTRASONIC IMAGING APPROACHES

In this section, crucial components of an imaging system that generates, processes, and displays the ultrasound signal into an image are discussed. It must be stressed that these images yield mostly information about the anatomy (structure) of the object being imaged. There are various modes of ultrasound imaging that have been used over the years; the most popular of which has been *B-mode imaging*.

4.3.1 A (Amplitude)-Mode and B (Brightness)-Mode Imaging

4.3.1.1 A-Mode Imaging

A-mode is the simplest and earliest mode of ultrasonic instrumentation. A block diagram for an analog A-mode instrument is shown in Figure 4.5. In modern instruments, several of these components may be replaced by digital devices, and the instrument is under the control of a computer. A *transducer* is an energy conversion device that converts electrical energy to acoustic energy and vice versa [4,7]. The simplest type is a single-element transducer, the construction of which is shown in Figure 4.6a along with a photograph in Figure 4.6b. The most important component of the device is a *piezoelectric* layer that facilitates the required energy conversion between electrical energy and acoustic energy. A number of piezoelectric materials have been used in medical ultrasonic transducers. Most notable are a ferroelectric ceramic, lead zirconate titanate (PZT), and a piezoelectric polymer, polyvinylidene difluoride (PVDF) [4,7]. There are also several supporting structures. A backing block is used to support the piezoelectric layer and to control the *bandwidth* of the device, which is the frequency band within which most of the energy concentrates. Matching layers, which act much like an electrical transformer in electrical circuits, step the acoustic impedance down, allowing a better acoustic match with a *loading medium* (the medium into which the ultrasonic wave is launched) such as water or biological tissues. This is because the acoustic impedance of piezoelectric materials like PZT is much higher than those of water and tissues. Better acoustic matching minimizes reflection at the interface between piezoelectric materials and the loading medium, and improves energy transmission to the loading medium. A lens typically made of silicon rubber helps to focus the ultrasonic beam.

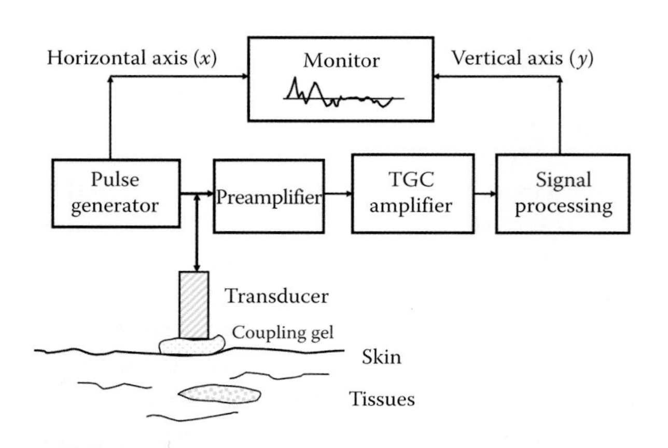

FIGURE 4.5 Block diagram of an A-mode instrument.

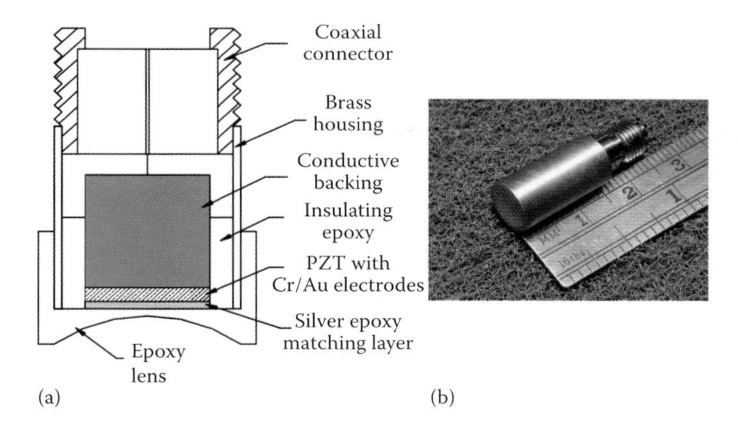

FIGURE 4.6 (a) Detailed construction of a single-element ultrasonic transducer with matching layers and the backing material. (b) Photograph of a transducer.

A signal generator (or pulser) that produces high-amplitude short pulses excites the transducer and causes emission of ultrasonic waves. The returned echoes from the tissues are sensed by the same transducer, amplified, and processed for display. A coupling medium in the form of an aqueous gel or oil is used to couple the transducer to the body because of the mismatch in acoustic impedance between air and the transducer. The echoes are first amplified by a low-noise, high-input-impedance preamplifier and then by the time gain compensation (TGC) amplifier, which compensates for the energy loss suffered by the beam as it penetrates deeper into the medium. The amplified signal is then demodulated and logarithmically compressed. Demodulation is a signal unmixing process that separates a carrier signal from the signal of interest. The carrier signal can be removed by low-pass filtering. Logarithmic compression is needed because the *dynamic range* of the received echoes, which is defined as the ratio of the largest echo to the smallest echo above the noise level detected by the transducer, is very large—on the order of 100 to 120 dB. Typical display units can only display signals with a dynamic range up to 40 dB at best. The horizontal axis of the monitor is synchronized or triggered by pulses generated by the pulser, whereas the vertical axis or vertical deflection of the electron beam is driven by the output of the signal processing unit, the video signal. Most of these components have been replaced by digital devices today, and their functions can be achieved digitally as well.

The echo data acquired by an A-mode instrument or an A-line correspond to the scattered signal received by the transducer as a function of time originating from a single line through the object. Scatterers closer to the transducer will produce an echo earlier than scatterers further away. Therefore, while the A-line is a plot of echo intensity versus time, it can be interpreted as echo intensity versus depth.

4.3.1.2 B-Mode Imaging

Rather than displaying the A-line graphically as shown in Figure 4.5, it can be displayed in a format where the echo amplitude as a function of time is encoded as a grayscale value, called B (brightness)-mode. The simplest way to obtain B-mode display is to use the echo amplitude to modulate the intensity of the electronic beam of the display by translating or scanning the ultrasonic beam, as described later. Different trajectories of the ultrasonic beam through the object are sampled, and a B-mode image can be constructed. It is sometimes desirable not to display the echo information in a linear manner. In a majority of the scanners, the echo amplitude or video signal versus gray level mapping is made adjustable.

An A-line or a single line of the B-mode display yields information about the position of the echo given by $d = (c \times t)/2$, where d is the distance from the transducer to the target, t is the time of flight or the time needed for the pulse to travel from the transducer to the target and return to the transducer, and c is the sound velocity in tissues, which is assumed to be a constant of 1540 m/s in commercial scanners because sound velocities in tissues do not vary significantly. This assumption, however, sometimes may cause errors in distance, area, or volume measurements and image distortion.

A majority of commercial scanners on the market today are 2-D B-mode scanners in which the beam position is also monitored. The positional information of the beam and the video signal representing echoes are converted into a format compatible with a display in a device called a scan converter, which is almost invariably digital today. Images can also be formed by superposition of multiples images after translating and rotating the transducer at a fixed position within a sector angle, called a compound B-scan. The advantages of doing so are making the image look smoother or suppressing the speckle pattern and averaging out the specular echoes due to flat interfaces. Modern B-mode scanners are capable of acquiring images faster than 30 frames per second to allow real-time monitoring of organ motion.

Rather than mechanically translating a single-element transducer, current real-time scanners almost exclusively utilize linear array transducers capable of electronic scanning. Array transducers consist of 128 or more rectangular piezoelectric elements, as shown in Figure 4.7a. A group of elements (32 or more) is fired simultaneously to form a beam. The beam is then swept from one end to the other, as illustrated in Figure 4.7b, to form an image. The array aperture may be linear or slightly curved (termed a curved linear array). An image of a liver and surrounding structures obtained using a curved linear array at 5 MHz is shown in Figure 4.7c. Another version of the linear array transducers is the phased linear array or simply phased array. The phased array has fewer elements, but all elements are fired at the same time to allow both dynamic focusing and beam steering (explained in Section 4.3.1.3 on beam forming). This type of array, which has a smaller "footprint" or aperture size, is better suited for imaging structures that have limited acoustic windows, regions of the body that do not contain bony structures or air, such as the heart.

The block diagram of a linear array B-mode imaging system is shown in Figure 4.8. A pulser is switched on to a group of transducer elements with or without delays. Delays are often used to better focus the beam. The returned echoes detected by the array elements are processed by the front-end analog beam former, consisting of a matrix of delay lines, transmit/receive (T/R) switches, and amplifiers. Several components in a B-mode scanner perform the same functions as those in the A-mode system. These include the TGC amplifier and signal processing units for signal compression, demodulation, and filtering. The timing and control signals are all generated by a central unit that may consist of multiple microprocessors. After signal processing, the signal is digitized by an analog-to-digital (A/D) converter. In high-end systems, the A/D conversion following preamplification is accomplished in the beam former. This makes the system more expensive since many more A/D converters of a higher sampling rate are required. An 8-bit A/D converter digitizes the signal into $2^8 = 256$ gray levels. For better contrast resolution, more bits are needed. The scan converter is a digital memory device that stores the data that have been

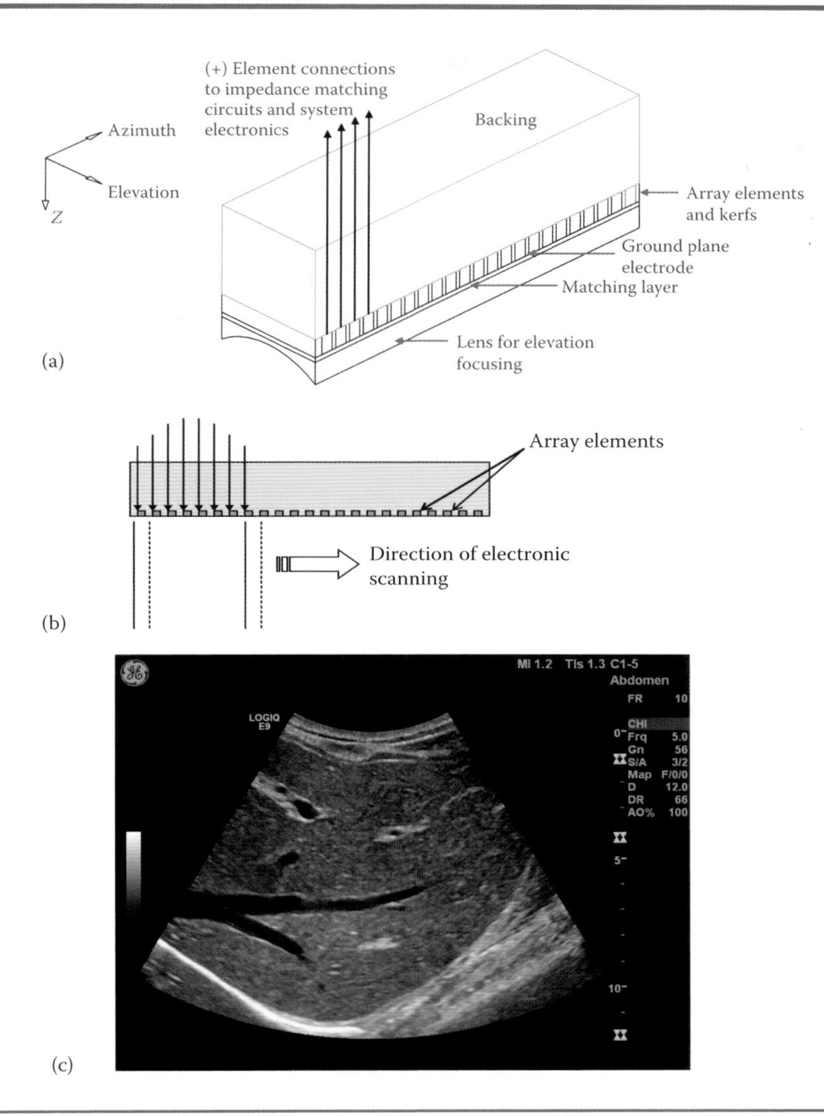

FIGURE 4.7 (a) Detailed construction of a linear array with two matching layers, a lens, and light backing. (b) An image is formed by a linear array by electronically sweeping the beam. A group of elements are fired simultaneously to form one beam. (c) B-mode image of liver and surrounding structures obtained with a curved linear array at 5 MHz. (Courtesy of GE Healthcare.)

converted from the format in which they were collected into a format that is displayable by a monitor. Before display, the video data may be processed again via band-pass filtering, high-pass filtering, low-pass filtering, grayscale mapping, and so forth. Signal processing performed before and after the scan converter is called preprocessing and postprocessing, respectively.

For most B-mode scanners, only one ultrasound pulse is being transmitted at any one instant of time. The time needed to form one image frame, t_f, can be readily calculated from the following equation.

$$t_f = \frac{2DN}{c}, \tag{4.22}$$

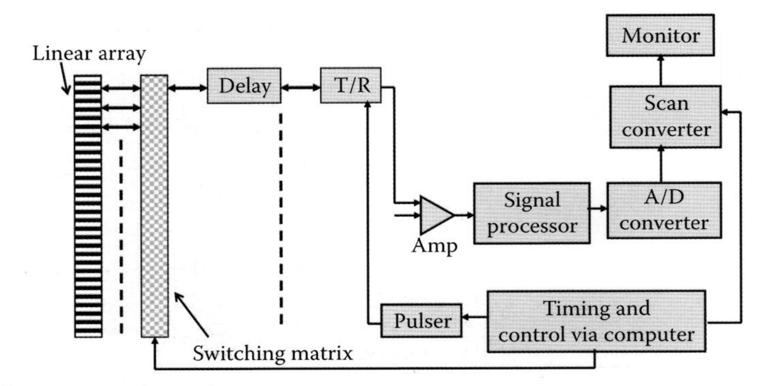

FIGURE 4.8 Block diagram of a B-mode scanner.

where D is the depth of penetration determined by the pulse repetition frequency of the pulser, N is the number of scan lines in the image, and c is the sound speed in tissue. Rearranging this equation,

$$FDN = \frac{c}{2}, \tag{4.23}$$

where $F = 1/t_f$ is the frame rate. The depth of penetration D is specified by the pulse repetition period of the scanner, which should be long enough to allow all the echoes of interest to be detected. Otherwise, *range ambiguity* can occur, that is, it is not clear which pulse causes the echo from the object, if the time of flight from an object of interest is longer than the pulse repetition period.

4.3.1.3 Beam Forming

In real-time imaging with phased arrays, the ultrasonic beam can be dynamically focused and steered by applying appropriate time delays to the transmitted pulses or/and received echoes. This is illustrated in Figure 4.9, which shows the top view of several elements of a linear array. The time delays can be obtained by considering that the difference in path length between the nth element and the center element is $\Delta r = r_n - r$. Hence,

$$\Delta t_n = \Delta r_n / c \approx \frac{x_n \sin\phi_x}{c} + \frac{x_n^2}{2cr}, \tag{4.24}$$

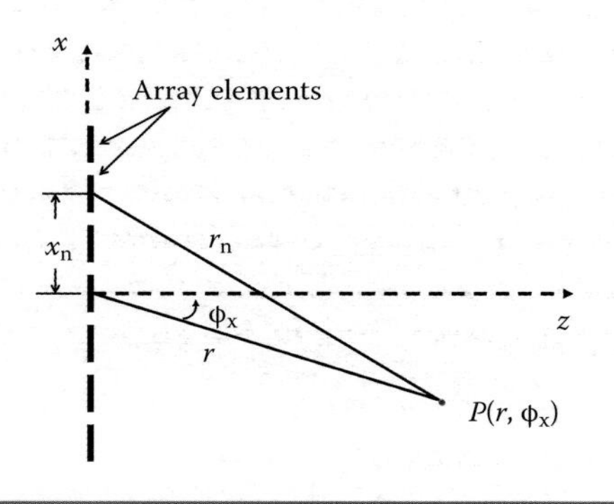

FIGURE 4.9 A 2-D coordinate system depicting the difference in path length between the center element of a linear array and the *n*th element.

where Δt_n and x_n denote the time delay to the nth element relative to the center element and the location of the nth element relative to the center element, respectively. The first and second terms in Equation 4.24 represent, respectively, the time delays needed for achieving beam steering and focusing. The same criteria can be applied to the receiving beam. Figure 4.10 illustrates how dynamic focusing is achieved.

This time delay capability is one of the functions provided by the beam former of the ultrasonic imaging system. Other functions of the beam former are weighting and apodization of the transmitted and received signals. Most high-end linear array systems also

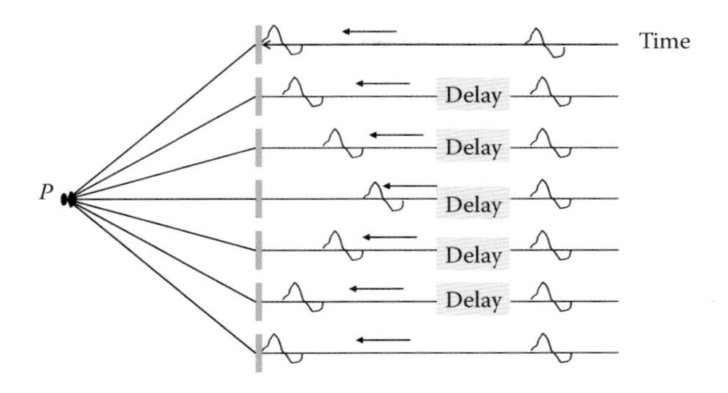

FIGURE 4.10 Illustration of dynamic transmit focusing. Transmitted signals are delayed to change the exiting time of each element to compensate for the path difference between the element and the point of interest.

have dynamic focusing and limited beam steering capability for Doppler blood flow applications. The steering capability is needed because oftentimes, the ultrasound beam is at 90° relative to the blood flow. No flow signal would be detected in this case. Slight steering of the beam can readily alleviate this problem.

4.3.1.4 Spatial Resolution of B-Mode Ultrasonic Imaging Systems

The resolution of a B-mode imaging system in the imaging plane is determined by the duration of the pulse in the depth direction (i.e., in the direction of the beam) and the width of the ultrasonic beam in the lateral direction (i.e., in the direction perpendicular to the beam). The slice thickness of the imaging plane or the beam width in the plane perpendicular to the imaging plane is fixed and determined by the lens in front of the linear transducer array.

The emitted pulse duration and the beam width of a transducer may be defined as the time duration and spatial distance between the −3 or −6 dB points because whether the echoes from two targets either in the axial or in the lateral direction can be separated or resolved is directly related to these parameters. This is graphically illustrated in Figure 4.11a, where it can be seen that the echoes from two targets can be clearly resolved if they are far apart. As the targets are moved closer, as shown in Figure 4.11b, they become more difficult to resolve.

The axial and lateral resolution of a transducer can be improved from an increase in the bandwidth by using backing and/or matching and focusing. The spectrum of an ultrasonic pulse varies as it penetrates into tissue because the attenuation of the tissues is frequency dependent. It is known that the center frequency and bandwidth of an ultrasonic pulse decrease as the ultrasound pulse penetrates deeper. Since axial resolution of a beam is determined by the center frequency and bandwidth, the axial resolution worsens as the beam penetrates deeper into the tissue. In commercial scanners, pulse shape and duration are maintained by TGC, which increases signal gain as a function of echo depth and some form of signal processing, for example, applying higher gain to higher-frequency components.

4.3.2 Doppler Flow Measurements

The Doppler effect provides a unique capability for ultrasound to noninvasively measure instantaneous blood flow in a human body. Blood flow measurements are frequently

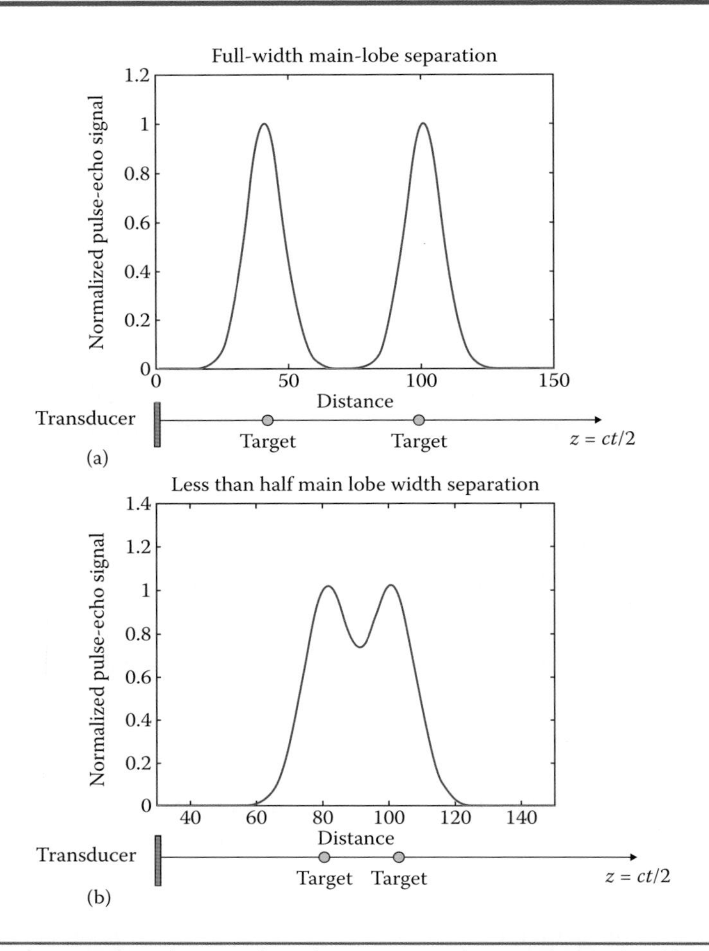

FIGURE 4.11 The spatial resolutions of an ultrasonic transducer in the axial and lateral directions are determined by the pulse duration and beam width. (a) When the targets are far apart, they can be readily resolved. (b) When they are closer, they can be more difficult to be resolved.

performed in a clinical environment to assess the state of blood vessels and functions of an organ. Combined with pulse-echo imaging, instantaneous flow rate in a blood vessel as a function of time and cardiac output can be measured noninvasively with ultrasound. Figure 4.12 shows an ultrasound beam at frequency f in sonifying a blood vessel, making an angle of θ relative to the blood flow velocity v, assuming that blood flows in a vessel with a uniform velocity v. The returned echoes are Doppler shifted (Section 4.2.9). The Doppler shift frequency f_d is related to the ultrasound frequency f by the following equation:

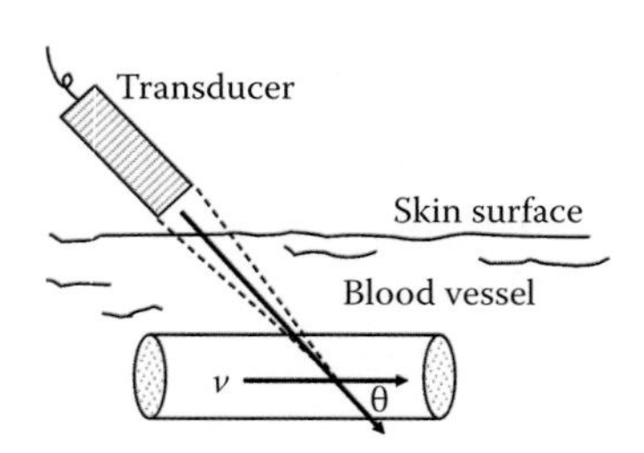

FIGURE 4.12 An ultrasound beam is incident upon a blood vessel and makes an angle of θ relative to the direction of blood flow.

$$f_d = \frac{2v\cos\theta}{c} f, \tag{4.25}$$

where c is the sound velocity in blood and may be assumed to be 1540 m/s. The Doppler-shifted frequencies happen to be in the audio range for blood flow velocities in the human body for an ultrasound frequency between 1 and 15 MHz. In reality, blood flow in a blood vessel is pulsatile, and the velocity varies over the lumen. As a result, the Doppler signal covers a spectrum of frequencies. Two different approaches have been used for ultrasonic Doppler flow measurements: continuous-wave (CW) and pulsed-wave (PW) Doppler. In the former, continuous sinusoidal ultrasound is generated by a transmitting piezoelectric element, and the returned echoes are received by another element. In the latter, the same element generates a burst of a few cycles and detects the returned echoes much like a pulse-echo imaging approach. Echoes from a certain depth are processed to extract the Doppler signal using a time-gate function. CW Doppler has a fixed sensitive volume, which is the overlapping region of the two piezoelectric elements, but does not have the signal aliasing problem. The pulsed Doppler overcomes the fixed-depth problem associated with CW Doppler but suffers from signal aliasing [2–4] caused by undersampling of the signal.

4.3.3 Color Doppler Imaging

Ultrasonic B-mode real-time imaging can be combined with Doppler in a scanner so that the scanner is capable of providing not only anatomical information but also blood flow data. Both sets of information are displayed simultaneously. A cursor line is typically superimposed on the B-mode image to indicate the direction of the Doppler beam. A fast Fourier transform (FFT) algorithm is used to compute the Doppler spectrum that is displayed in real time. This type of scanner is called a duplex scanner. Alternatively, blood flow data can be displayed in real time and superimposed with the B-mode image if the data acquisition rate and image processing algorithms are sufficiently fast to allow that.

In advanced ultrasonic imaging systems, color Doppler flow imaging is an important function. In these systems, grayscale B-mode and color blood flow information extracted from Doppler frequencies is displayed simultaneously in real time [2–4]. Conventionally, the color red is assigned to indicate flow toward the transducer, and the color blue is assigned to indicate flow away from the transducer. The magnitude of the velocity is represented by different shades of the color. Typically, the lighter the color, the higher the velocity. A color Doppler image of blood vessels on the umbilical cord in the uterus is shown in Figure 4.13.

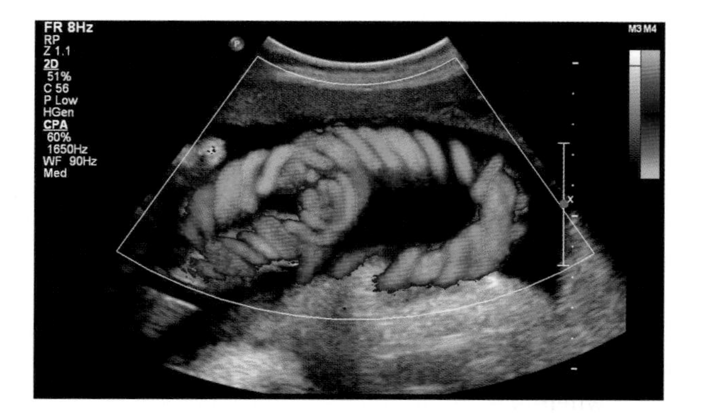

FIGURE 4.13 Color Doppler image of blood vessels on umbilical cord in uterus. (Courtesy of Philips.)

The fundamental principle involved in color Doppler is fairly straightforward. Assuming a plane wave propagation in the z-direction given by Equation 4.3, the phase of the wave is $\phi = 2\pi ft - kz$. The time derivative of the phase is then

$$\frac{d\phi}{dt} = 2\pi f. \tag{4.26}$$

This equation can be rearranged to become

$$f = \frac{1}{2\pi}\frac{d\phi}{dt}. \tag{4.27}$$

Clearly, from the phase of the returned echoes, the Doppler frequency shift can be extracted, whereas the amplitude of the echoes yields the grayscale B-mode image.

The basic instrumentation for a color Doppler is similar to that of a pulsed Doppler instrument that extracts the mean Doppler shift frequency from a sample volume defined by the beam width and the width of the time gate that specifies the depth. The exception is that color Doppler instruments are capable of estimating the mean Doppler shifts of many sample volumes along a scan line in a very short period of time, on the order of 30 to 50 ms. To be able to do so, fast algorithms have to be developed [5,13,14]. The current approach takes advantage of the autocorrelation function, which, with the power spectrum, forms a Fourier Transform pair. Through approximations and simplifications, it can be shown that for a complex autocorrelation function $H(t) = R(t) + jI(t)$ where the phase $\gamma = \tan^{-1}[I(t)/R(t)]$, the mean frequency can be roughly estimated from the slope of the phase of the autocorrelation function at two points in time γ and γ' separated by Δt, that is,

$$\langle f \rangle = \frac{1}{2\pi}\frac{\gamma - \gamma'}{\Delta t}. \tag{4.28}$$

In practice, this relationship is achieved first by computing the autocorrelation function of the time waveform $f(t)$, and then, from the complex correlation function, the phase of $f(t)$ is extracted after each transmission of the Doppler burst, that is, $\Delta t = T$, the burst repetition period.

The accuracy of the estimated mean frequency is ultimately determined by the time duration in which the estimation is performed. The longer the time duration, the better the accuracy. This requirement must be compromised in real-time ultrasonic imaging. Typically, the estimation is accomplished with 8 to 10 bursts, resulting in a reduction in the frame rate of color Doppler scanners (15 frames per second or less).

In the heart, the myocardium is in motion during a cardiac cycle, and tissue color Doppler images of this motion can be acquired with color Doppler methods previously described as well. The difference is that myocardial motion is slower than blood flow and myocardial echoes are stronger than blood. The spurious Doppler signals from blood in this case can be eliminated by thresholding the echoes.

Many clinical applications have been found for color Doppler flow imaging, including diagnosing tiny shunts in the heart wall and valvular stenosis. This method considerably reduces the examination time in many diseases associated with flow disturbance. Problematic regions can be quickly identified first from the flow mapping, and more quantitative conventional Doppler measurements are then made on these areas.

4.3.4 Color Doppler Power Imaging

Color Doppler information may be displayed differently in the so-called power mode (sometimes called *energy mode*) [5,15]. Instead of the mean Doppler shift, the power or energy scattered by blood contained in the Doppler signal is displayed in this approach, which has several advantages over color Doppler: (1) A threshold can be set to minimize the effect of noise; (2) the data can be averaged to achieve a better signal-to-noise ratio, while this cannot be done in traditional color Doppler, in which Doppler frequency shift is estimated; (3) the images are less dependent upon the Doppler angle (θ in Figure 4.12); and (4) aliasing is no longer a problem since only the power is detected. As a result, signals from blood flowing in much smaller vessels can be detected. The images so produced have an appearance similar to that of x-ray angiography. The disadvantages of this approach are that it is more susceptible to motion artifacts because of frame averaging and the image contains no information on flow velocity and direction. In a color Doppler "power" image, it is important to note that the color region indicates the presence of blood flow. The shade of the color delineates the strength of the Doppler signal and has nothing to do with the flow velocity.

4.3.5 Elasticity Imaging

Ultrasound has been widely used to differentiate cysts from solid tumors in tumor imaging because liquid-containing cysts are typically echo poor. However, certain solid tumors, harder than surrounding tissues, are frequently missed by ultrasound because their echogenicity is similar to that of surrounding tissues. These tumors or harder tissues are identifiable if their elastic properties can be imaged. At present, several ultrasound approaches have been developed to image tissue elasticity, and these can be classified into two categories: static and dynamic methods [16–20].

4.3.5.1 Elastography

In the static approach, dubbed elastography [17], a flat plate, which may be the ultrasonic probe, is used to compress the tissue by a distance Δz. The displacement of the tissue is estimated via a mathematical technique known as cross-correlation by comparing the returned RF echoes before and after compression. Under idealistic conditions, assuming a 1-D case, that is, the compression is transmitted only in the z-direction, the stress, defined as tensile force per unit area, and strain, defined as displacement per unit distance, can be estimated from the force applied and area of the plate or transducer and Δz, respectively. The *Young's modulus* of the tissue is then given by the ratio of longitudinal stress to longitudinal strain. This idea can be illustrated by the following simple example where mediums I and II have different elastic properties (Figure 4.14a). Medium II is harder than medium I. When a force is applied to the surface, then along z, the medium displacement will be different in mediums I and II. This is plotted in Figure 4.14b, where the strains in mediums I and II are given respectively by $\Delta z/d$ and $\Delta z'/(L - d)$. The symbols Δz and $\Delta z'$ represent, respectively, the displacements produced by the compression in mediums I and II.

This approach has been the focus of attention in cancer imaging in the breast, prostate, and other organs. It has been incorporated into several commercial scanners. An example is shown in Figure 4.15, where the elastogram (an image obtained by elastography) and B-mode image of a breast lesion are compared. Although elastography may be capable of imaging elastic properties of tissues, which cannot be achieved with standard B-mode sonography, it suffers from similar problems and limitations. The stress propagating into

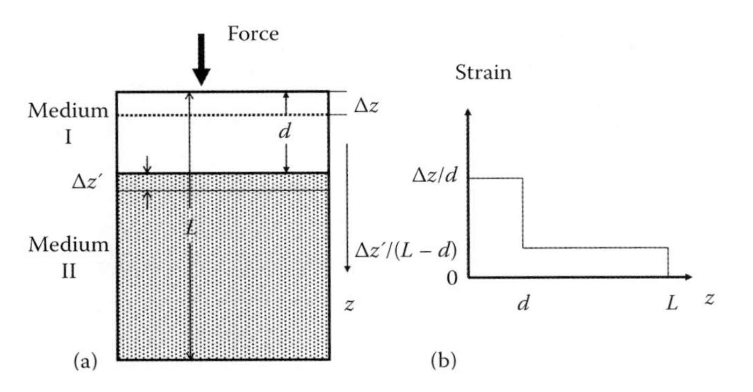

FIGURE 4.14 Diagrams illustrating the concept of elastography. (a) A force applied to medium I produces displacements in medium I and II. (b) Strain in medium I and II is plotted as a function depth.

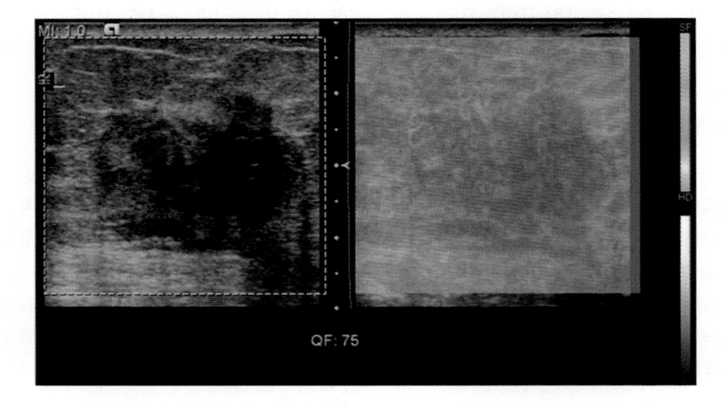

FIGURE 4.15 Elastogram and B-mode sonogram of a lesion in breast. The elasticity is delineated in color scale in the elastogram on the right. (Courtesy of Siemens.)

the tissue of interest is attenuated by other tissues, spreads into other directions from the primary incident direction, and interacts with the boundary between two media of different elastic properties. The effect of boundaries on an elastogram can sometimes be distracting.

4.3.5.2 Sonoelasticity Imaging

Sonoelasticity imaging [18] is a dynamic approach in which a motion monitoring device, for example, a color Doppler flow mapping system, is used to measure the motion of tissues produced by the vibration of a source inserted into one of the cavities of the body or placed externally. The source typically vibrates at a frequency of a few hundred hertz to a few kilohertz so that a conventional Doppler device can be used to monitor the motion with little modification. This approach has been used to assess prostate cancer with an intrarectal vibrating source. Sonoelasticity imaging suffers from similar problems to those mentioned previously. The vibration produced by the source is nonuniform and is attenuated by tissues. The energy of the ultrasound generated by the scanner to probe the vibration is also affected by attenuation.

4.3.5.3 Acoustic Radiation Force Imaging

A body force called radiation force is generated from a transfer of momentum by an ultrasonic wave on a medium caused by attenuation [7,20]. If the ultrasound intensity is

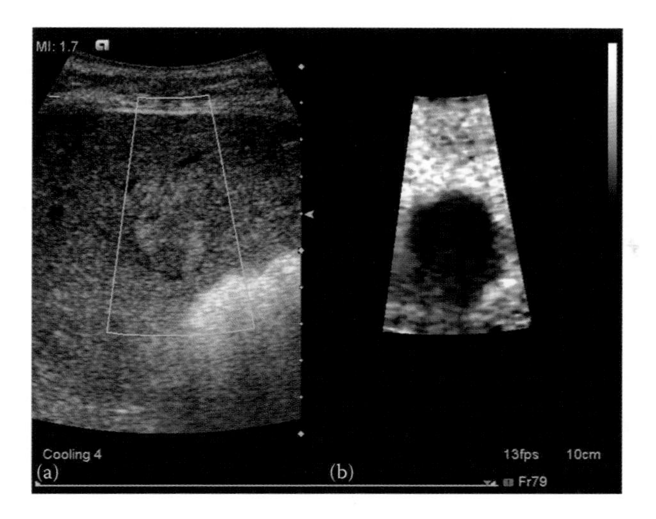

FIGURE 4.16 Elastogram acquired by ARFI (b) and B-mode sonogram (a) of a lesion in liver. The lesion is poorly distinguished from the surrounding tissues, whereas it is quite apparent in the elastogram. (Courtesy of Siemens.)

sufficiently high, this force may be used to produce a displacement of the tissue, which can be subsequently estimated from an RF signal or video (demodulated RF) signals before envelope detection by cross-correlating these signals before and after displacement. In the initial experiment carried out by Nightingale [7] and Palmeri and Nightingale [20], a commercial 7.2 MHz linear array was modified to consecutively fire a pushing beam (to apply the radiation force) with a duration of a fraction of a millisecond and an imaging beam with a pulse duration of a fraction of a microsecond at a pulse repetition frequency of a few kilohertz. Image acquisition time was less than 350 ms. Several commercial scanners now have acoustic radiation force imaging (ARFI) as an option. Figure 4.16 shows an elastic image acquired with ARFI.

4.4 ULTRASONIC CONTRAST AGENTS AND HARMONIC IMAGING

4.4.1 Contrast Agents

Contrast media have been used widely in radiology, cardiology, and other medical disciplines to enhance the contrast of anatomic structures or functional properties in an image. Iodinated compounds have been used for many years in x-ray angiography (Section 2.3.4.3) to better visualize the coronary vasculature in the heart by injecting agents into the coronary artery. Ultrasonic agents have also been successfully developed [21,22], and many applications have been found.

These ultrasonic contrast agents consist of small particles with acoustic properties very different from blood. These particles should be (1) nontoxic, (2) more echogenic than tissues, (3) capable of traversing the pulmonary circulation, (4) stable, and (5) of narrow size distribution. Requirement (3) is dictated by the need for intravenous injection, since this is much less risky than intra-arterial injection. To allow the applications of these agents to imaging left cardiac structures and other parts of the body, for example, the liver and kidney, via intravenous injection, they must be smaller than a few microns, stable, and able to persist at least for a few seconds. To satisfy requirements (2) and (3), a majority of these agents utilize microscopic air bubbles encapsulated in a shell, which are extremely strong ultrasound scatterers because of the acoustic impedance mismatch between air and blood.

These bubbles resonate when disturbed. As a result, the echoes from the bubbles can be further enhanced if the incident wave is tuned to the resonant frequency of the bubbles [23]. The resonant frequency for a free air bubble, a bubble without a shell, f_r, is related to the radius of the bubble, a, by

$$f_r = \frac{1}{2\pi a}\sqrt{\frac{3\gamma P_0}{\rho_w}}, \quad (4.29)$$

where γ is the ratio of the specific heats at constant pressure and constant volume of a gas and equals 1.4 for air, P_0 is the hydrostatic ambient pressure and equals 1.013×10^5 Pa (equal to 1.013×10^6 dynes/cm^2) at 1 atm, and ρ_w is the density of the surrounding medium, for example, water. The resonating frequency for a bubble with a radius of 1.7 μm is 2 MHz. When the bubble is encapsulated by a shell, for example, a layer of lipid, the resonance response is damped. To produce a resonance at 2 MHz, the encapsulated bubble radius has to be increased to 2.4 μm.

4.4.2 Harmonic Imaging

The nonlinear interaction of a bubble with an impinging ultrasonic wave in an incompressible fluid has been studied extensively [24,25]. These studies showed that harmonic signals at $2f$, $3f$, ... and subharmonic signals at $1/2f$, $1/4f$, ..., where f denotes the resonating frequency of the bubble, could be generated by the bubble. These results were modified later [26] to include the effect of a shell for encapsulated contrast agents. A very exciting application of this phenomenon is harmonic imaging, where the effect of the surrounding stationary structures on the image is minimized. Harmonic imaging and Doppler measurements following the injection of a gas-containing contrast agent are possible because only microbubbles resonate when impinged upon by ultrasound of appropriate frequency and emit ultrasound at harmonic frequencies. Ultrasound at harmonic frequencies is therefore only produced within anatomic structures that contain these agents. Tissues that do not contain gaseous contrast agents presumably do not normally produce harmonic signals. A good example is blood flowing in a blood vessel. Blood containing the contrast agent produces harmonic signals, but the blood vessel does not. The contrast between the blood and the blood vessel is therefore significantly improved in the resultant harmonic image if the transducer detects only the echoes at the appropriate harmonic frequency. The simplest approach that may be taken to produce harmonic images is to generate a wide-band acoustic signal. Only the harmonic signals are received for image formation. An array or transducer is excited with a wide-band pulse with the center frequency being the first harmonic or fundamental frequency signal, and the echoes received by the transducer are then filtered to collect the data only at the harmonic frequency. A major problem for such an approach is that it is impossible to completely suppress echoes at the fundamental frequency due to spectral leakage, which means that signals in the fundamental band spill into the harmonic band. To overcome this, several approaches have been studied. The most common is the pulse inversion method [5,27], shown in Figure 4.17. In this approach, two pulses that are 180° out of phase are transmitted sequentially. The returned echoes are summed up. The

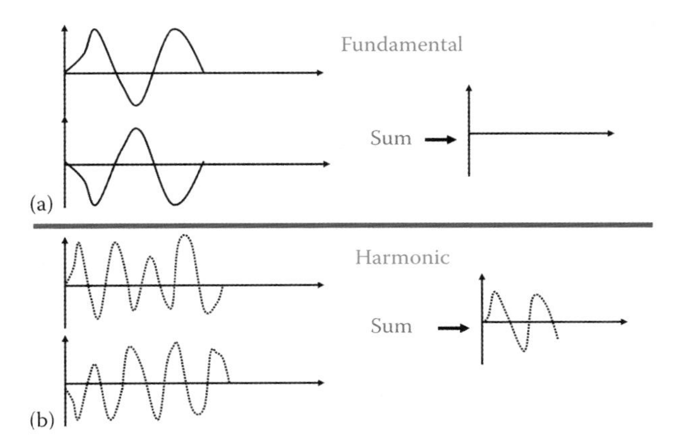

FIGURE 4.17 In pulse inversion harmonic imaging, two pulses 180° out of phase are transmitted. The returned echoes at fundamental frequency cancel each other out upon summation (a), whereas the echoes at harmonic frequency do not (b).

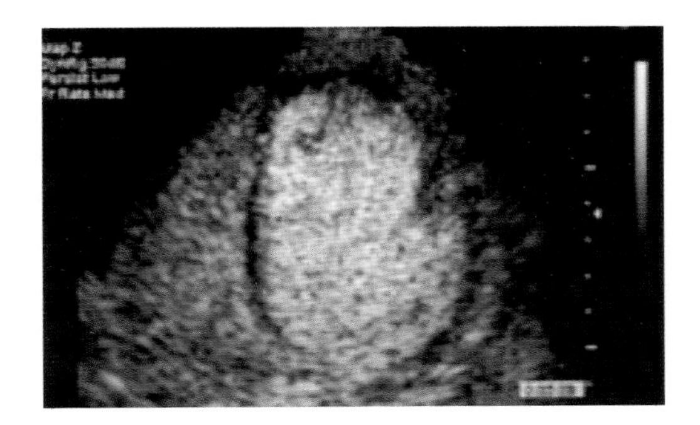

FIGURE 4.18 Ultrasonic harmonic image of the left ventricle of a human heart following intravenous injection of a contrast agent. (Courtesy of Philips.)

echoes at the fundamental frequency are cancelled out, whereas those at the harmonic frequencies are not. The price to pay, however, is a lower frame rate.

Figure 4.18 shows a harmonic image of the left ventricle of a human heart following intravenous injection of a contrast agent in which the echogenicity of the myocardium is lower than the blood pool that contains the contrast agent.

4.4.3 Native Tissue Harmonic Imaging

Harmonic imaging has also been performed on tissues without the injection of a contrast agent. As previously mentioned, harmonic and subharmonic signals are produced as the ultrasound pulse penetrates into the body because of the nonlinear interaction between the tissues and ultrasound energy. Energy at the fundamental frequency is partly absorbed, partly scattered, and partly converted into harmonic and subharmonic signals. The harmonic signals increase initially, reach a plateau, and then decrease as a function of depth.

Native tissue harmonic imaging has a major advantage over conventional B-mode imaging in that it is capable of penetrating deeper into tissues since more harmonic signals are generated due to nonlinear interactions between the tissues and the beam until they are offset by the increased attenuation. Consequently, obese patients frequently are better imaged with harmonic imaging.

4.5 OTHER DEVELOPMENTS

A majority of ultrasonic scanners today are color Doppler B-mode devices that are capable of acquiring 2-D tomograms of a slice of the body, for example, a slice of the heart, in real time. These devices utilize 1-D linear arrays or phased arrays, and 3-D imaging may be obtained by mechanically scanning the linear arrays. However, to achieve 4-D imaging, that is, 3-D imaging in real time, 2-D transducer arrays must be used [4,28,29]. Recently 2-D arrays consisting of more than 80 × 80 elements with 256 electronic channels have been developed and shown to be capable of performing 4-D color Doppler imaging. Clinical applications of these devices in diagnosing a number of cardiac disorders including valvular stenosis have been demonstrated.

Another significant development is in high-frequency (>30 MHz) ultrasonic imaging, which offers improved spatial resolution by sacrificing depth of penetration [4,30]. Mechanical high-frequency scanners have long been used for eye, skin, small animal, and IVUS imaging [4,31,32]. A major obstacle for high-frequency ultrasonic imaging in real time with an image quality and performance similar to low-frequency clinical scanners has been the lack of available high-frequency linear arrays; this has now been overcome [33,34]. A high-frequency linear array scanner is commercially available for small animal or preclinical imaging in the frequency range from 30 to 50 MHz [34]. This capability combined with microbubble contrast agents that can be conjugated with different molecules to allow them to bind with proteins and cells has made it possible for ultrasound to migrate into the arena of molecular imaging [34]. Figure 4.19 shows two images obtained at 50 MHz of a melanoma implanted on a nude mouse following the injection of a microbubble contrast agent conjugated with a molecule that targets endothelial cells. In the image on the right, the echoes from the bubble are encoded in green. Ultrasound, along with micro-CT, micro-MR and micro-PET, micro-SPECT, and optical imaging, has become an essential tool for preclinical imaging.

As a final note, the trend of ultrasound in recent years has been in the miniaturization of ultrasonic scanners. A number of small, lightweight portable scanners have been introduced by commercial vendors. Figure 4.20 shows an iPod-size scanner developed by GE that weighs only 390 g and has an 8.9 cm display. It is equipped with a 3 MHz phased array capable of color Doppler. The anticipated applications of these devices are in emergency rooms or primary physicians' offices for initial screening.

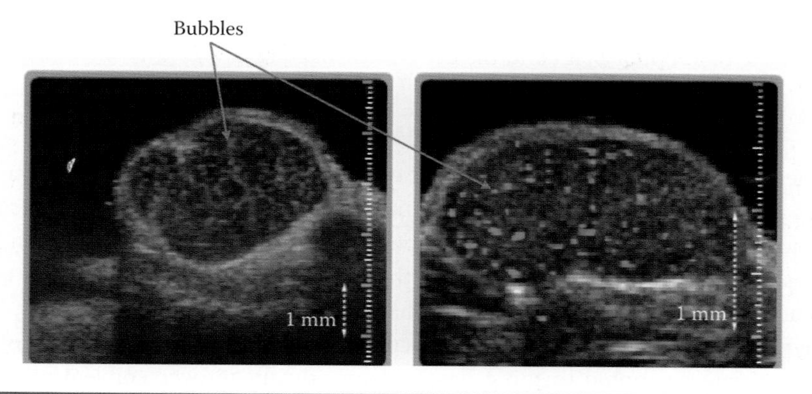

FIGURE 4.19 Two images obtained at 50 MHz of a melanoma implanted on a nude mouse following the injection of a microbubble contrast agent conjugated with a molecule that targets endothelial cells. In the image on the right, the echoes from the bubble are encoded in green. (Courtesy of Fuji Visualsonics.)

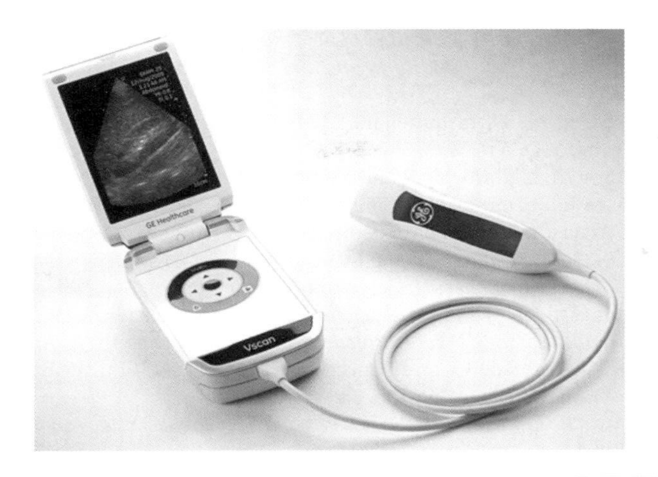

FIGURE 4.20 A miniature lightweight portable ultrasonic scanner. (Courtesy of GE Healthcare.)

4.6 ULTRASOUND BIOEFFECTS

Ultrasound is a form of nonionizing radiation, meaning that the energy carried by it in diagnostic ultrasound is not sufficient to ionize molecules or atoms. Possible ultrasound bioeffects have been classified into thermal effects and nonthermal effects (such as those caused by cavitation, the generation of air bubbles). However, these effects are usually difficult to separate in real experiments. A large body of data has been accumulated over the years from both water tank studies and animal studies in an effort to establish whether diagnostic ultrasound produces any biological effects. These data suggest that in the low megahertz frequency range, no adverse bioeffects are observed on nonhuman biological tissues exposed *in vivo* under the following experimental ultrasound conditions: (1) if, when a thermal mechanism is involved and an unfocused beam is used, the spatial peak temporal averaged intensity (I_{SPTA}) is below 100 mW/cm^2; (2) if, when a thermal mechanism is involved and a focused beam is used, the I_{SPTA} is below 1 W/cm^2; and (3) if, when a nonthermal mechanism is involved, the *in situ* peak rarefactional pressure (negative peak pressure) is below approximately 0.3 MPa.

In summary, it is safe to say that the current consensus on possible bioeffects of ultrasound is that as long as the energy imparted to the tissues is smaller than 50 J/cm^2, no adverse bioeffects will be produced.

The output levels of a number of diagnostic ultrasound instruments have been reported [35], ranging from 20 mW/cm^2 (I_{SPTA}) for B-mode scanners to 234 mW/cm^2 (I_{SPTA}) for pulsed Doppler flowmeters. Finally, it is worthwhile to point out that the bioeffects produced by ultrasound can also be used to its advantage for therapeutic applications, for example, hyperthermia, high-intensity focused ultrasound (HIFU) surgery, drug delivery, and gene transfection [35–38].

REFERENCES

1. Goldberg BB and Kimmelman BA. *Medical Diagnostic Ultrasound: A Retrospective on Its 40th Anniversary.* Laurel, MD: AIUM, 1988.
2. Szabo T. *Diagnostic Ultrasound Imaging: Inside Out.* Amsterdam: Elsevier Academic Press, 2004.

3. Cobbold RSC. *Foundations of Biomedical Ultrasound*. New York: Cambridge Press, 2007.
4. Shung KK. *Diagnostic Ultrasound: Imaging and Blood Flow Measurements*. Boca Raton, FL: CRC Press, 2005.
5. American Institute of Ultrasound in Medicine. *Safety Considerations for Diagnostic Ultrasound*. Laurel, MD: AIUM, 1984.
6. Nightingale K, Soo M, Nightingale R and Trahey G. Acoustic radiation force impulse imaging: In vivo demonstration of clinical feasibility. *Ultrasound Med Biol* 2002; 28: 227–235.
7. Nightingale K. Acoustic radiation force impulse (ARFI) imaging: A review. *Curr Med Imaging Rev* 2011; 7: 328–339.
8. Hchwan HP. *Biological Engineering*. New York: Wiley, 1969.
9. Morse PM and Ingard KU. *Theoretical Acoustics*. New York: McGraw Hill, 1968.
10. Hamilton MF and Blackstock DT. *Nonlinear Acoustics*. San Diego, CA: Academic Press, 1998.
11. Tranquart F, Grenier N, Eder V and Pourcelot L. Clinical use of ultrasound tissue harmonic imaging. *Ultrasound Med Biol* 1999; 25: 889–894.
12. Pierce A. *Acoustics: An Introduction to Physical Principles and Applications*. New York: McGraw Hill, 1986.
13. Kasai C, Namekawa K, Koyano A and Omoto R. Real-time two-dimensional blood flow imaging using an autocorrelation technique. *IEEE Trans Sonics Ultrason* 1985; 32: 458–463.
14. Jensen JA. *Estimation of Blood Velocities Using Ultrasound*. Cambridge, UK: Cambridge Press, 1996.
15. Rubin M, Bude RO, Carson PL, Bree RL and Adler RS. Power Doppler US: A potentially useful alternative to mean frequency-based color Doppler US. *Radiology* 1994; 190: 853–856.
16. Lerner RM, Huang SR and Parker K. Sonoelasticity images derived from ultrasound signals in mechanically vibrated tissues. *Ultrasound Med Biol* 1990; 16: 231–239.
17. Ophir J, Cespedes I, Ponnekanti H, Yazdi Y and Li X. Elastography: A quantitative method for imaging the elasticity of biological tissues. *Ultrason Imaging* 1991; 13: 111–134.
18. Park KJ, Doyley MM and Rubens DJ. Imaging the elastic properties of tissue: The 20-year perspective. *Phys Med Biol* 2011; 56: R1–R29.
19. Treece G, Lindop J, Cehn L, Housden J, Prager R and, Gee A. Real-time quasi-static ultrasound elastography. *Interface Focus* 2011; 1: 540–552.
20. Palmeri MA and Nightingale KR. Acoustic radiation force-based elasticity imaging methods. *Interface Focus* 2011; 1: 553–564.
21. Goldberg BB, Liu JB and Forsberg F. Ultrasound contrast agents: A review. *Ultrasound Med Biol* 1994; 20: 319–333.
22. De Jong N. Improvements in ultrasound contrast agents. *Eng Med Biol Mag* 1996; 15: 72–82.
23. Leighton TG. *The Acoustic Bubble*. San Diego, CA: Academic Pres, 1994.
24. Rayleigh L. On the pressure developed in a liquid during the collapse of a spherical cavity. *Phil Mag* 1917; 34: 94–98.
25. Plesset MS. The dynamics of cavitation bubbles. *J Appl Mech* 1949; 16: 277–282.
26. De Jong N, Cornet R and Lancee CT. High harmonics of vibrating gas-filled microspheres. Part one: Simulations. *Ultrasonics* 1994; 32: 447–453.
27. Averukiou MA. Tissue harmonic imaging. *2000 IEEE Ultrasonics Symp Proc* 2000; 1563–1572.
28. Savord B and Solomon R. Fully sampled matrix transducer for real-time 3D ultrasonic imaging. *2003 IEEE Ultrasonics Symp Proc* 2003; 945–953.
29. Powers J and Kremaku F. Medical ultrasound systems. *Interface Focus* 2011; 1: 477–489.
30. Shung KK. High frequency ultrasonic imaging. *J Med Ultrasound* 2009; 17: 25–30.
31. Pavlin CJ and Foster FS. *Ultrasound Microscopy of the Eye*. New York: Springer-Verlag, 1995.
32. Saijo Y and van der Steen AFW. *Vascular Ultrasound*. Tokyo: Springer, 2003.
33. Ritter TA, Shrout TR, Tutwiler R and Shung KK. A 30 MHz composite ultrasound array for medical imaging applications. *IEEE Trans Ultrason Ferroelectr Freq Control* 2002; 49: 217–230.
34. Foster FS, Mehi J and Lukacs MA. A new 15–50 MHz based micro-ultrasound scanner for preclinical imaging. *Ultrasound Med Biol* 2009; 35: 1700–1708.
35. Patton CA, Harris GR and Philips RA. Output levels and bioeffect indices from diagnostic ultrasound exposure data reported to FDA. *IEEE Trans Ultrason Ferroelectr Freq Control* 1994; 41: 353–359.

36. Miller MW. Gene transfection and drug delivery. *Ultrasound Med Biol* 2000; 26: S59–S63.
37. Ter Haar G. Intervention and therapy. *Ultrasound Med Biol* 2000; 26: S51–S54.
38. Ter Haar G. Ultrasonic imaging: Safety considerations. *Interface Focus* 2011; 1: 686–697.

FURTHER READINGS

The following texts provide more detailed information on many of the topics covered in this chapter:

Cobbold RSC. *Foundations of Biomedical Ultrasound*. New York: Cambridge Press, 2007.
Shung KK. *Diagnostic Ultrasound: Imaging and Blood Flow Measurements*. Boca Raton, FL: CRC Press, 2005.
Szabo T. *Diagnostic Ultrasound Imaging: Inside Out*. Amsterdam: Elsevier Academic Press, 2004.

5

Optical and Optoacoustic Imaging

Adrian Taruttis and Vasilis Ntziachristos

5.1 INTRODUCTION

This chapter describes *in vivo* optical and optoacoustic imaging techniques. The focus is on methods that use light to provide molecular imaging of living organisms.

Optical imaging has a number of attractive general characteristics. It does not involve the use of ionizing radiation, so the safety concerns for patients and practitioners associated with x-ray and nuclear imaging are not present. As will be explained in more detail in this chapter, optical imaging can provide highly sensitive detection of wide-ranging contrast. For example, the oxygenation states of hemoglobin can be separately identified because oxyhemoglobin and deoxyhemoglobin absorb light differently. Fluorescence gives optical imaging a vast toolbox for monitoring biological processes through fluorescent proteins, agents, and cell labeling. The use of multiple wavelengths of light allows visualization of several different channels, or colors, simultaneously. Another key reason for the widespread use of optical imaging seen in biological research laboratories today is the relative convenience and simplicity of many optical imaging systems, which are often little more than boxes containing lights and cameras. Naturally, such devices often cost significantly less than other modalities.

A large proportion of optical biomedical imaging techniques rely on cameras to capture images, and sometimes, only minimal processing is performed after acquisition. Cameras are very similar to the human eye: One or more lenses form an image on the sensor, or in the case of the eye, the retina. Photographic approaches to optical imaging relate directly to the human field of view, augmenting it with additional information, for example, fluorescence distributions captured with appropriate filters. The advantages of these approaches are clear: Users are able to interpret the images just the same as something they see with their own eyes.

Despite these attractive properties, optical imaging approaches suffer from one key limitation. Tissue puts up a barrier to the progress of light, as evidenced by the observation that we are not transparent. Red light can typically travel on the order of a centimeter in tissue, while blue light typically only travels a fraction of a millimeter. Thus, the drawbacks of optical imaging primarily relate to the limited penetration depth of light in tissue. The very strong depth dependence of the optical signal also makes absolute quantification challenging.

Light microscopy has been leading biological discovery for centuries. Already, in the 17th century, Antonie van Leeuwenhoek and Robert Hooke were using microscopes to see individual cells and discover bacteria, spermatozoa, and more. Fluorescence microscopy has become a standard tool for biologists to observe specific targets. More recent advances in light microscopy like confocal and multiphoton techniques have greatly improved contrast, resolution, and depth penetration. Still, these methods are limited to imaging close to the tissue surface, with maximum penetration depths in the range of hundreds of micrometers.

Because of all the advantages of using light for biological imaging, it is highly desirable to extend its use to deeper layers of tissue for *in vivo* investigations. As described here, light in the near-infrared region of the spectrum is absorbed far less than visible light and allows penetration depths sufficient for whole-body imaging of mice and several important targets, like the breast and extremities in humans. Many other tissues in humans also are accessible to visible light during endoscopic or surgical interventions. However, light is strongly scattered in tissue, degrading spatial resolution and complicating optical imaging at any depth below the surface.

This chapter discusses the use of light for *in vivo* imaging, concentrating on techniques that are able to reach below superficial tissue layers. The optical imaging techniques described here are relatively new when compared to the modalities discussed in other chapters of this book. Methods are constantly evolving, and in many cases, there is no scientific consensus on which of the approaches to solving problems, such as the modeling of light propagation through tissue that may have highly heterogeneous optical properties, is the best. This chapter aims to present an instructive overview of the field that will allow the reader to gain insight into optical imaging methods. We do not present an exhaustive review of all advances and possible approaches.

The chapter starts by introducing how light interacts with tissue to the degree of detail required for a general understanding of optical imaging. This forms a basis for the subsequent section on the sources of contrast in optical imaging. An overview of different optical imaging techniques, presented roughly in order of complexity, follows. Optoacoustic imaging, a more recent development that attains superior spatial resolution in deeper tissue layers, is described in its own section. The remainder of the chapter is devoted to preclinical and clinical applications of optical and optoacoustic imaging.

5.2 LIGHT AND TISSUE

The photon model of light is convenient for describing interactions of light with biological tissue. Photons are elementary particles representing a quantum of electromagnetic radiation, including light. For the purposes of this chapter, when photons interact with matter, we first restrict ourselves to considering two cases: (1) absorption of the photon and (2) elastic scattering, where the direction of the photon changes but not the energy. The energy of a photon is given by

$$E = h\nu, \tag{5.1}$$

where h is Planck's constant (6.626×10^{-34} m^2 kg/s) and ν is the frequency of the light. In optical imaging, light is typically characterized by its wavelength measured in nanometers (nm), which is $\lambda = c/\nu$, where c is the speed of light in empty space. Wavelengths of light that are visible to us range from roughly 380 to 750 nm, as shown in Figure 5.1.

5.2.1 Absorption

Absorption of light can be considered as matter taking up the energy of a photon. For our purposes, this energy is taken up by the electrons in molecules within tissue. One important property of tissue is the *optical absorption coefficient*, which essentially describes how quickly light traveling through tissue is absorbed. In the case that absorption is the only mechanism by which a beam of light is attenuated, the absorption can be defined by the *Beer–Lambert law*:

$$\frac{I}{I_0} = e^{-\mu_a l}, \tag{5.2}$$

where I_0 is the incident light intensity, I is the resultant light intensity after the beam has traveled through a length l in the tissue in question, and μ_a is the optical absorption coefficient of that tissue. In practice, an optical absorption coefficient measured in or assigned to tissue is a bulk property describing the overall absorption resulting from a mixture of tissue components. The absorption properties of individual tissue molecules are often described in terms of molar absorption (or extinction) coefficients. This is a measure of how strongly

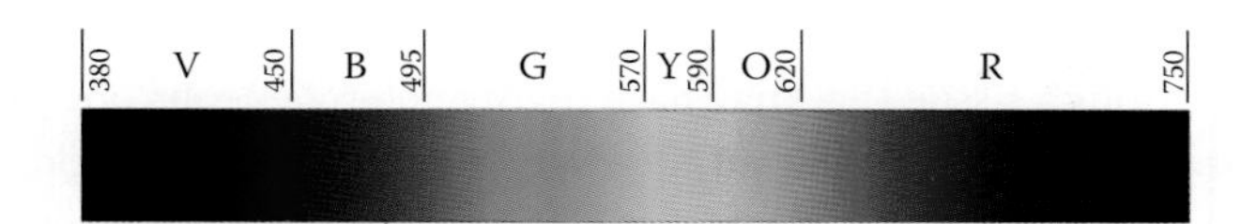

FIGURE 5.1 The visible spectrum. The numbers refer to wavelength in nm. The near-infrared region starts at wavelengths longer than 750 nm. Ultraviolet light has wavelengths shorter than 380 nm. V, violet; B, blue; G, green; Y, yellow; O, orange; R, red.

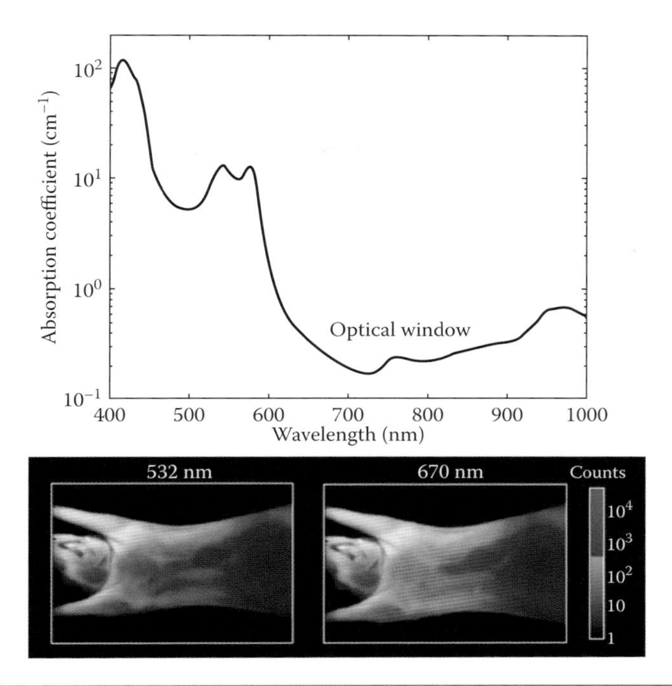

FIGURE 5.2 Optical absorption in tissue. The graph shows the variation of tissue absorption with wavelength. The absorption numbers are calculated assuming a realistic combination of hemoglobin and water. At shorter wavelengths, the overall absorption is dominated by the contribution from hemoglobin, which sharply decreases in the red/near-infrared region. At wavelengths longer than 800 nm, water absorption becomes significant and increases with wavelength. Overall, the wavelength region where hemoglobin and water absorption is low, from around 650 to 900 nm; offers an opportunity for deep optical tissue penetration; and is commonly referred to as the "optical window." The mouse images at the bottom show experimentally measured photon counts through the body of a nude mouse at 532 nm (left) and 670 nm (right). The excitation source was a point of light placed on the chest wall. Signal in the near-infrared range is orders of magnitude stronger compared with illumination with green light under otherwise identical conditions. (Reproduced from Weissleder, R. and V. Ntziachristos, *Nat. Med.*, 9, 2003. With permission.)

a particular chemical absorbs light. The usual units are M^{-1} cm^{-1}, which is absorption per unit of molar concentration.

The absorption properties of tissue constituents generally vary with the optical wavelength. Overall, the absorption properties of tissue are usually dominated by hemoglobin in the wavelength range from about 400 to 900 nm. The absorption of hemoglobin, and therefore of most tissues, drops steeply, by orders of magnitude, after about 600 nm, that is, at red and near-infrared wavelengths (Figure 5.2) [1]. This can be easily verified: If you hold a torch against the palm of your hand in the dark, you will see red light leaving your hand at the other side. This is because blue and green wavelengths are absorbed in the tissue. In simple terms, red and near-infrared light penetrates tissue, while other colors such as blue and green do not.

5.2.2 Scattering

While near-infrared light is capable of penetrating through centimeters of tissue, advanced light microscopy methods cannot produce high-resolution images deeper than a few hundred micrometers from the tissue surface. The problem is *scattering*. Scattering in this context

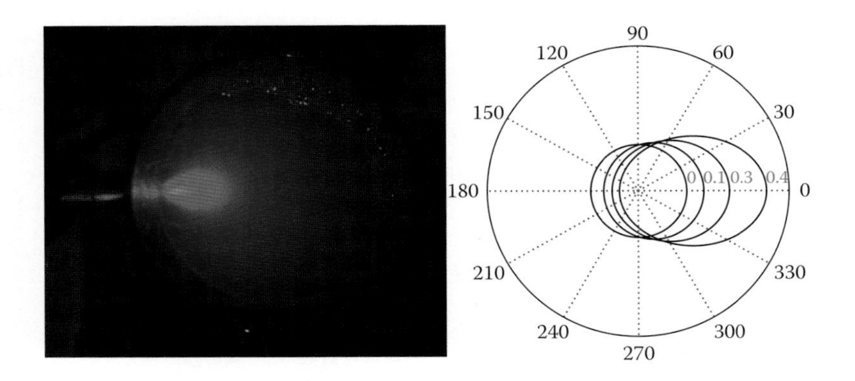

FIGURE 5.3 Scattering of light. The photograph shows a top view of a red laser pointer shining a beam of light into a glass of water mixed with 12 drops of yogurt. The beam rapidly loses its shape and direction due to scattering. The plot on the right illustrates how the directionality of scattering changes with increasing g. Note that $g = 0$ is isotropic scattering, where every scatter direction has an equal probability.

means that the direction of travel of photons is changed as they interact with tissue. In the dominant form of scattering in tissue, while the direction changes, the energy of the photons does not change—this is called *elastic scattering*. Scattering can be described by the scattering coefficient, μ_s, which is the probability of a scattering event per unit length. The *scattering mean free path*, which is the mean distance a photon travels between two scattering events, is then given by $1/\mu_s$. Photon scattering in tissue is generally not isotropic: Photons scatter with a preference for the direction that they were originally traveling (i.e., the forward direction). The amount of angular variation in scattering is often represented by a parameter g, called the *anisotropy* factor. The closer g is to 1, the more forward scattering the tissue is, whereas 0 represents isotropic scattering. Typical values of g for tissue are in the range of 0.8–0.99.

Ballistic photons are those that are not yet scattered off course by the medium and therefore travel in straight lines. These photons can be used to produce optical images with a high resolution limited only by diffraction. However, since the likelihood of a photon being scattered is quite high in tissue, the ballistic regime, where sufficient unscattered photons can be detected, is commonly limited to the first few hundred micrometers from the surface. This ballistic regime is where microscopic techniques operate. Beyond this regime, light rapidly loses its direction and, therefore, information on where it came from (Figure 5.3). This light cannot be focused by a lens and is therefore beyond the reach of microscopy.

5.2.3 Radiative Transfer and the Diffusion Approximation

The propagation of light in tissue can be described by the *radiative transfer equation* (or Boltzmann equation), which represents the way energy is transferred by absorption and scattering:

$$\frac{1}{c}\frac{\partial L(\vec{r},\hat{s},t)}{\partial t} = -\nabla\cdot[L(\vec{r},\hat{s},t)\hat{s}] - (\mu_a + \mu_s)L(\vec{r},\hat{s},t) + \mu_s\int_{4\pi} L(\vec{r},\hat{s},t)P(\hat{s}'\cdot\hat{s})d\Omega' + Q(\vec{r},\hat{s},t), \quad (5.3)$$

where $L(\vec{r},\hat{s},t)$ is the radiance, which is the energy flow per unit area and solid angle (W/m^2 sr) at position $\vec{r}$ along the direction of unit vector $\hat{s}$ at time t; $P(\hat{s}'\cdot\hat{s})$ is a phase function, which represents the probability of a change in photon propagation angle from $\hat{s}'$ to $\hat{s}$; $d\Omega'$ is a solid angle element around $\hat{s}'$; and $Q(\vec{r},\hat{s},t)$ represents an illumination source. The term $\nabla\cdot[L(\vec{r},\hat{s},t)\hat{s}]$ represents the divergence of the photon beam as it propagates, $(\mu_a+\mu_s)[L(\vec{r},\hat{s},t)]$ represents energy loss by absorption and scattering, and $\mu_s\int_{4\pi}L(\vec{r},\hat{s},t)P(\hat{s}'\cdot\hat{s})d\Omega'$ represents photons scattered into the path under consideration. This equation is difficult to solve because, apart from the three independent spatial dimensions, it also contains dependencies on the angle of photon propagation. However, the scattering of light in tissue is so strong that light propagation in layers deeper than the ballistic regime closely follows the physical phenomenon of diffusion. Diffusion equations are easier to solve than the radiative transfer equation. In the so-called diffusive regime, scattering is approximated by an isotropic model, represented by a reduced scattering coefficient $\mu_s'=(1-g)\mu_s$, where g is the anisotropy defined in Section 5.2.2. The applicable diffusion equation is

$$\frac{\partial\phi(\vec{r},t)}{c\,\partial t}+\mu_a\phi(\vec{r},t)-\nabla\cdot[D\nabla\phi(\vec{r},t)]=S(\vec{r},t), \tag{5.4}$$

where $\phi(\hat{r},t)$ is the fluence rate (W m^{-2}), $D=1/3(\mu_a+\mu_s')$ is the diffusion coefficient, and $S(\vec{r},t)$ is an isotropic illumination source. Note that the fluence rate does not have the angular dependency of the radiance. The majority of the imaging techniques introduced in this chapter involve diffusive light, and many of these methods rely on a form of the diffusion equation for image reconstruction.

5.3 SOURCES OF IMAGE CONTRAST

5.3.1 Fluorescence

Fluorescence is considered to be the single most powerful source of contrast for molecular optical imaging. Fluorescence refers to the short-lived emission of photons of a lower energy (longer wavelength) after molecules are excited by higher-energy (shorter wavelength) photons. Fluorescence emission typically occurs within approximately 10^{-9} s of excitation. The reason for the energy difference between absorbed and emitted photons in fluorescence is that the absorbing molecule quickly releases some of the absorbed energy to its surroundings by nonradiative (thermal) relaxation, as illustrated in Figure 5.4.

If the fluorescent substance has a ground energy state S_0 and an excited energy state S_1, then we have

$$S_0+\frac{hc}{\lambda_{ex}}\rightarrow S_1\text{ (excitation)} \tag{5.5}$$

$$S_1\rightarrow S_0+\frac{hc}{\lambda_{em}}+\text{vibrational relaxation energy (emission)}, \tag{5.6}$$

where the energy of the photons is represented in terms of h, Planck's constant, and c, the speed of light in free space, and λ_{ex} *and* λ_{em} are the excitation and emission wavelength, respectively.

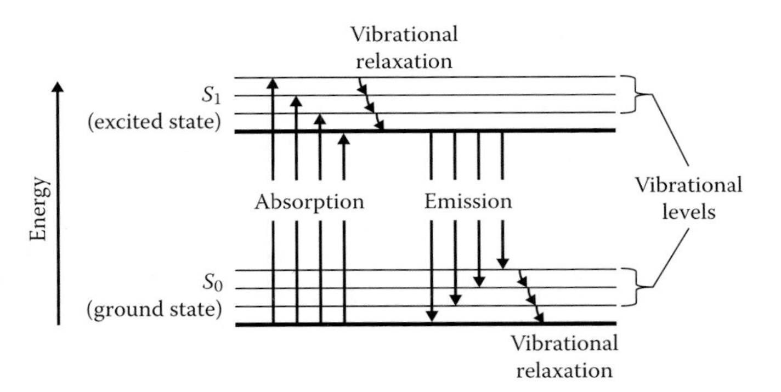

FIGURE 5.4 A Jablonski diagram illustrating the process of fluorescence. Photons are absorbed and excite a molecule into various vibrational levels of an excited state (S_1). Nonradiative vibrational relaxation into the surroundings quickly brings the molecule into the lowest vibrational level of the excited state. From there, it can return to the ground state (S_0) by emitting a fluorescence photon. Since some of the absorbed energy is given off by vibrational relaxation, the emitted photon has less energy and thus a longer wavelength than the absorbed photon.

Certain materials display high fluorescence when excited at suitable wavelengths. These materials can be used as *fluorophores*, that is, to make something visible by means of fluorescence. Figure 5.5 shows the absorption and emission spectra of a common fluorescent dye, Cy5.5. Such spectra give crucial information for the selection of appropriate fluorophores. This wavelength difference observed between the excitation (absorption) peak and emission peak of a fluorescent agent is the *Stokes shift*. This shift allows highly efficient isolation of the emitted light from the excitation, which is one of the reasons why fluorescence imaging is very sensitive. A fluorescent agent is further characterized by its

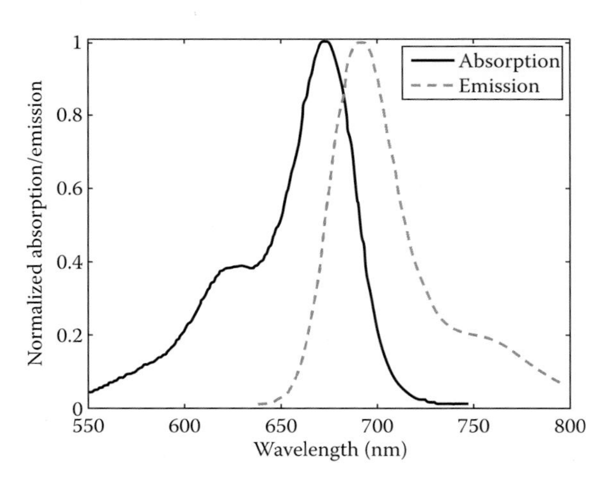

FIGURE 5.5 The absorption and emission spectra of a common cyanine far-red fluorescent dye, Cy5.5. Note that the emission maximum around 690 nm is at a longer wavelength than the excitation maximum, which is around 670 nm. This shift allows the separation of fluorescence photons from the excitation light. For example, if a subject containing Cy5.5 is excited with laser light at 670 nm, then a long-pass filter that only allows light of higher wavelengths than 680 nm could be used to visualize the fluorescence.

quantum yield, which describes the proportion of the absorbed energy released as fluorescence, that is,

$$\text{Quantum yield} = \frac{\text{photons emitted}}{\text{photons absorbed}}. \tag{5.7}$$

The time taken for excited fluorophores to emit light is referred to as their fluorescence *lifetime*. This time depends on the fluorophore as well as its molecular microenvironment and can thus be exploited as an additional source of contrast, provided that the measurements are time resolved (see Section 5.4.6). Sections 5.3.1.1 through 5.3.1.4 present different approaches for producing fluorescence contrast in biomedical imaging.

5.3.1.1 Exogenous Dyes

The most suitable exogenous fluorophores for deep-tissue imaging are those with excitation and emission wavelengths in the "optical window" of the far red and near infrared (see Section 5.2.1). Cyanine dyes, of which there are many varieties and derivatives, are among the most commonly utilized fluorophores in this context.

The use of fluorescent dyes for applications in biomedical imaging can be categorized as follows:

1. Free dyes with no specificity for particular biological targets can be employed to visualize blood vessels or accumulation in specific organs/tissues resulting from physiological processes.
2. Fluorophores can be used to tag ligands that bind specifically to biological targets of interest, thereby enabling fluorescence imaging of those targets (for example, specific receptors that are overexpressed in cancer).
3. Dye molecules can be arranged in such a way that they only display significant fluorescence after being activated by a biological process.
4. Fluorescent labels can be added to cells or nanoparticles, which can then be monitored *in vivo*.

The most prominent example of a dye for optical imaging without specificity is indocyanine green (ICG). ICG has absorption and emission maxima in the near infrared, at around 800 nm. It binds to plasma proteins and remains confined to blood vessels while it circulates, which makes it useful for visualizing the vasculature. ICG is rapidly removed from the circulation by the liver (circulation half-life in the range of a few minutes in small animals and humans) and excreted through bile. It is clinically approved and used for a number of diagnostic purposes, including assessment of hepatic function and cardiac output. Applications of ICG as a fluorescence imaging agent are the subject of an increasing number of investigations, particularly in intraoperative scenarios, such as lymphatic mapping [2].

Targeted fluorescence imaging agents are intended to highlight specific biological targets to make them visible via fluorescence. This can be understood as the *in vivo* equivalent of immunofluorescence techniques used in histology. The usual method is to inject the targeted agent intravenously, allow time for it to reach its target, allow further time for unbound agent to be cleared and excreted, and then perform imaging of the subject. It follows that targeted agents should be able to reach their targets and unbound agent should subsequently be rapidly eliminated from the tissue to allow for fast imaging with a low background signal. Commonly used ligands include small molecules, peptides, proteins, and antibodies [3]. Because of their slow clearance from the circulation and longer retention in tissue when unbound, antibodies are not as ideal for targeted *in vivo* imaging as they are for histological immunofluorescence. However, therapeutic antibodies approved for clinical use (e.g., Bevacizumab, Trastuzumab) offer a unique potential to tumor-targeted imaging in humans by adding fluorescent tags [4]. The ability to visualize specific biological targets *in vivo* has a range of applications, from making tumors visible to monitoring their response to treatment. Common imaging targets include integrins and growth factor receptors; however, a large part of the power of the targeted fluorescence imaging approach is the vast range of potential biological targets and available agents.

A particularly powerful tool in fluorescence imaging is the ability to activate agents within the body, that is, to design them to display fluorescence only after activation by some biological environmental condition or process. A prominent example of this is the imaging of protease activity *in vivo* [5]. This is achieved by using agents that combine a fluorescent dye with a quencher (something that stops the fluorescence, which could be more of the same dye) in close proximity and that release the fluorescent dye when the agent is cleaved by the specific protease it is designed to detect (Figure 5.6).

Cell imaging via fluorescence is often performed using fluorescent protein expression (see Section 5.3.1.2), but there are also methods for labeling cells with exogenous dyes, including the use of targeted agents *in vitro* prior to administration of the cells. This has the advantage of a wider range of available dyes and emission wavelengths but is hampered by diminishing signals as cells divide. Similarly, fluorescence labels have been added to nanoparticles for monitoring their pharmacokinetics and biodistributions *in vivo*. This technique could find increasing use in the emerging field of nanomedicine to characterize novel therapeutic nanocarriers, ranging from liposomes to carbon nanotubes.

Since a multitude of near-infrared fluorescent dyes with distinct emission peaks are available, as are filter sets to discriminate them, the approaches described in this section can be combined to achieve simultaneous imaging of multiple signals in one organism. For example, a nonspecific dye for lymphatic mapping could be combined with a targeted agent for tumor identification in intraoperative imaging.

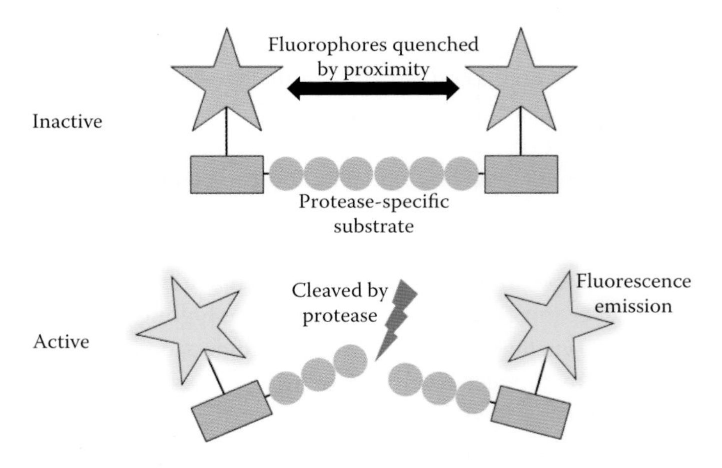

FIGURE 5.6 Protease-activatable fluorescent imaging agents. In its inactive state, the fluorophores are quenched by close proximity to one another. After being cleaved by a specific protease, the fluorophores are released (dequenched), resulting in fluorescence emission.

5.3.1.2 Fluorescent Proteins

The use of fluorescent proteins as reporter genes has become a powerful and widely used tool for the study of molecular biology [6]. The gene encoding a fluorescent protein is typically introduced to cells (by viral or other means) along with the gene to be studied, allowing coexpression of the fluorescent protein together with the protein of interest. Another favored application for fluorescent proteins is the *in vitro* labeling of specific cell populations, to allow them to be monitored over time *in vivo*. One example of this is the tracking of monocytes expressing green fluorescent protein as they travel through the bloodstream. While discoveries using fluorescent proteins have, until now, been dominated by microscopic imaging techniques, the development of fluorescent proteins with excitation and emission wavelengths in the far red and near infrared, where tissue absorption is much reduced, now provides a method for studies in whole organisms as large as mice [7].

5.3.1.3 Fluorescence Resonance Energy Transfer

Fluorescence (or Förster) *resonance energy transfer* (FRET) is the nonradiative transfer (no photon) of energy between two fluorescent molecules in close proximity. FRET efficiency is dependent on the inverse of the sixth power of the distance between the molecules, making it an extremely sensitive method to determine close proximity, as well as the integral of the spectral overlap between the emission spectrum of the donor molecule (the molecule giving up energy) and the absorption spectrum of the acceptor molecule (the molecule accepting the energy). FRET using either fluorescent proteins or exogenous fluorescent dyes can be used to make protein interactions visible because it is highly sensitive to the distance between fluorescent molecules. The power of FRET is the ability to distinguish between fluorescence from donor molecules in the non-FRET state when the molecules are not sufficiently close and fluorescence from acceptor molecules when donor and acceptor are in close proximity (Figure 5.7). The distances involved are in the vicinity of 10 nm, suitable for dimensions of biomolecules.

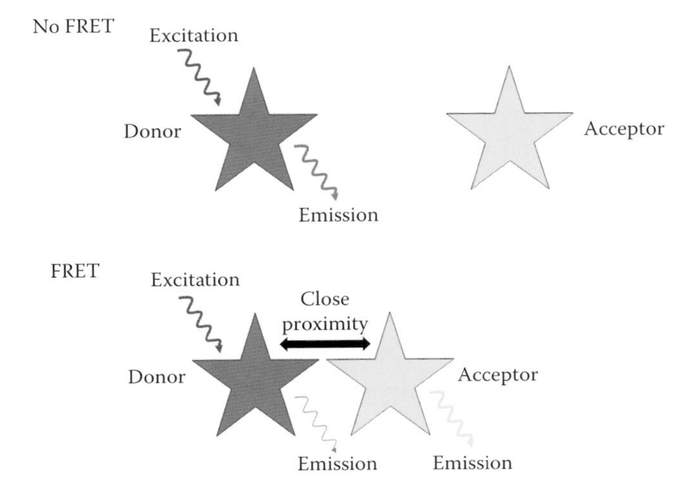

FIGURE 5.7 Fluorescence resonance energy transfer (FRET). For FRET to produce a significant signal, the donor and acceptor fluorophores have to be brought into close proximity, on the order of 10 nm. The energy transfer from donor to acceptor results in a decrease in fluorescence emission from the donor and a corresponding increase from the acceptor. This can be observed as a change in color of the fluorescence.

Examples of how FRET can be applied include measurements of distances between domains on a protein, by encoding a FRET pair such as cyan and yellow fluorescent protein on different positions of the protein, and studying protein–protein interaction by placing a donor on one protein and an acceptor on the other [8]. Note that FRET is a well-established method in microscopy, but macroscopic use of FRET *in vivo* is still in its infancy.

5.3.1.4 Autofluorescence

Autofluorescence is a term used to describe the natural fluorescence displayed by tissue without the addition of exogenous fluorophores or fluorescent proteins. Tissue autofluorescence is commonly regarded as a problematic source of background fluorescence signal when performing optical imaging. While tissue autofluorescence is dominated by molecules excited in the ultraviolet to blue wavelength range, the increased penetration depth of far-red and near-infrared light allows autofluorescence in that region to also gain significance. Multispectral approaches that employ multiple excitation and/or emission wavelengths can achieve improved rejection of autofluorescence by more completely identifying the spectral signature of the fluorophore under study. Some diagnostic applications for measuring and utilizing tissue autofluorescence exist, examples of which include imaging of lipofuscin deposits in the retina, visualizing cancer, and readouts of advanced glycation end products in the skin as a marker of diabetes.

5.3.2 Bioluminescence

Bioluminescence is the emission of light by a chemical reaction within an organism, a well-known example of which is the light emitted by fireflies. In this particular case, a substrate called *luciferin* is oxidized in a reaction catalyzed by an enzyme, *luciferase*, produced by the firefly. For use as a genetic reporter *in vivo*, this luciferase gene (luc) is introduced into the cells of interest. Prior to imaging, luciferin is administered to the animal—often injected intraperintoneally in mice. Luciferin, which is a small molecule, quickly distributes throughout the body, and wherever it encounters cells expressing the luciferase gene, bioluminescence is produced. This emitted light has a relatively broad spectrum, peaking at around 560 nm at physiological temperatures. Although other bioluminescence systems exist, the firefly luciferase/luciferin system is by far the most widely used in molecular imaging. Because of the relatively low light yield of bioluminescence, and its nonoptimal emission wavelength for tissue penetration, the light that escapes the subject and can be detected noninvasively is typically very weak and therefore cannot be observed by eye but, rather, requires long integration times. Bioluminescence imaging has the advantage of relative simplicity: Animals are typically placed in a light-tight box and imaged with a sensitive camera. No excitation light or filters are required. Common applications include monitoring of tumor growth in laboratory mice during treatment, *in vivo* tracking of other cells labeled *in vitro*, and *in vivo* profiling of gene expression in general [9].

5.3.3 Endogenous Tissue Contrast

Endogenous optical tissue contrast is another factor that contributes to the attractiveness of optical imaging. The primary endogenous optical contrast source is the absorption of hemoglobin, which is exploited by several techniques and can provide functional information based on the different absorption spectra of oxyhemoglobin and deoxyhemoglobin (Figure 5.8). While the absorption of hemoglobin is orders of magnitude lower in the near-infrared than in the visible-light region, it remains the dominant contribution to overall absorption in many tissue imaging scenarios.

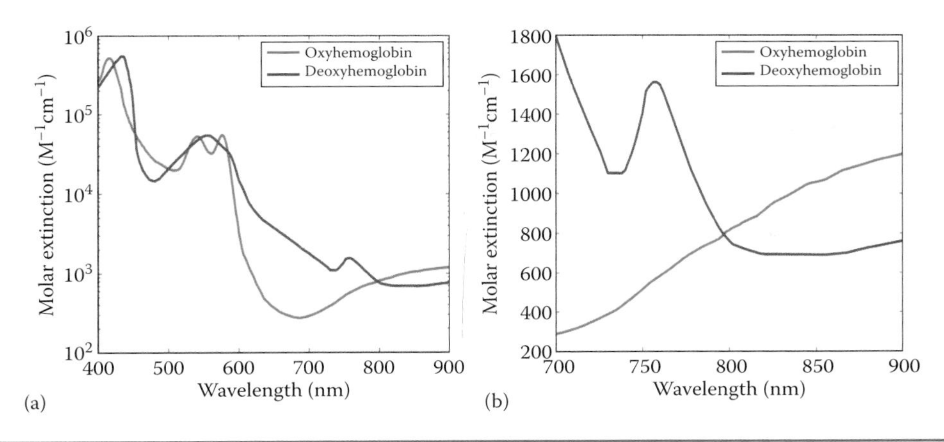

FIGURE 5.8 The absorption spectra of oxygenated and deoxygenated hemoglobin. (a) Spectra plotted on a logarithmic scale from 400 to 900 nm. Hemoglobin absorption in the near infrared is orders of magnitude lower than in the visible region. (b) Spectra plotted on linear scale from 700 to 900 nm. (Data compiled by Scott Prahl from multiple sources, Oregon Medical Laser Center, http://omlc.ogi.edu/spectra.)

Further tissue absorbers can also contribute to image contrast provided that the illumination wavelength range is chosen appropriately to exploit a particular feature in the absorption spectrum of the substance. Examples of absorbers targeted by optical imaging include water, melanin, and lipids.

5.3.4 Exogenous Absorption Contrast

Optical absorption can be visualized by several techniques, including diffuse optical tomography and optoacoustic imaging, which are explained in more detail later in this chapter. In many cases, optical absorption can even be observed by the naked eye. In addition to relying on endogenous optical absorbers like hemoglobin and melanin, absorption contrast can be provided by administering an agent that is a strong absorber. These agents include dyes that also produce fluorescence but are administered to provide additional absorption of the incident light. Perhaps the most common such example is the use of blue dyes (e.g., Patent Blue V) for lymphangiography or the sentinel lymph node procedure used during surgery, where the dye is injected around a tumor to be resected and subsequently colors the draining lymph vessels and nodes blue, which surgeons can then visually identify. On the other end of the complexity scale, new light-absorbing nanoparticles such as carbon nanotubes and gold nanorods have been applied for preclinical molecular imaging studies, particularly in connection with optoacoustic imaging [10].

5.4 TECHNIQUES FOR OPTICAL IMAGING

This section introduces the principles of the more prominent optical imaging techniques in use and under investigation today. We start with brief descriptions of intravital microscopy and optical projection tomography, techniques that can overcome the scattering barrier only by brute force. A discussion of photographic or planar approaches to fluorescence imaging follows. Tomographic techniques that enable quantitative volumetric imaging of fluorescence distributions are introduced, followed by related methods that involve the use of time or phase information to gain more information from optical measurements through tissue.

5.4.1 Intravital Microscopy

Advanced optical microscopy techniques like confocal and multiphoton microscopy allow high-resolution imaging with light within the first few hundred microns from the tissue surface. Beyond those depths, photon scattering makes the required focusing of light impossible. Intravital microscopy refers to methods where superficial tissue layers are removed or prepared so that optical microscopy techniques can be applied to image deeper layers, which are not usually accessible. In many cases, this is achieved by introducing a transparent window on mice through which skin flaps or the brain can be imaged in a repeatable way over longer periods of time. These techniques have seen wide adoption in neuroscience and cancer biology, in applications where wide fields of view are unnecessary [11].

5.4.2 Optical Projection Tomography

Scattering of light in tissue prevents us from accurately determining by which path light detected at the surface has traveled through the tissue. This results in a degradation of resolution with increasing penetration depth. The most direct approach to overcoming this obstacle is to image tissue that displays minimal scattering. This can be achieved either by the selection of transparent model organisms or by immersing a scattering specimen in a solvent in an optical clearing process, which essentially makes the tissue transparent. The result is that light travels through the specimen in straight lines, and a technique analogous to x-ray computed tomography, known in the optical case as optical projection tomography, can be used to reconstruct high-resolution images through specimens such as whole mouse embryos [12]. Since it is not possible to apply optical clearing to living tissue, subjects to be imaged *in vivo* are restricted to those that do not display significant optical scattering in their natural state.

5.4.3 Planar Fluorescence Imaging

One of the most widely adopted macroscopic optical techniques in biomedical research laboratories is photographic or planar fluorescence imaging. Its primary advantage is simplicity. As shown in Figure 5.9, planar fluorescence imaging requires only three major components: a light source to excite the fluorophore, an optical filter that only allows light emitted by the fluorophore through, and a camera to record an image of the fluorescence. (Bioluminescence is very similar, requiring only a camera.) Several vendors exist for such systems. It can be used for quick experiments—several small animals can be imaged at a time, and the exposure times involved can be less than a second. No complicated image reconstruction algorithms are necessary, and the results can be observed immediately. The application of near-infrared fluorescence in particular allows the visualization of subsurface tissue fluorescence. However, this simple approach does have distinct disadvantages that limit its applicability. In the strict sense, the images resulting from planar fluorescence imaging (and bioluminescence) cannot be relied upon for quantitative accuracy. The absorption and scattering of both the excitation and emission light means that signals originating near the illuminated skin surface are much stronger when compared to signals originating from the same fluorophore concentration deeper beneath the surface. This effect is referred to as *surface weighting*. In addition, because of scattering, while the surface displays

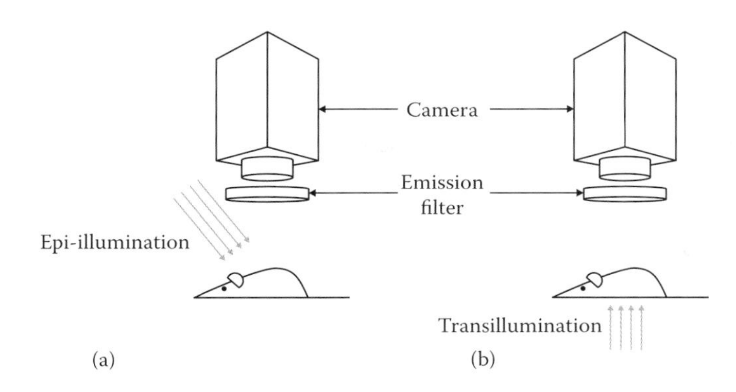

FIGURE 5.9 Planar fluorescence imaging. (a) Reflectance mode with epi-illumination, where the excitation light illuminates the same surface of the tissue visible in the camera image. (b) Transillumination mode, where the excitation light is incident on the opposite tissue surface to that viewed by the camera.

a high spatial resolution, deeper signals are blurred. Another often-neglected problem with simple planar fluorescence imaging is the heterogeneous nature of optical properties, particularly absorption. Regions of high absorption, such as tissues with high concentrations of blood, will allow less excitation light to reach the fluorophore and less emission light to escape than less absorbing tissue regions, thus skewing the intensity distribution in the resulting images. While approaches to estimate the optical absorption and correct for it in different tissue regions based on color images and other information exist, they are not universally adopted by system vendors, and some care should be taken when interpreting results. Overall, planar fluorescence imaging is quick, easy, and well suited for imaging fluorophores that are on or near the surface, such as in targeted agent accumulation in subcutaneous tumors in mice (Figure 5.10a and b) [13,14].

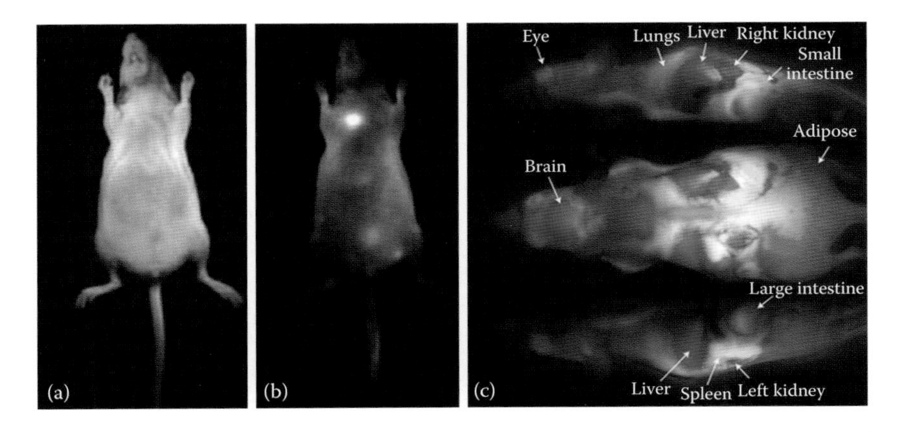

FIGURE 5.10 Examples of epi-illumination fluorescence imaging. (a and b) Imaging of protease activity in a tumor-bearing mouse by means of a protease-activated near-infrared fluorescent agent. (a) White light image. (b) Corresponding fluorescence image showing protease activity in a tumor on the chest. (c) Epi-illumination fluorescence imaging of indocyanine green over time and subsequent separation of the temporal profiles allow identification of different organs by means of their fluorescence uptake. ([a and b] Reproduced from Weissleder, R. et al., *Nat. Biotechnol.*, 17, 1999. With permission. [c] Reproduced from Hillman, E.M. and A. Moore, *Nat. Photonics,* 1, 2007. With permission.)

The geometry where the illumination source is on the same side of the tissue as the detector (camera), as shown in Figure 5.9a, is referred to as epi-illumination or fluorescence reflection imaging. An alternative to this geometry is transillumination, where the light source is positioned on the opposite side of the tissue as the detector (Figure 5.9b). This configuration requires a sufficiently thin object so that the resulting fluorescence light can still be detected on the opposite side. Using near-infrared light, whole mice can be imaged in this geometry. Transillumination fluorescence imaging offers some advantages in accuracy. The primary advantage is that the fluorescence images are not surface weighted, because light needs to travel through the whole tissue layer before being captured by the camera. Other related problems arising in reflectance imaging are also reduced: Surface autofluorescence is much less of a problem in transillumination, as is excitation light that travels through the emission filter, so-called bleed-through. However, transillumination adds a difficulty: Variations in the thickness of the transilluminated layer will lead to vast differences in detected light signals, since light is rapidly attenuated as it passes through tissue. In other words, thinner regions, like the outside edges of a mouse, will let through a lot more light. There are several ways to overcome this problem, from restricting imaging to subjects with uniform thickness to the normalization approaches discussed in Section 5.4.4.

5.4.4 Normalized Fluorescence

Planar fluorescence imaging in its simplest form, whether in reflectance or transillumination mode, completely ignores the influence of heterogeneity in tissue optical properties. Optically absorbing regions, such as organs with high blood content, strongly absorb both excitation and emitted fluorescence light, thus skewing the detected intensity. A common approach to partially correcting these inaccuracies is the use of images of the same field of view that record data in spectral bands other than the fluorescence emission. One of the simplest approaches is the use of an image recording the excitation light itself. In both the reflectance and transillumination cases, dividing the fluorescence image by the image of the excitation light results in an image that is less sensitive to variations in optical properties. This can be expressed as

$$I_{\text{norm}} = \frac{I_{\text{fluo}}}{I_{\text{ex}}}, \tag{5.8}$$

where I_{norm} is the normalized image displaying reduced sensitivity to optical property heterogeneity, I_{fluo} is the fluorescence image captured by the camera using an appropriate emission filter for the fluorophore of interest, and I_{ex} is the excitation light image, which is the image captured by the camera without the emission filter. Regions with highly absorbing tissues will appear darker in the excitation image, and the fluorescence from those regions will therefore be divided by a lower value than less absorbing regions (Figure 5.11) [15]. This normalization only makes sense where the tissue optical properties at the excitation and emission wavelengths are similar, which is a reasonable assumption in practice. This normalization also provides correction for spatial variations in excitation light on the tissue surface.

5.4.5 Fluorescence Tomography

Just like CT was developed as a spatially more informative technique than planar x-ray images (see Chapter 2), the acquisition of multiple optical projections allows for a

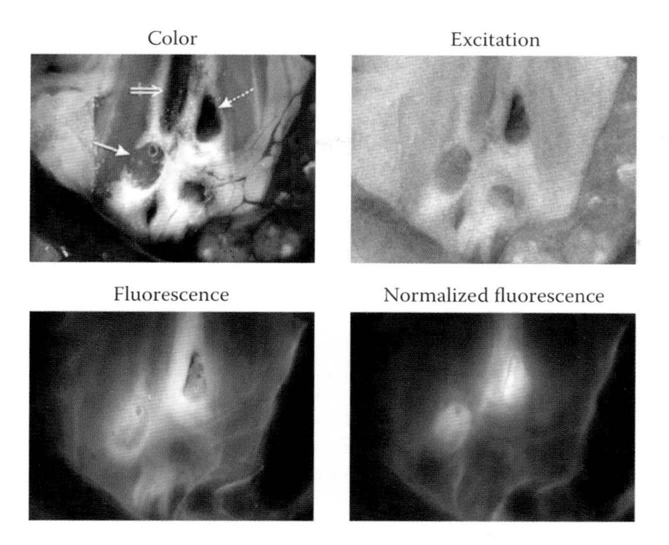

FIGURE 5.11 An example of normalized fluorescence. For the illustration of normalization, two lymph nodes in a mouse (postmortem) were injected with the same amount of a fluorescent dye but different amounts of light-absorbing ink. The images show lymph nodes (single-line arrows) around the inferior vena cava (double arrow). The image obtained at the excitation wavelength shows absorption differences between the two lymph nodes due to the differential injection of ink. The recorded fluorescence image shows low signal intensity from the lymph nodes compared to bright background signals. The normalized image shows fluorescence originating from the absorbing lymph nodes. (Reproduced from Themelis, G. et al., *J. Biomed. Opt.*, 14, 2009. With permission.)

better-quantified representation of fluorescence information in 3-D space. This technique is referred to as fluorescence tomography or *fluorescence molecular tomography* (FMT) to indicate that it is a molecular imaging by means of fluorescence. The similarity to x-ray CT is unfortunately rather limited: While x-rays, to a good approximation, travel through tissue in straight lines, allowing simple analytical image reconstruction approaches, light, as has been discussed, undergoes much scattering. Careful design of detection geometries and reconstruction algorithms is necessary in the optical case. The basic principle of operation is to combine varying source illumination positions with detector measurements at multiple projections around the tissue being imaged. Sources are usually points of light on the skin surface originating from either a laser beam in free space or an optical fiber output. Detector positions are generally optical fiber ends that collect light or individual pixels on a camera chip (all the pixels together represent a 2-D array of detectors). An example schematic of such a system is shown in Figure 5.12. FMT implementations where projections are acquired from different angles around the animal or tissue being imaged, either by rotating the instrumentation around the subject or rotating the subject itself, represent the most complete form of data acquisition. Alternatively, several implementations utilize data acquired from one plane only, for more convenient and rapid imaging (Figure 5.13).

5.4.5.1 Instrumentation

The hardware involved in modern FMT systems is not very different from planar fluorescence imaging. Generally, laser sources are used for excitation. This allows convenient point source illumination on the tissue surface by collimating or loosely focusing the laser beam. Robust and

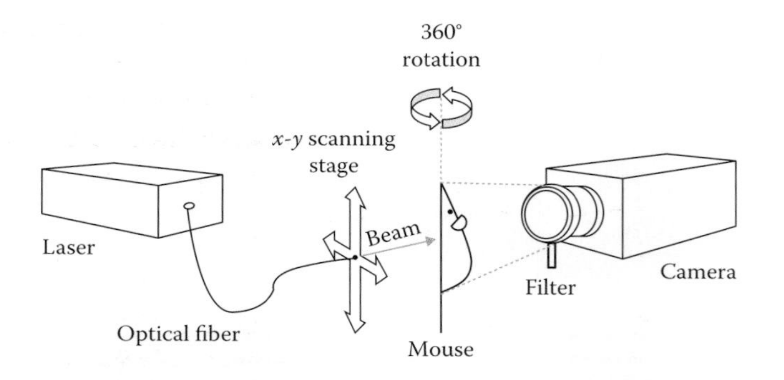

FIGURE 5.12 A small animal FMT imaging system with a rotational, free-space geometry. For each angular projection, the laser beam is scanned to a number of different source positions in transillumination mode, producing point illumination on the skin. For each source position, the camera acquires a fluorescence image and an image of the excitation light for normalization. The mouse is then rotated to the next angular projection.

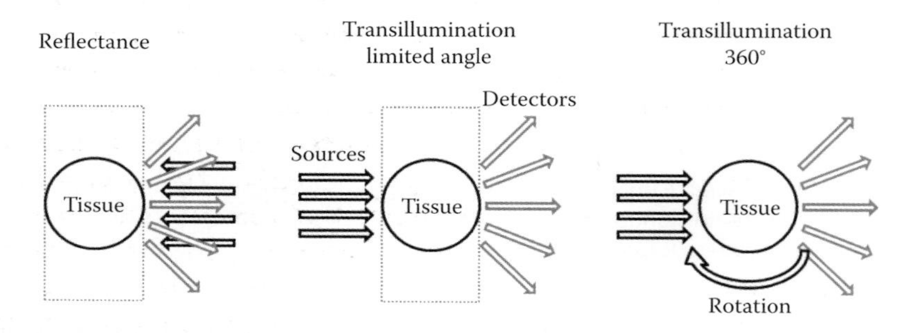

FIGURE 5.13 Tomographic geometries found in FMT implementations, illustrated in 2-D. Excitation is by approximate point sources on the tissue surface, which are activated/scanned sequentially. Detection is generally parallel. Limited angle geometries, whether in reflectance or transillumination mode, are especially common in small animal systems in combination with imaging chambers or cassettes, represented by dotted lines in the diagram. Detection from all angles around 360° results in improved spatial accuracy. This can be achieved in free-space implementations by rotation of the animal or instrumentation.

inexpensive continuous-wave (CW) diode lasers are generally employed for time-independent tomography. Sensitive charge-coupled device (CCD) cameras are used to capture images. The need for excitation light images as well as fluorescence, possibly in multiple wavelength bands, requires the use of several filters, which can be changed automatically by a filter wheel or similar device. Variation of the excitation source position requires a translation stage to move the laser beam. If multiple angular projections are required, an additional stage is used to rotate the mouse or, alternatively, the instrumentation. In an alternative configuration, the CCD camera can be replaced with multiple detection fibers, each of which is then fed into separate photomultiplier tubes or avalanche photodiode devices for photon counting.

5.4.5.2 Model-Based Reconstruction

We have already mentioned that the strong scattering of light as it propagates through tissue makes image reconstruction in optical tomography challenging. Light detected as it emerges from the tissue surface could have been influenced by any parts of the overall volume being imaged, instead of just a straight line through the volume, as in x-ray CT.

The problem of optical tomographic image reconstruction is commonly split into two parts: modeling the forward problem and inversion. The forward problem refers to modeling the way light propagates through tissue from sources to detectors. In general, modeling the forward problem can be achieved by discretizing a suitable differential equation that describes the propagation of both excitation and emission light through tissue. By discretization, we mean that the volume being imaged is split up into small parts and the relevant equations are then solved for each part in a simplified form. Examples of well-established numerical methods for such problems include the *finite element method* (FEM) and the *finite difference method* (FDM). Of special interest in fluorescence tomography is the mathematical description of light propagation applied in the forward model. The diffusion equation, in which scattering is assumed to be isotropic, is key to practical forward modeling in optical tomography because it allows far simpler computation than the radiative transfer equation of which it is a first-order approximation (see Section 5.2.3).

The diffusion equation can be solved using a Green's function approach, a well-known technique for solving differential equations, to the boundary value problem that represents the known point illumination on the surface. This point illumination is provided by scanning the light source along the tissue surface. The Green's functions then provide the probabilities of photons traveling through each voxel in the imaged volume from the point source to other points, such as the detectors (Figure 5.14). The relevant optical properties of the tissue, the absorption coefficient and the reduced scattering coefficient, are critical parameters in the equation but typically are not known for a given subject. Therefore, they are commonly assigned to known average values for the tissue type involved, often assigning the same values to the whole volume of interest or, where possible, segmenting the volume into different tissue regions and assigning different properties to each region. Segmentation into different regions can be accomplished using a hybrid imaging approach where FMT is combined with a modality that provides anatomical information, such as x-ray CT (see Section 5.4.5).

Modeling the overall propagation of light in fluorescence tomography can be achieved by using one diffusion equation each for excitation and fluorescence emission. A highly useful approach applied in FMT is the use of the *normalized* Born approximation, in which the fluorescence emission measurements are divided by the excitation measurements in transmission mode. This is similar to the normalization approach described earlier for planar fluorescence imaging, and it reduces the sensitivity of the reconstructed image to theoretical inaccuracies of the model, variations in the optical properties of the tissue being imaged, and unequal gain factors of sources and detectors in the system [16].

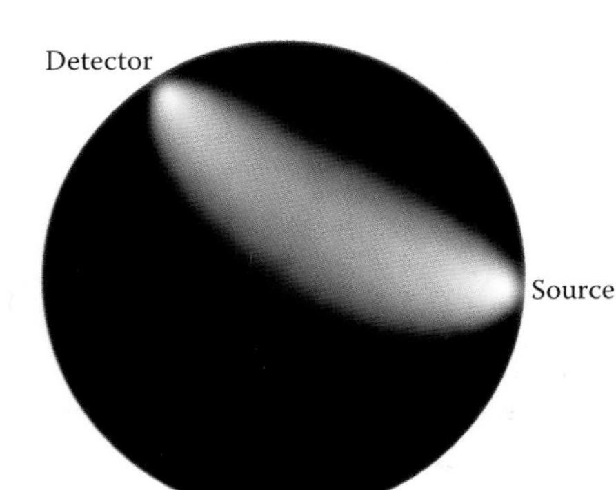

FIGURE 5.14 Green's function for a source–detector pair in the diffusive regime. The function describes the paths taken by photons traveling from a point source to a detector. These photon distributions are used to generate the forward matrix for optical tomography.

The forward model describes mathematically what the (normalized) measured image data would be for a given fluorescence distribution, thus mimicking exactly what the imaging system does experimentally. The matrix equation is of the form

$$y = Wx, \tag{5.9}$$

where y is a vector containing the measurement data (e.g., normalized CCD pixel values), W is the model matrix (sometimes called *weight* matrix) generated from the relevant Green's functions in the discretized volume, and x is the vectorized fluorescence distribution to be solved for.

To obtain the fluorescence distribution in the volume of interest from the measured data, which is what we want, the forward model (Equation 5.9) must be inverted. This is nontrivial. The linear forward model obtained from fluorescence tomography is large: The model matrix to be inverted has dimensions of $N_{vox} \times N_{src} \times N_{proj} \times N_{det}$, where N_{vox} is the number of elements or voxels that the volume of interest is split into, N_{src} is the number of excitation source positions, N_{proj} is the number of angular projections, and N_{det} is the number of detectors utilized (e.g., CCD pixels). In a typical FMT system, the number of matrix elements in the model can be as high as 20 million. More significantly, though, the inverse problem is *ill posed* or *ill conditioned*, meaning that the solution is very sensitive to small variations in the measurements (such as inevitable noise in the detector or background from autofluorescence), and there are some scenarios in which there is no unique solution for the fluorescence distribution given the measurements obtained. The fact that the problem is ill posed is a direct result of the scattering of light, which makes it difficult to tell where photons measured on the tissue surface originate. Regularization approaches, which amount to a spatial smoothing of the problem, are required to obtain robust reconstruction of fluorescence distributions. Overall, these limitations caused by photon scattering result in a reconstructed spatial resolution that is vastly inferior to methods using *ballistic photons*, and that degrades rapidly with increasing depth. Typical spatial resolutions obtained through whole mice are on the order of 1 mm or worse. However, the ability to extract quantitative volumetric fluorescence distributions from whole mice, which is not possible with planar imaging, can be valuable.

5.4.5.3 Hybrid Approaches

FMT provides information about the spatial distribution of fluorescent agents within the subject. However, it lacks anatomical information that can be useful for determining which organs the signals originate from. As is the case with other hybrid imaging modalities such as PET/CT, FMT and other optical imaging methods can benefit from combination with a second modality that provides anatomical imaging. This consideration has led to the development of multimodal FMT systems. One approach to this is the use of imaging cassettes inside of which small animals can be positioned and imaged in multiple systems, for example, FMT and MRI or x-ray CT. The cassettes include *fiducial markers* (markers that can be seen in the images from each modality) to aid coregistration between modalities. An alternative and more recent approach is to combine two modalities into one system for simultaneous dual-modality imaging, as shown in Figure 5.15 [17].

Coregistered fluorescence (molecular) and anatomical images are clearly a great advantage over stand-alone optical imaging for determining exactly where in the body the signals come from (see Figure 5.16 for an example of this) [18]. However, there are further advantages to the hybrid approach. Anatomical images allow a more refined approach to assigning optical properties to the imaged tissue volume than the average values assumed for all tissues in most stand-alone FMT approaches. Specific organs or tissue types (e.g., bones, lungs, heart) can be automatically segmented from an x-ray CT volume and assigned different optical properties for a more accurate forward model. Further, regularization can also be driven by the anatomical data to provide more accurate inversion.

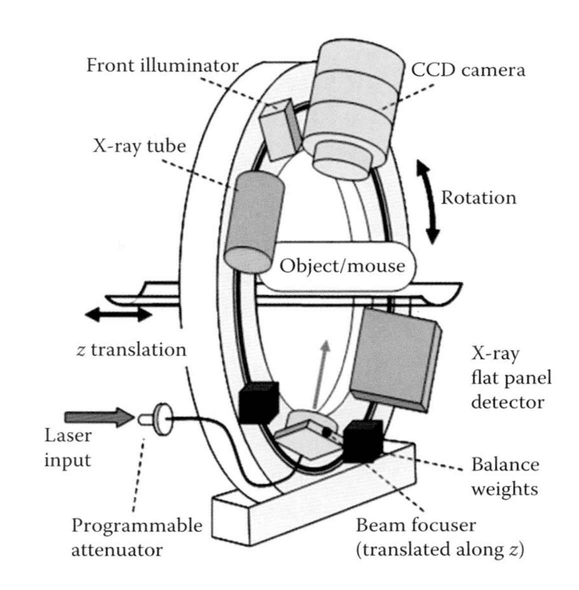

FIGURE 5.15 Illustration of a system for hybrid FMT and x-ray CT imaging of small animals. (Reproduced from Ale, A. et al., *Med. Phys.*, 37, 2010.)

Fluorescence tomography techniques have the advantage over simple planar fluorescence imaging in that they produce 3-D fluorescence distributions with much better quantitative accuracy. There are, however, some reasons why planar fluorescence is still widely used. Tomography requires more elaborate and costly hardware, including the mechanical systems required for rotation and linear translation. Image acquisition time is longer because of the multiple source positions and projections required. Typical acquisition times for FMT are in the low tens of minutes. Dynamics in shorter time ranges cannot be captured. Image reconstruction is also not immediate: A typical processing time is 10 min. Clearly, FMT is not suitable for applications where rapid imaging information is required. Nevertheless, the vast range of available fluorescent molecular imaging agents and the ease with which almost anything can be tagged with a fluorophore, combined with the quantitative, volumetric information obtained using FMT and the sensitivity of fluorescence imaging in general, make it a powerful choice for many molecular imaging applications, particularly those involving deep-seated signals.

5.4.6 Time-Dependent Imaging

Fluorescence tomography as described up to this point relies on measurements that are independent of time and applies time-independent models for image reconstruction. This is the simplest case and commonly involves the use of CW laser illumination to generate steady-state photon distributions that are imaged with cameras using sufficiently long integration times to achieve suitable signal levels. Such schemes can resolve fluorescence

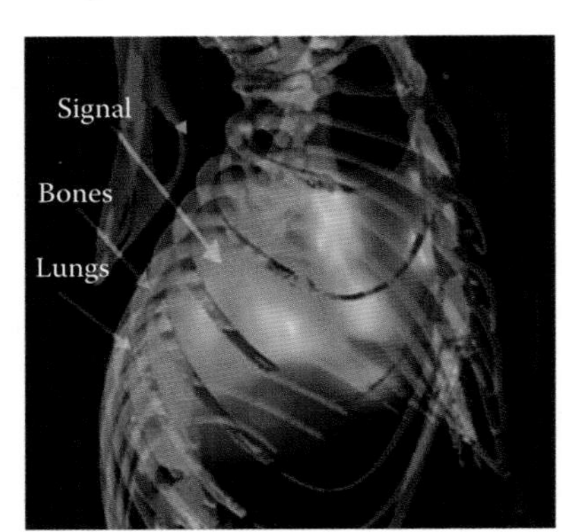

FIGURE 5.16 An example of 3-D imaging results from hybrid FMT/CT of a transgenic mouse model that spontaneously develops lung tumors. A near-infrared fluorescent agent targeted to $\alpha_V\beta_3$ integrin was injected intravenously 24 h prior to imaging. Fluorescence signal originating from tumors in the lungs is shown in orange. The lungs, as segmented from the x-ray CT volume, are shown in blue. (Reproduced from Ale, A. et al., *Nat. Methods*, 9, 2012.)

distributions in tissues assigned with average optical properties. In cases where more information is necessary, *frequency-domain* and *time-domain* optical imaging provide increasing amounts of information.

5.4.6.1 Gating and Time-Varying Light Sources

Obtaining a useful time axis in optical readouts is particularly challenging due to the extremely high speed of light (~3×10^8 ms^{-1}). Light detection devices such as photomultiplier tubes and avalanche photodiodes can, however, achieve very high temporal resolution. A useful approach in conjunction with CCDs is the image intensifier. Image intensifiers are instruments used in low-light imaging and night-vision applications to increase the intensity of incoming light. This is achieved by employing a photocathode to produce photoelectrons from the incoming photons, a microchannel plate to multiply those electrons, and a phosphor screen to convert the electrons back to photons. Besides the intensification of light, image intensifiers offer fast gating (activation of signal). In general, when combined with a light source of short pulse duration, gating can be applied with a varying time delay after the light pulse to sequentially measure the signal during different time windows. By putting together the sequence of measurements obtained with different delays, the characteristics of the emission as a function of time can be determined. This has several potential applications. The ultimate examples of time-varying illumination are ultrashort laser pulses. This term refers to pulses of light of duration in the range of picoseconds (10^{-12} s) or femtoseconds (10^{-15} s). Mode-locked lasers, in which the phase shifts between different longitudinal modes are kept constant, causing trains of destructive and constructive interference that result in ultrashort pulses, are commercially available and used in a range of applications, including two-photon microscopy. An alternative time-varying illumination, primarily used for frequency-domain imaging, is to modulate the intensity of a laser diode at a specific frequency. Together, these devices provide the necessary tools for time-dependent imaging.

5.4.6.2 Early Photon Tomography

As discussed in the context of FMT, the strong scattering of light in tissue makes the inverse problem associated with tomographic reconstruction for optical imaging ill posed. This is due to the fact that photons traveling between a source–detector pair may interact with a relatively large volume of tissue (Figure 5.14), compared to x-ray photons, which travel in approximately straight lines. Time-resolved detection of light offers a simple way around this problem: By capturing only the earliest photons to arrive at the detector, the later-arriving photons, which have been scattered away from the straight trajectory between source and detector and therefore travel a longer path, are disregarded. This improves both the spatial resolution of reconstructed images and the fidelity of the reconstructions, particularly in cases of distributed fluorescence, because the inverse problem is better posed (Figure 5.17). Clearly, discarding the vast majority of photons that do not travel in straight lines has a negative impact on the sensitivity of fluorescence detection. However, the loss of sensitivity is not as significant as the proportion of photons discarded might suggest, since photons that have been scattered off course do not contribute as much useful information about the fluorescence distribution as early photons. Early photon tomography requires ultrashort laser pulses and a gated intensified CCD camera to generate almost instantaneous pulses of light and record only the earliest arriving photons, respectively [19]. This increases the complexity and cost of the instrumentation required.

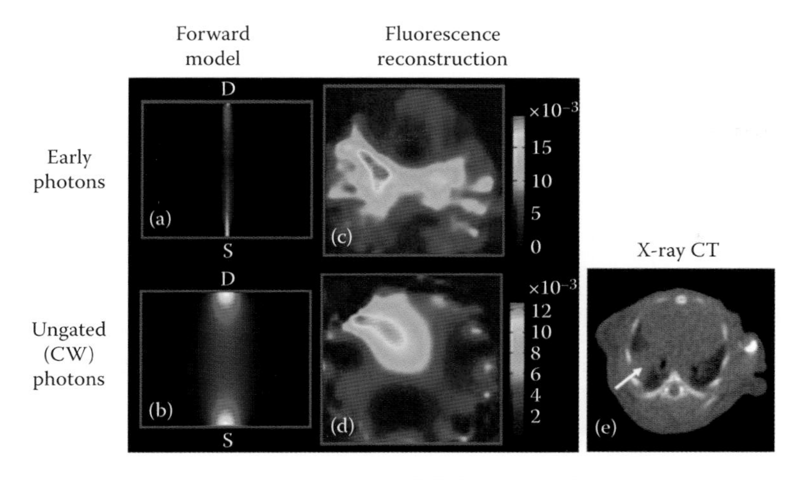

FIGURE 5.17 A comparison of early photon tomography to imaging with time-invariant illumination (CW laser). Green's functions for (a) early-arriving versus (b) continuous-wave photons. The early photons are scattered far less from the straight-line path between source and detector. (c and d) Resulting tomographic images of fluorescence resulting from a cathepsin-activated fluorescent agent in a mouse lung tumor model. (c) The early photon image displays higher spatial resolution and accuracy than (d) the ungated approach. (e) CT slice showing the location of the lung tumor. (Reproduced from Niedre, M.J. et al., *Proc. Natl. Acad. Sci. USA,* 105, 2008, Copyright 2008 National Academy of Sciences, U.S.A. With permission.)

5.4.6.3 Frequency-Domain Optical Tomography

As discussed previously, time-independent measurements using CW lasers, as used in FMT, can be used to determine fluorescence distributions in tissues, where normalization results in a reduced sensitivity to tissue optical property variations. There are, however, cases where this is not sufficient, for example, where the objective is to image the distribution of optical properties in the tissue. Imaging the optical properties of tissue can be useful for detecting endogenous changes, such as absorption increases due to increased vascularity, which can be used to diagnose such diseases as cancer and arthritis. The general approach is to use a source–detector configuration and a relevant forward model to reconstruct the absorption and scattering coefficients in the volume of interest. However, time-independent imaging typically does not provide enough information to robustly separate absorption and scattering properties. In these cases, frequency-domain imaging provides additional information over time-independent imaging that can add robustness to the quantification results.

The common approach to frequency-domain optical tomography in the diffusive regime is to modulate (vary) the intensity of the laser source illuminating the tissue and record the amplitude and phase shift of the light at detector positions by either photomultiplier tubes or gated intensified CCD cameras. As is the case with time-independent optical tomography, point sources of illumination are distributed across the tissue surface to provide many source–detector pair combinations. The forward model is then a frequency-domain representation of the diffusion equation or possibly higher-order approximations to the radiative transfer equation.

Inversion results in images of the absorption and scattering coefficients in the volume of interest. If imaging is performed at multiple wavelengths, these images can be combined in a spectral unmixing algorithm to obtain images of specific chromophore concentrations, such as oxyhemoglobin and deoxyhemoglobin. These approaches are particularly attractive for clinical imaging because they provide potentially useful information without the need for administration of exogenous contrast agents. This avoids issues of possible allergic reactions as well as circumventing the current lack of clinical approval for optical agents with molecular specificity. Further information on frequency-domain optical tomography for applications in breast cancer and arthritis imaging can be found in the scientific literature [20,21] and in Section 5.7.

5.4.6.4 Fluorescence Lifetime Imaging

Fluorescence lifetime imaging is best known in the context of microscopy as fluorescence lifetime imaging microscopy (FLIM; see Section 5.3.1) [22]. Instead of producing images based on fluorescence intensity, fluorescence lifetime images essentially give information on how long after the excitation pulse fluorescence emission occurs. The fluorescence lifetime is not only a property of the fluorophore but also depends on its environment. It has been used to detect FRET interactions by means of the timing of fluorescence emission in a way that is independent of the excitation intensity and can therefore yield accurate quantitative results. Fluorescence lifetime imaging *in vivo* in the diffusive regime has only recently been developed. The approach applies ultrashort laser pulses and gated intensified CCD cameras to record optical signals in the time domain. *In vivo* imaging of a FRET pair, where a free donor and acceptor can be distinguished from a linked donor–acceptor by differing fluorescence lifetimes, has been demonstrated in mice [23].

5.5 OPTOACOUSTIC IMAGING

One of the central themes of this chapter is that the scattering of light after the first few hundred microns of tissue penetration prevents deep-tissue high-resolution optical imaging. Optoacoustic (or photoacoustic) imaging uniquely provides optical contrast at high spatial resolution independently of whether the excitation light is scattered. It is based on the photoacoustic effect: A short pulse of light causes transient thermal expansion where it is absorbed in tissue, resulting in broadband pressure (ultrasound) waves that propagate outward and can be detected noninvasively. The propagation of ultrasound in tissue and its detection is discussed in Chapter 4. Because ultrasound waves scatter orders of magnitude less in tissue than light, the initial pressure distribution caused by light absorption can be reconstructed with high spatial resolution. As denoted in Equation 5.10, the initial pressure distribution (p_0) is proportional to the local optical absorption coefficient (μ_a) of the tissue, and optoacoustic imaging therefore measures optical absorption contrast.

$$p_0 = \Gamma \phi \mu_a. \tag{5.10}$$

Γ describes the thermodynamic properties of the absorber, and ϕ is the local light fluence in J cm^{-1}.

Because hemoglobin is the dominant absorber of light in tissue at visible and near-infrared wavelengths, optoacoustic imaging is naturally sensitive to blood vessels and organs with a high concentration of blood. The value for the absorption coefficient of whole blood in a blood vessel at 800 nm in the near-infrared spectrum is μ_a = 2.3 cm^{-1}.

A typical value for surrounding tissue is $\mu_a \sim 0.2\ cm^{-1}$. This means that blood vessels provide an order of magnitude more optoacoustic signal per volume than the tissue background, providing strong image contrast (later in the chapter, see Figures 5.25 and 5.28 for examples).

Optoacoustic signals encode the spatial dimensions of the absorbers that generate them, as can be seen in Figure 5.18. As a consequence, signals from small features, such as microvasculature, have higher frequency content than signals from larger features. Since ultrasound detectors have a limited bandwidth (range of detectable ultrasound frequencies), this implies that higher-frequency detectors should be selected to detect smaller features. Ultrasound attenuation in tissue increases with frequency, meaning that the attainable spatial resolution gets worse with increasing depth. (It should be noted that optoacoustic imaging is nevertheless able to obtain much higher resolution beyond the scattering barrier than pure optical imaging.) As a result of this relationship between depth and resolution, optoacoustic imaging is referred to as multiscale. Close to the surface, very-high-frequency ultrasound transducers can be used to obtain high-resolution images of a microscopic field of view. Within the ballistic regime, excitation light can be focused to provide optical resolution. On the other hand, whole small animals can be imaged with lower resolution by lower-frequency ultrasound detectors.

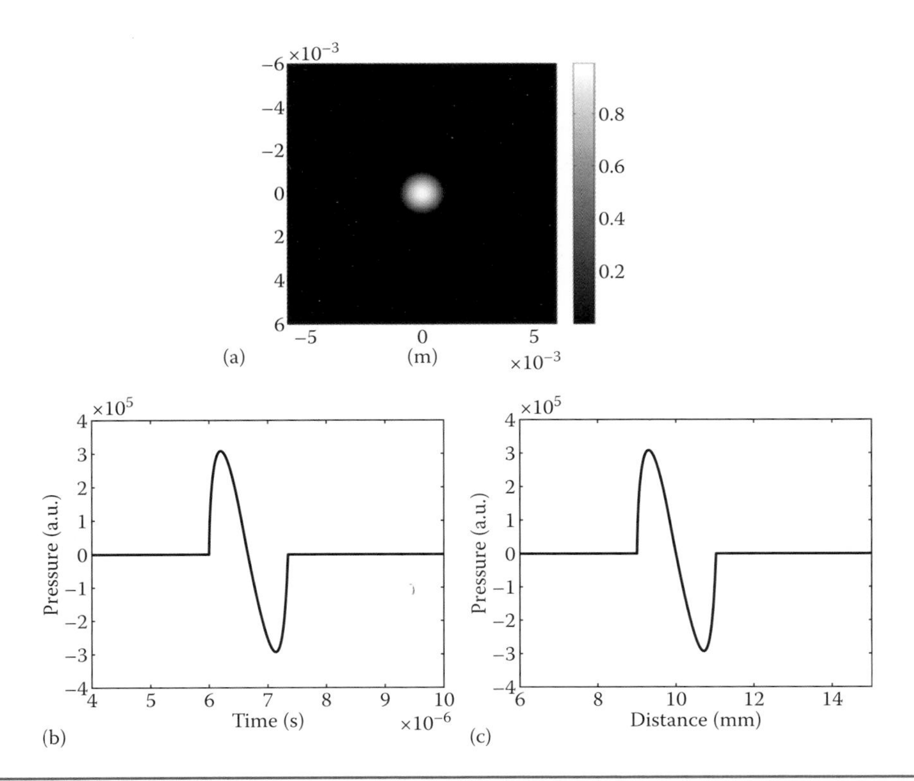

FIGURE 5.18 A simulation showing how optoacoustic signals encode the spatial characteristics of absorbers. (a) A circular absorber with a diameter of 2 mm. (b) The time-resolved optoacoustic signal detected from the absorber by a detector placed 10 mm from the center of the object. (c) The time axis of the signal can be replaced with distance by multiplying by the speed of sound in the medium. The signal now clearly displays a width equal to that of the absorber that generated it (2 mm).

Optoacoustic imaging is an emerging technology represented by a wide range of different implementations tailored for a variety of imaging applications. A large fraction of these approaches can be categorized as either tomographic imaging systems, where relatively large volumes are imaged, or microscopy systems using focused light and/or ultrasound detection to image one point at a time (Figure 5.19). Regardless of geometry, the basic principle remains the same: Absorption contrast is measured via ultrasound waves generated by the absorption of light. The use of ultrasound waves means that signal sources can be pinpointed at a high spatial resolution, even if the optical illumination is diffuse.

Instrumentation for optoacoustic imaging varies according to the implementation, but there are some common features. In many cases, illumination is provided by nanosecond pulsed lasers. These deliver enough energy to generate detectable optoacoustic signals within a sufficiently short time that the energy can be considered to be deposited instantaneously. For reasons that will be described later, the ability to vary the excitation wavelength is a common requirement. Popular illumination sources include *Q*-switched Nd:YAG (neodymium-doped yttrium aluminium garnet) lasers pumping either an optical parametric oscillator (OPO) or a Ti:sapphire laser to provide wavelength tuning. Optoacoustic imaging relies on time-resolved ultrasound detection, which can be achieved using piezoelectric transducers similar to those applied in diagnostic ultrasound imaging (see Chapter 4). As in ultrasound imaging, acoustic coupling between the tissue and transducer is necessary; that is, air gaps in the ultrasound signal path must be prevented. This is commonly achieved using ultrasound gel or water. Scanning imaging systems can be implemented using a single ultrasound transducer, but tomographic systems can take advantage of parallel ultrasound detection using multielement transducer arrays. The use of such arrays along with multichannel digitizers allows multiple projections to be acquired simultaneously to provide complete

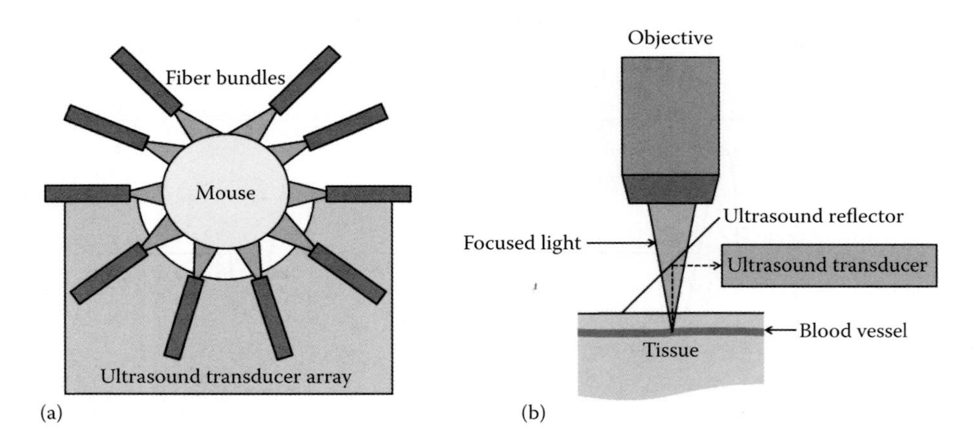

FIGURE 5.19 Optoacoustic imaging systems. (a) An example of optoacoustic tomography, where the entire surface around the region of interest is illuminated evenly, to generate diffusive light within the subject. An ultrasound detection array is employed, here in a semicircular geometry, to acquire multiple projection angles simultaneously. In such a system, imaging through an entire mouse is feasible. (b) An example of an optoacoustic microscopy system using focused light to image one point at a time.

images from each laser pulse, resulting in an imaging frame rate governed by the laser pulse repetition rate.

Optoacoustic images are commonly reconstructed using algorithms that neglect ultrasound scattering. Signals detected at each time point are assumed to come from points on a circle or sphere (in 2-D or 3-D imaging systems, respectively) of a radius corresponding to the distance that the ultrasound waves have traveled in the time since the excitation pulse. Because of this, model-based reconstruction methods, in contrast to pure optical tomography, rely on matrices that are sparse and generally well conditioned. However, the resulting images are reconstructions of the initial pressure distribution p_0, which depends not only on the local optical absorption properties but also on the local light fluence, as can be seen in Equation 5.10. Recovering fluence-independent information on tissue properties requires additional computation and is an active topic of current research.

Clearly, the use of ultrasound detection brings some of the limitations of ultrasound imaging to optoacoustics. The assumption that ultrasound waves do not scatter breaks down if bones or air is in the propagation path. Imaging through the skull in humans is therefore not a likely application, although this is possible through the much thinner skulls of small animals. Imaging of the lungs is also of highly doubtful feasibility because of the ultrasound scattering caused by tissue–air interfaces.

5.5.1 Multispectral Optoacoustic Imaging

Optoacoustic imaging provides optical absorption contrast at the high spatial resolution provided by ultrasound detection. The image contrast can originate from endogenous tissue absorbers, including hemoglobin, or exogenous contrast agents, which can be anything that absorbs light, ranging from organic dyes like ICG to plasmonic nanoparticles such as gold nanorods. However, this combination of endogenous and exogenous contrast results in difficulty in determining the source of any particular signal. Optoacoustic experiments often overcome this by acquiring a baseline image and then observing the change in contrast over time after an exogenous contrast agent is administered. This method has promise in scenarios involving a short period of time, for example, contrast-enhanced lymphatic mapping. But measurements over longer times are highly challenging because movement of the subject and potential changes in endogenous contrast could mask exogenous contrast enhancement. This challenge can be overcome by multispectral optoacoustic imaging.

Multispectral optoacoustic imaging is the application of multiple excitation wavelengths and subsequent spectral unmixing algorithms to obtain separated images of specific chromophores (absorbers) of interest. Any absorption source that provides significant contrast and has a unique absorption spectrum can be imaged by this method. This includes the different oxygenation states of hemoglobin (see absorption spectra in Figure 5.8 and the imaging example in Figure 5.20) and contrast agents with unique absorption peaks, thus allowing both functional and molecular imaging of tissue.

Ideally, contrast agents for use in multispectral optoacoustic imaging should provide high optical absorption and a narrow absorption peak to allow their optoacoustic spectra to be separated from background tissue absorption. As mentioned previously, single-wavelength optoacoustic imaging can also be used in cases where comparable baseline measurements without presence of the agent are possible.

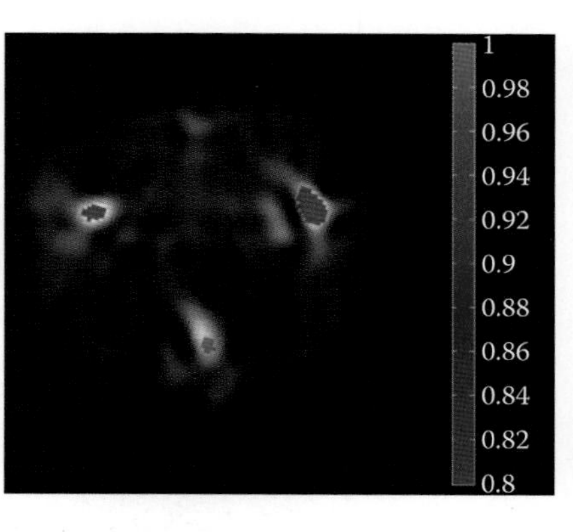

FIGURE 5.20 Multispectral optoacoustic tomography showing hemoglobin oxygen saturation through the tail of a mouse. Optoacoustic imaging was performed at six different wavelengths in the near infrared. Linear spectral unmixing was applied to obtain separated oxyhemoglobin and deoxyhemoglobin images, which were then recombined using the formula $HbO_2/(Hb + HbO_2)$. The tail veins and artery are clearly visible.

5.5.2 Sources of Contrast for Optoacoustic Molecular Imaging

5.5.2.1 Fluorescent Dyes

Multispectral optoacoustic imaging can be applied to image molecular fluorescent agents. Optoacoustic imaging detects the part of absorbed optical energy that is released by thermal relaxation, not fluorescence. In other words, fluorescent agents with a low fluorescence quantum yield have a higher optoacoustic efficiency if everything else is equal. The primary advantage of using multispectral optoacoustic imaging instead of conventional optical fluorescence to resolve fluorescent dyes is the improved spatial resolution at depth. Furthermore, while quantitative volumetric fluorescence imaging by tomographic approaches like FMT requires long acquisition times, as illumination source positions and, sometimes, detector projections are scanned sequentially, optoacoustic tomography can provide much-improved temporal resolution by illuminating the entire volume of interest and applying parallel ultrasound detection. These faster imaging rates can be crucial for experiments where fast dynamic processes are to be captured. However, there is a price to pay in terms of detection sensitivity compared to fluorescence-based methods (actual sensitivity numbers depend on the particular systems used), which can detect extremely low levels of light. Where detection sensitivity is critical, conventional fluorescence imaging remains the primary method for visualizing fluorescent agents.

5.5.2.2 Light-Absorbing Nanoparticles

The absorption contrast of optoacoustic imaging provides a unique method for detection of light-absorbing nanoparticles, which include carbon nanotubes and a wide variety of gold and silver nanoparticles. A particularly interesting example of molecular imaging with nanoparticles is enabled by the shift in the absorption spectrum by plasmon resonance coupling when nanoparticles are brought close together, for example, by binding to specific receptors. This phenomenon has been demonstrated for gold nanoparticles that are targeted to epidermal growth factor receptors [24].

Carbon nanotubes are a subject of much research related to possible applications as drug delivery vehicles. Because they absorb light, they can be detected by optoacoustic imaging, as shown in Figure 5.21, where they are investigated as tumor-targeting imaging agents [25].

Overall, optoacoustic imaging provides a method for imaging a wide range of nanoparticles *in vivo* without the need for additional labeling by fluorescence or radioisotopes. By providing high spatial and temporal resolution, it may establish itself as a powerful method for small animal imaging in biomedical nanotechnology research.

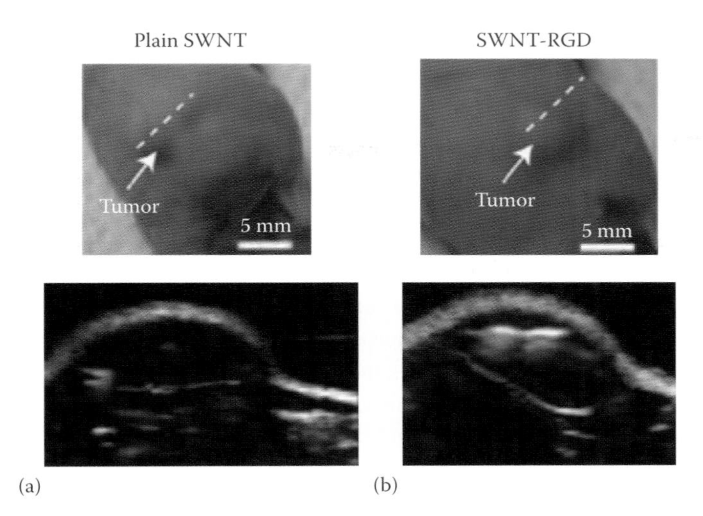

FIGURE 5.21 Optoacoustic imaging of carbon nanotubes in living mice. Mice with subcutaneous tumors were injected intravenously with single-walled carbon nanotubes (SWNT). (a) Images from control SWNT. (b) Images from tumor-targeted SWNT (with Arg-Gly-Asp [RGD] peptides). The green overlay on the gray ultrasound images shows the increase in optoacoustic signal after 4 h. (Reproduced from De la Zerda, A. et al., *Nat. Nanotechnol.*, 3, 2008. With permission.)

5.5.2.3 Fluorescent Proteins and Other Reporter Genes

As described in Section 5.3.1, fluorescent proteins have become a standard tool in biological research, and recent efforts to shift their excitation and emission peaks into the red and near-infrared wavelength regions have improved their suitability for *in vivo* deep-tissue imaging. Optoacoustic imaging also can provide high-resolution visualization of these fluorescent proteins via their absorption. Multispectral optoacoustic imaging of red-shifted and near-infrared fluorescent proteins has been demonstrated in model organisms such as zebra fish and in the setting of tumor imaging, as shown in Figure 5.22 [26,27].

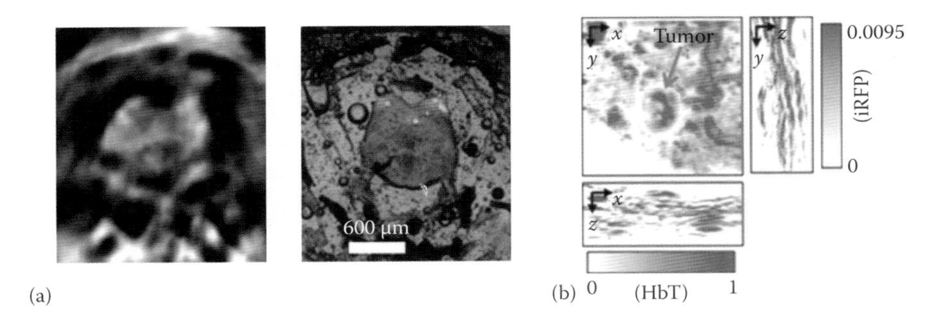

FIGURE 5.22 Multispectral optoacoustic imaging of fluorescent proteins. (a) Imaging of mCherry expression in the hindbrain of a zebra fish. (Left) Multispectral optoacoustic tomography. (Right) Corresponding fluorescence microscopy of dissected fish. (b) Multispectral optoacoustic microscopy of subcutaneous tumor expressing iRFP, a near-infrared fluorescent protein, in a mouse. The iRFP signal is shown in blue and the total hemoglobin (HbT) signal in red. ([a] Reproduced from Razansky, D. et al., *Nat. Photonics*, 3, 2009. With permission. [b] Reproduced from Filonov, G.S. et al., *Angew. Chem. Int. Ed. Engl.*, 51, 2012. With permission.)

In addition to imaging fluorescent proteins, optoacoustic imaging of reporter genes with more efficient optoacoustic signal generation, that is, increased optical absorption, has been investigated. Of particular interest has been the use of tyrosinase expression, which results in increased production of eumelanin, a strongly absorbing endogenous pigment [28].

5.6 PRECLINICAL APPLICATIONS

Optical imaging in the general sense is a familiar tool in biomedical research laboratories because of the ubiquity of fluorescence microscopy. As this chapter has shown, near-infrared fluorescence expands those capabilities to *in vivo* macroscopic deep-tissue imaging. Countless examples of the use of bioluminescence and planar near-infrared fluorescence can be found in the literature. The main advantage of these methods is their simplicity and versatility. Figure 5.23 shows an example of bioluminescence imaging to track the progression of Alzheimer's disease *in vivo* [29].

In deep tissues, obtaining quantitative information from simplistic planar methods is challenging, and more advanced techniques like FMT can be advantageous. There are several recent examples of FMT/CT imaging being applied for cardiovascular disease research, where the accurate quantification of deep-seated signals is essential. In one such study, FMT imaging of a protease-activatable near-infrared fluorescent agent, combined with CT for improved anatomical registration, provided key evidence that atherosclerosis is accelerated by prior myocardial infarction (Figure 5.24) [30].

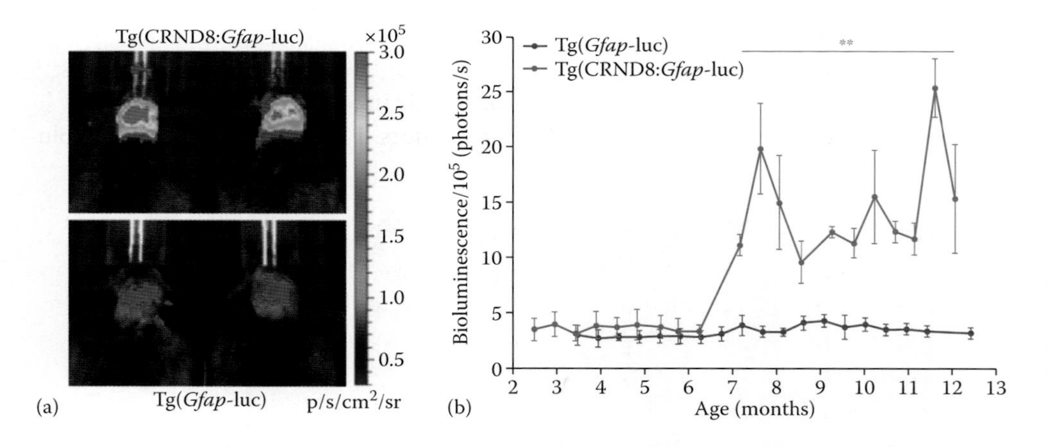

FIGURE 5.23 Bioluminescence imaging of beta-amyloid deposition during the progression of Alzheimer's disease. (a) Bioluminescence signals from the brains of 12-month-old mice. The mice in the top row are a transgenic model of Alzheimer's disease, while the bottom row are controls. Both the disease models and the controls express firefly luciferase under control of the glial fibrillary acidic protein (GFAP) promoter. (b) Mean bioluminescence signals of Alzheimer's mice (red) versus control mice (blue). **, statistically significant difference. (Reproduced from Watts, J.C. et al., *Proc. Natl. Acad. Sci. USA*, 108, 2011, Copyright 2011 National Academy of Sciences, U.S.A. With permission.)

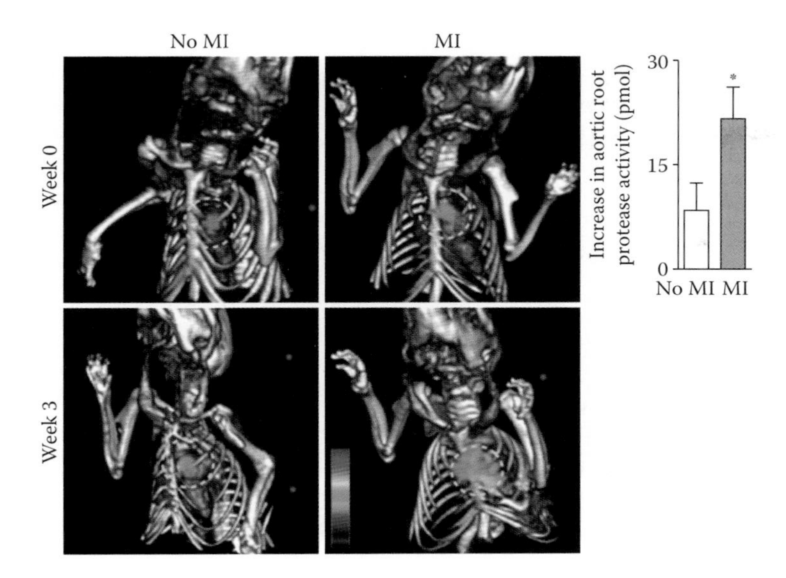

FIGURE 5.24 FMT/CT imaging of protease activity provides evidence that myocardial infarction (MI) accelerates atherosclerosis. Longitudinal FMT/CT imaging combined with a protease-activatable near-infrared fluorescent agent allows a quantitative *in vivo* comparison of inflammation in the aortic root between mice with previous MI and a control group. *P < .05. (Reproduced from Dutta, P. et al., *Nature,* 487, 2012. With permission.)

FMT is a valuable tool for imaging fluorescence distributions through whole mice, but the achievable spatial resolution on the order of 1 mm means that it has limited applicability to imaging heterogeneous features in smaller tissue volumes. For such applications, the multiscale capabilities of optoacoustic imaging can play a valuable role. A prominent example is the study of individual tumors. FMT lacks the spatial resolution to distinguish heterogeneous features within tumors of a few millimeters in diameter. Intravital microscopy techniques, which are commonly applied to studies of tumor biology, are limited to microscopic fields of view near the accessible tumor surface. In contrast, optoacoustic imaging at appropriately selected ultrasound frequencies allows noninvasive imaging of tumor heterogeneity. The natural sensitivity of optoacoustic imaging to hemoglobin, as well as the ability to distinguish between different oxygenation states of hemoglobin by multispectral optoacoustic imaging, allows visualization of functional aspects of the developing tumor vasculature as well as response to therapy (Figure 5.25) [31,32].

Apart from the ability to obtain high-resolution images of optical contrast, the parallel detection of ultrasound by multielement arrays can provide a far better temporal resolution than FMT. This allows the pharmacokinetic profiles of light-absorbing agents to be characterized *in vivo,* as shown for a dye being filtered by the kidneys in Figure 5.26 [33]. This capability can be applied to characterize new imaging agents and appropriately tagged therapeutics, as well as assess organ function (e.g., kidney and liver) in connection with novel agents by using organ-specific dyes.

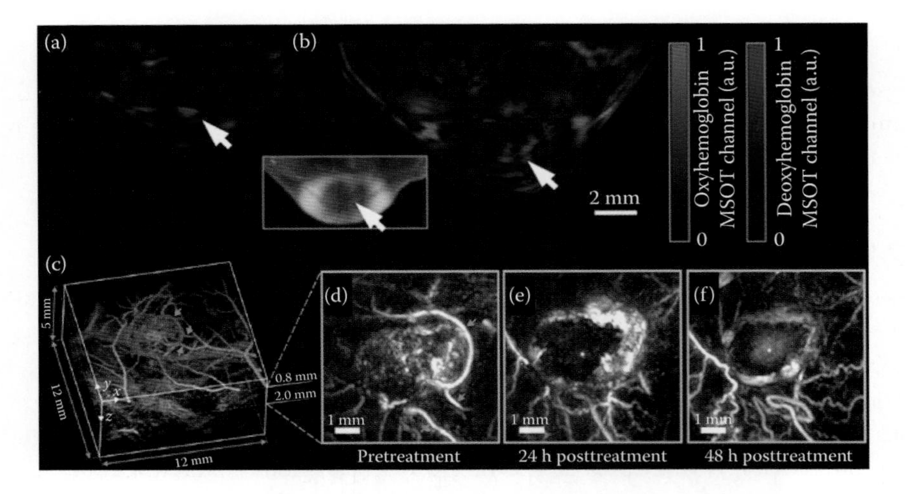

FIGURE 5.25 Preclinical optoacoustic imaging of tumors. Multispectral optoacoustic tomography (MSOT) to visualize oxyhemoglobin and deoxyhemoglobin distributions in a developing tumor in a mouse at (a) 6 days and (b) 13 days after cancer cell inoculation. The inset shows an *ex vivo* cryosection of the tumor, indicating an accumulation of deoxygenated blood in the core, as imaged by MSOT. (c through f) 3-D optoacoustic imaging of tumor in a mouse in response to treatment with a vasculature-disrupting agent. Green arrows indicate features recognizable throughout the images for reference. Optoacoustic imaging is able to visualize the change in tumor vasculature over time. ([a and b] Reproduced from Herzog, E. et al., *Radiology*, 263, 2012. With permission. [c through f] Reproduced from Laufer, J. et al., *J. Biomed. Opt.*, 17, 2012. With permission.)

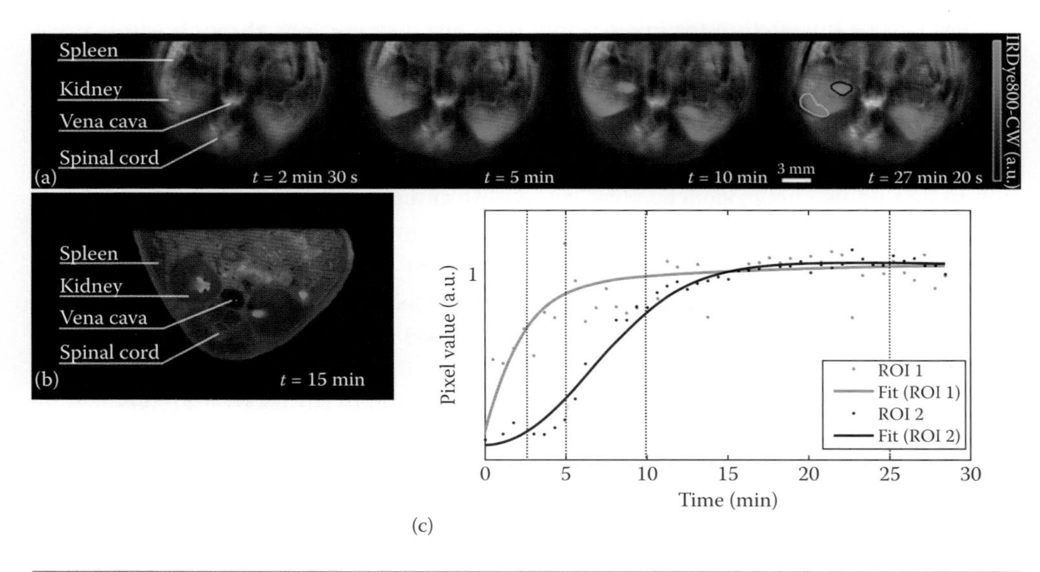

FIGURE 5.26 Multispectral optoacoustic tomography visualizing kidney filtration of a near-infrared fluorescent dye. (a) Green spectrally resolved signal from the near-infrared dye superimposed on grayscale images from hemoglobin absorption. The images are selected from different time points during an imaging session, showing the evolving dye distribution as it is filtered by the kidneys. (b) *Ex vivo* planar fluorescence/color image of a cryosection for comparison. (c) Quantities extracted from the multispectral optoacoustic images of the dye in two selected regions of interest (ROIs) over time. (Reproduced from Taruttis, A. et al., *PLoS One*, 7, 2012. With permission.)

5.7 CLINICAL APPLICATIONS

The sections that follow describe major emerging applications of optical and optoacoustic imaging in the clinical domain, without any attempt to provide an exhaustive review. Because molecular optical imaging is a relatively recent development, none of the applications highlighted here are in routine clinical use.

5.7.1 Fluorescence-Guided Surgery

Surgeons have relied on their eyes and hands (visual and tactile information) for centuries to discriminate between healthy and unhealthy tissue, for example, in tumor resections. Unfortunately, the visual contrast in their field of view is quite low because of the dominant role that hemoglobin plays in the overall optical properties of tissue. Here, fluorescence can come to the rescue. For example, by using fluorescent agents that are targeted to receptors that are overexpressed in cancerous cells, regions containing tumor cells can be made to emit far more fluorescence than the surrounding tissue, thus increasing the contrast. Progress in the field of fluorescence-guided surgery depends to a large extent on the approval of fluorescent imaging agents for experimental clinical studies. While ICG has been widely studied for fluorescence-guided lymphatic mapping and also can accumulate in tumors passively through leaky vasculature, targeted agents are generally considered to be a prerequisite for widespread clinical application. A recent study investigating the use of tumor-targeting folate–fluorescein isothiocyanate (FITC) for fluorescence-guided surgery in the context of ovarian cancer was the first in-human study of this kind (Figure 5.27) [34]. It is expected that further agents will be investigated for intraoperative and endoscopic fluorescence imaging, in particular, taking advantage of the superior penetration depth of near-infrared fluorescence.

5.7.2 Intravascular Fluorescence

Atherosclerosis is a disease that progresses slowly but can eventually cause myocardial infarction, stroke, and sudden death. Current diagnostic imaging for atherosclerosis focuses on the morphology of plaques, especially the degree of stenosis, which is not a good predictor for later events. Imaging of plaque biology *in vivo*, such as processes related to inflammation, could provide clinicians with far more valuable information for risk stratification of individual plaques. By means of targeted and especially protease-activatable fluorescent agents, plaques in locations that are challenging to image, such as the coronary arteries, can be highlighted for optical imaging. Fiber-based catheters for delivering excitation light and detecting emitted fluorescence can be inserted into the arteries alongside established modalities like intravascular ultrasound or emerging methods like optical coherence tomography, to provide combined imaging of morphology and molecular biology. The reduced optical absorption of blood in the near-infrared wavelength range makes such imaging feasible without flushing, that is, in the presence of blood. While this technique is a relatively recent development that, so far, has not been used on humans, results from studies on atherosclerotic rabbit models have been highly promising [35,36].

5.7.3 Breast Imaging

Near-infrared light has been investigated for decades in connection with breast cancer diagnosis because of the high prevalence of the disease and difficulties in diagnosis, particularly

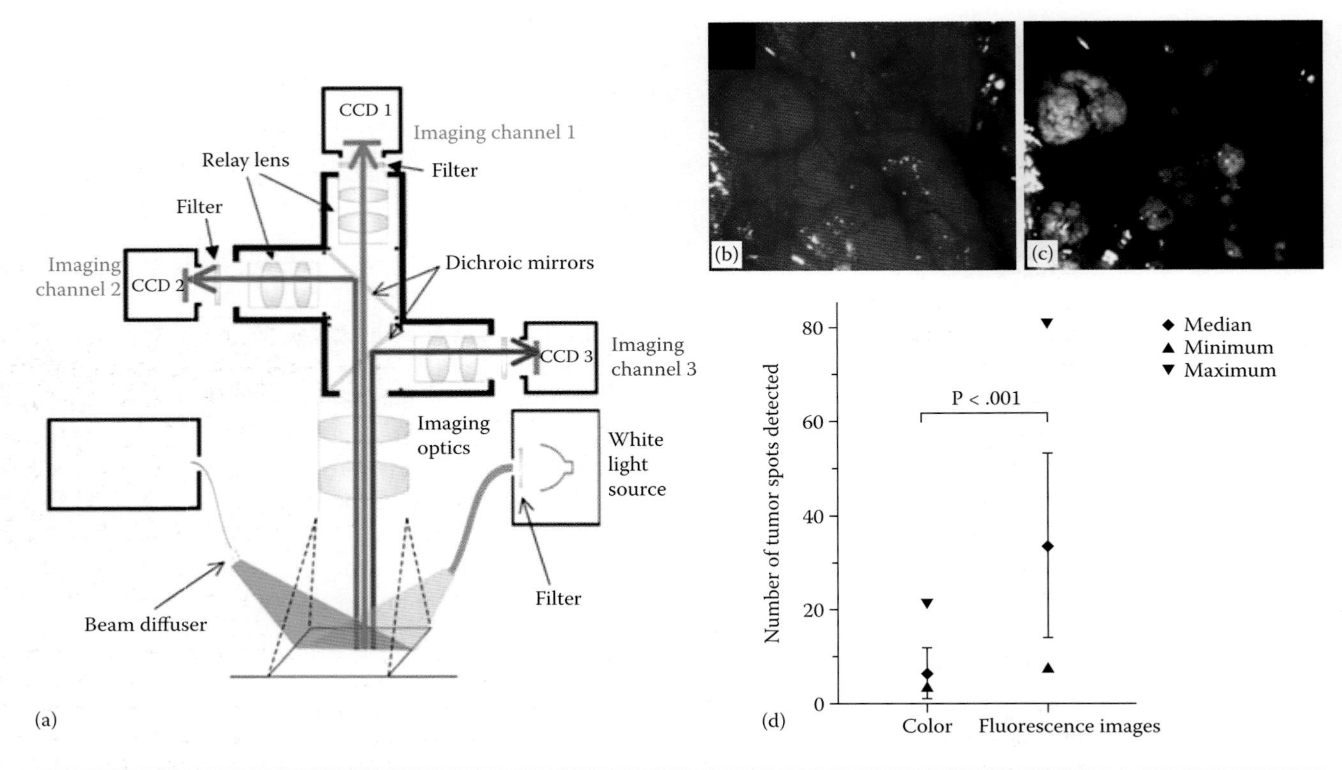

FIGURE 5.27 Fluorescence-guided surgery for enhanced tumor detection. Folate-FITC was used as a tumor-targeted fluorescent agent for patients with ovarian cancer. (a) Schematic of multispectral intraoperative fluorescence imaging system that simultaneously captures epi-illumination fluorescence, excitation wavelength, and color images. Excitation images can be used to produce images of normalized fluorescence. (b) Color image of tumor spots. (c) Corresponding fluorescence image highlighting the tumors. (d) Similar images were independently analyzed by surgeons asked to count the number of visible tumor spots in color and fluorescence images. A far larger number of tumors were visualized in the fluorescence images. (Reproduced from van Dam, G.M. et al., *Nat. Med.*, 17, 2011. With permission.)

in dense breast tissue. The ability of near-infrared light to penetrate the entire breast and the fact that it is nonionizing are particular motivations in this pursuit. A wide variety of optical methods have been proposed, ranging from early attempts at simple transillumination to identify dark lesions to advanced multimodal tomographic approaches with or without contrast enhancement. Approaches with diffuse optical tomography have been most successful in combination with MRI. As demonstrated in Figure 5.28a through c, this approach is able to multispectrally map endogenous chromophore concentrations through the breast [20]. Early attempts at diffuse optical tomography with exogenous contrast enhancement also demonstrated promising results [37]; however, the lack of clinically approved optical agents with molecular specificity has slowed progress in this area. Interest is expected to accelerate in the near future, when the number of agents approved for exploratory studies increases. More recently, optoacoustic breast imaging has been investigated for visualization of endogenous hemoglobin contrast (Figure 5.28d and e). Optoacoustic imaging offers higher spatial resolution in this application and provides sufficient anatomical contrast to

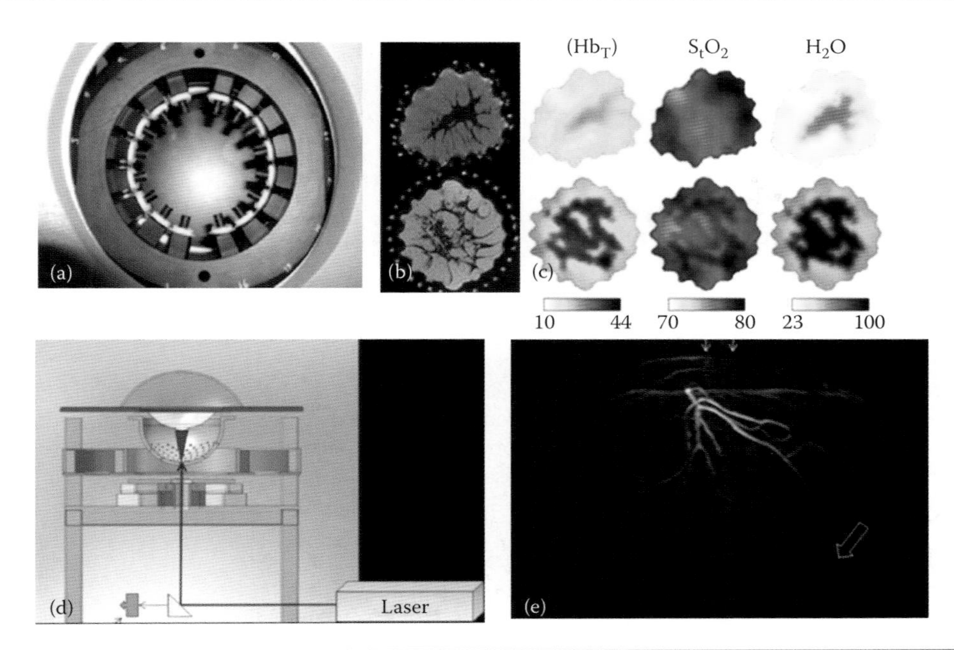

FIGURE 5.28 Optical and optoacoustic imaging of the breast. (a through c) Frequency-domain diffuse optical tomography of breasts from healthy volunteers in a hybrid implementation with MRI. (a) Photograph of source–detector fiber setup, which is positioned around the breast within the MRI system. (b) T_1-weighted MRI images of coronal slices corresponding to the optical tomography plane. (c) Images from diffuse optical tomography using MRI volumes as priors. Multiwavelength excitation was applied to produce images of hemoglobin, tissue hemoglobin oxygen saturation, and water. (d and e) Optoacoustic imaging of a breast of a healthy volunteer. (d) Optoacoustic breast imaging system. The breast is positioned in a cup within a cup-shaped, rotating, multielement ultrasound transducer array. (e) Resulting optoacoustic image of blood vessels in the breast, illuminated at 800 nm. The image is a maximum-intensity projection in lateral projection. ([a through c] Reproduced from Brooksby, B. et al., *Proc. Natl. Acad. Sci. USA*, 103, 2006, Copyright 2006 National Academy of Sciences, U.S.A. [d and e] Reproduced from Kruger, R.A. et al., *Med. Phys.*, 37, 2010. With permission.)

make it feasible as a stand-alone method. Investigations of patients are currently in very early stages [38,39].

5.7.4 Neuroimaging

The natural sensitivity of optical imaging to hemoglobin lends itself to imaging of the brain. Dynamic imaging made possible by parallel detection approaches provide information on hemodynamics from which brain function can be elucidated. Several experimental systems have been developed for diffuse optical tomography of the brain, with particular attention being paid to bedside imaging of infants. The systems consist of an array of optical source/detector fibers attached to the head to allow imaging of dynamics. The aim is to detect functional deficits and brain injury, with a view to providing prognostic information on future development of infants (Figure 5.29) [40].

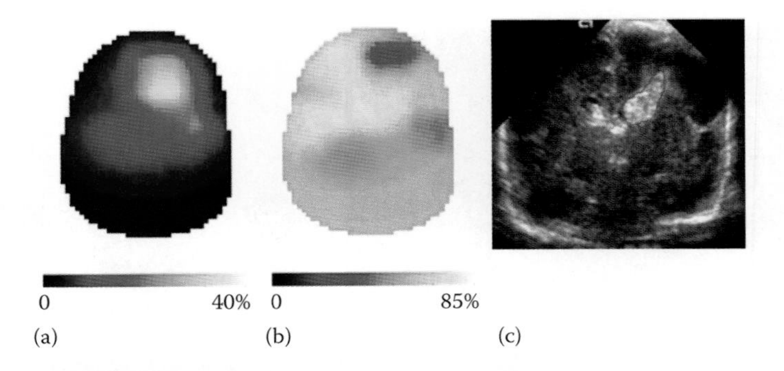

FIGURE 5.29 Diffuse optical imaging of the brain in infants. (a) A coronal section through the imaged volume showing regional blood volume. (b) Regional hemoglobin oxygen saturation in the same section. (c) Corresponding cranial ultrasound scan showing a lesion outlined in red. The detected lesion is a hemorrhage and is characterized by (a) increased blood volume and (b) decreased oxygenation. (Reproduced from Austin, T. et al., *Neuroimage,* 31, 2006. With permission.)

5.7.5 Arthritis

Optical imaging for simpler and more accurate early diagnosis of arthritis has been attempted using a wide range of approaches. The focus has been on imaging hands and fingers. These efforts have largely been based on detecting the increased optical absorption in inflamed tissue due to higher hemoglobin concentration. Both optical [21] and optoacoustic [41] imaging has been investigated. The most advanced clinical study so far has applied frequency-domain optical tomography to image 99 finger joint of patients, where potential values for sensitivity and specificity of over 85% for diagnosis of rheumatoid arthritis were reported [21].

5.7.6 Skin Examinations

The ability to scale optoacoustic imaging to microscopic resolutions as well as its natural sensitivity to hemoglobin makes the modality a promising candidate for imaging of skin lesions for cancer screening. Multispectral optoacoustic imaging is able to distinguish between oxygenation states of hemoglobin, allowing the extraction of measures of oxygen metabolism, which, alongside high-resolution imaging of microvasculature and melanin, could be used to identify cancerous lesions. Figure 5.30 shows an example of optoacoustic microscopy of the human skin [42].

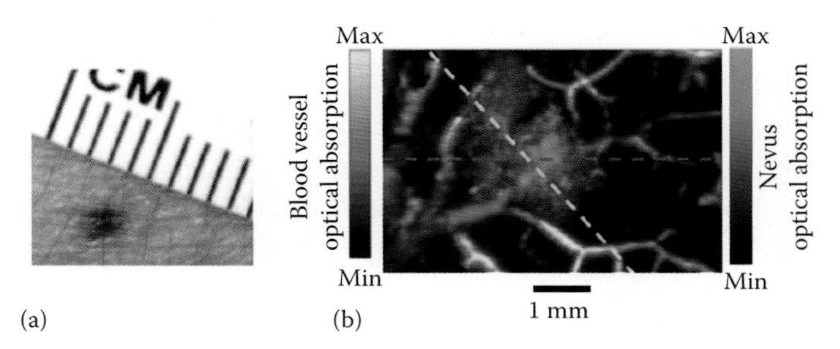

FIGURE 5.30 (a) Imaging of a benign skin lesion (nevus) by acoustic resolution optoacoustic microscopy. (b) Blood vessels and the nevus itself are visible on the resulting image. (Reproduced from Favazza, C.P. et al., *J. Biomed. Opt.,* 16, 2011. With permission.)

REFERENCES

1. Weissleder R, Ntziachristos V. Shedding light onto live molecular targets. *Nat Med.* 2003; 9(1):123–8.
2. Schaafsma BE, Mieog JS, Hutteman M, van der Vorst JR, Kuppen PJ, Löwik CW, Frangioni JV, van de Velde CJ, Vahrmeijer AL. The clinical use of indocyanine green as a near-infrared fluorescent contrast agent for image-guided oncologic surgery. *J Surg Oncol.* 2011;104(3):323–32.
3. Luo S, Zhang E, Su Y, Cheng T, Shi C. A review of NIR dyes in cancer targeting and imaging. *Biomaterials.* 2011;32(29):7127–38.
4. Scheuer W, van Dam GM, Dobosz M, Schwaiger M, Ntziachristos V. Drug-based optical agents: Infiltrating clinics at lower risk. *Sci Transl Med.* 2012;4(134):134ps11.
5. Ntziachristos V, Tung CH, Bremer C, Weissleder R. Fluorescence molecular tomography resolves protease activity *in vivo*. *Nat Med.* 2002;8(7):757–60.
6. Tsien RY. Constructing and exploiting the fluorescent protein paintbox (Nobel Lecture). *Angew Chem Int Ed Engl.* 2009;48(31):5612–26.
7. Filonov GS, Piatkevich KD, Ting LM, Zhang J, Kim K, Verkhusha VV. Bright and stable near-infrared fluorescent protein for *in vivo* imaging. *Nat Biotechnol.* 2011;29(8):757–61.
8. Jares-Erijman EA, Jovin TM. FRET imaging. *Nat Biotechnol.* 2003;21(11):1387–95.
9. Contag CH, Bachmann MH. Advances in *in vivo* bioluminescence imaging of gene expression. *Annu Rev Biomed Eng.* 2002;4:235–60.
10. Ntziachristos V, Razansky D. Molecular imaging by means of multispectral optoacoustic tomography (MSOT). *Chem Rev.* 2010;110(5):2783–94.
11. Jain RK, Munn LL, Fukumura D. Dissecting tumour pathophysiology using intravital microscopy. *Nat Rev Cancer.* 2002;2(4):266–76.
12. Sharpe J, Ahlgren U, Perry P, Hill B, Ross A, Hecksher-Sørensen J, Baldock R, Davidson D. Optical projection tomography as a tool for 3D microscopy and gene expression studies. *Science.* 2002;296(5567):541–5.
13. Weissleder R, Tung CH, Mahmood U, Bogdanov A Jr. *In vivo* imaging of tumors with protease-activated near-infrared fluorescent probes. *Nat Biotechnol.* 1999;17(4):375–8.
14. Hillman EM, Moore A. All-optical anatomical co-registration for molecular imaging of small animals using dynamic contrast. *Nat Photon.* 2007;1(9):526–30.
15. Themelis G, Yoo JS, Soh KS, Schulz R, Ntziachristos V. Real-time intraoperative fluorescence imaging system using light-absorption correction. *J Biomed Opt.* 2009;14(6):064012.
16. Ntziachristos V, Weissleder R. Experimental three-dimensional fluorescence reconstruction of diffuse media by use of a normalized Born approximation. *Opt Lett.* 2001;26(12):893–5.
17. Ale A, Schulz RB, Sarantopoulos A, Ntziachristos V. Imaging performance of a hybrid x-ray computed tomography-fluorescence molecular tomography system using priors. *Med Phys.* 2010;37(5):1976–86.
18. Ale A, Ermolayev V, Herzog E, Cohrs C, de Angelis MH, Ntziachristos V. FMT-XCT: *In vivo* animal studies with hybrid fluorescence molecular tomography-X-ray computed tomography. *Nat Methods.* 2012;9(6):615–20.
19. Niedre MJ, de Kleine RH, Aikawa E, Kirsch DG, Weissleder R, Ntziachristos V. Early photon tomography allows fluorescence detection of lung carcinomas and disease progression in mice *in vivo*. *Proc Natl Acad Sci U S A.* 2008;105(49):19126–31.
20. Brooksby B, Pogue BW, Jiang S, Dehghani H, Srinivasan S, Kogel C, Tosteson TD, Weaver J, Poplack SP, Paulsen KD. Imaging breast adipose and fibroglandular tissue molecular signatures by using hybrid MRI-guided near-infrared spectral tomography. *Proc Natl Acad Sci U S A.* 2006;103(23):8828–33.
21. Hielscher AH, Kim HK, Montejo LD, Blaschke S, Netz UJ, Zwaka PA, Illing G, Muller GA, Beuthan J. Frequency-domain optical tomographic imaging of arthritic finger joints. *IEEE Trans Med Imaging.* 2011;30(10):1725–36.
22. Lakowicz JR, Szmacinski H, Nowaczyk K, Johnson ML. Fluorescence lifetime imaging of free and protein-bound NADH. *Proc Natl Acad Sci U S A.* 1992;89(4):1271–5.
23. Nothdurft RE, Patwardhan SV, Akers W, Ye Y, Achilefu S, Culver JP. *In vivo* fluorescence lifetime tomography. *J Biomed Opt.* 2009;14(2):024004.

24. Mallidi S, Larson T, Tam J, Joshi PP, Karpiouk A, Sokolov K, Emelianov S. Multiwavelength photoacoustic imaging and plasmon resonance coupling of gold nanoparticles for selective detection of cancer. *Nano Lett.* 2009;9(8):2825–31.
25. De la Zerda A, Zavaleta C, Keren S, Vaithilingam S, Bodapati S, Liu Z, Levi J, Smith BR, Ma TJ, Oralkan O, Cheng Z, Chen X, Dai H, Khuri-Yakub BT, Gambhir SS. Carbon nanotubes as photoacoustic molecular imaging agents in living mice. *Nat Nanotechnol.* 2008;3(9):557–62.
26. Razansky D, Distel M, Vinegoni C, Ma R, Perrimoon N, Köster R, Ntziachristos V. Multi-spectral optoacoustic tomography of deep-seated fluorescent proteins *in vivo*. *Nat Photon.* 2009;3(7):412–417.
27. Filonov GS, Krumholz A, Xia J, Yao J, Wang LV, Verkhusha VV. Deep-tissue photoacoustic tomography of a genetically encoded near-infrared fluorescent probe. *Angew Chem Int Ed Engl.* 2012;51(6):1448–51.
28. Krumholz A, Vanvickle-Chavez SJ, Yao J, Fleming TP, Gillanders WE, Wang LV. Photoacoustic microscopy of tyrosinase reporter gene *in vivo*. *J Biomed Opt.* 2011;16(8):080503.
29. Watts JC, Giles K, Grillo SK, Lemus A, DeArmond SJ, Prusiner SB. Bioluminescence imaging of Abeta deposition in bigenic mouse models of Alzheimer's disease. *Proc Natl Acad Sci U S A.* 2011;108(6):2528–33.
30. Dutta P, Courties G, Wei Y, Leuschner F, Gorbatov R, Robbins CS, Iwamoto Y, Thompson B, Carlson AL, Heidt T, Majmudar MD, Lasitschka F, Etzrodt M, Waterman P, Waring MT, Chicoine AT, van der Laan AM, Niessen HW, Piek JJ, Rubin BB, Butany J, Stone JR, Katus HA, Murphy SA, Morrow DA, Sabatine MS, Vinegoni C, Moskowitz MA, Pittet MJ, Libby P, Lin CP, Swirski FK, Weissleder R, Nahrendorf M. Myocardial infarction accelerates atherosclerosis. *Nature.* 2012;487(7407):325–9.
31. Herzog E, Taruttis A, Beziere N, Lutich AA, Razansky D, Ntziachristos V. Optical imaging of cancer heterogeneity with multispectral optoacoustic tomography. *Radiology.* 2012;263(2):461–8.
32. Laufer J, Johnson P, Zhang E, Treeby B, Cox B, Pedley B, Beard P. *In vivo* preclinical photoacoustic imaging of tumor vasculature development and therapy. *J Biomed Opt.* 2012;17(5):056016.
33. Taruttis A, Morscher S, Burton NC, Razansky D, Ntziachristos V. Fast multispectral optoacoustic tomography (MSOT) for dynamic imaging of pharmacokinetics and biodistribution in multiple organs. *PLoS One.* 2012;7(1):e30491.
34. van Dam GM, Themelis G, Crane LM, Harlaar NJ, Pleijhuis RG, Kelder W, Sarantopoulos A, de Jong JS, Arts HJ, van der Zee AG, Bart J, Low PS, Ntziachristos V. Intraoperative tumor-specific fluorescence imaging in ovarian cancer by folate receptor-α targeting: First in-human results. *Nat Med.* 2011;17(10):1315–9.
35. Jaffer FA, Calfon MA, Rosenthal A, Mallas G, Razansky RN, Mauskapf A, Weissleder R, Libby P, Ntziachristos V. Two-dimensional intravascular near-infrared fluorescence molecular imaging of inflammation in atherosclerosis and stent-induced vascular injury. *J Am Coll Cardiol.* 2011;57(25):2516–26.
36. Yoo H, Kim JW, Shishkov M, Namati E, Morse T, Shubochkin R, McCarthy JR, Ntziachristos V, Bouma BE, Jaffer FA, Tearney GJ. Intra-arterial catheter for simultaneous microstructural and molecular imaging *in vivo*. *Nat Med.* 2011;17(12):1680–4.
37. Ntziachristos V, Yodh AG, Schnall M, Chance B. Concurrent MRI and diffuse optical tomography of breast after indocyanine green enhancement. *Proc Natl Acad Sci U S A.* 2000;97(6):2767–72.
38. Heijblom M, Piras D, Xia W, van Hespen JC, Klaase JM, van den Engh FM, van Leeuwen TG, Steenbergen W, Manohar S. Visualizing breast cancer using the Twente photoacoustic mammoscope: What do we learn from twelve new patient measurements? *Opt Express.* 2012;20(11):11582–97.
39. Kruger RA, Lam RB, Reinecke DR, Del Rio SP, Doyle RP. Photoacoustic angiography of the breast. *Med Phys.* 2010;37(11):6096–100.
40. Austin T, Gibson AP, Branco G, Yusof RM, Arridge SR, Meek JH, Wyatt JS, Delpy DT, Hebden JC. Three dimensional optical imaging of blood volume and oxygenation in the neonatal brain. *Neuroimage.* 2006;31(4):1426–33.
41. Xiao J, Yao L, Sun Y, Sobel ES, He J, Jiang H. Quantitative two-dimensional photoacoustic tomography of osteoarthritis in the finger joints. *Opt Express.* 2010;18(14):14359–65.
42. Favazza CP, Jassim O, Cornelius LA, Wang LV. *In vivo* photoacoustic microscopy of human cutaneous microvasculature and a nevus. *J Biomed Opt.* 2011;16(1):016015.

6

Radionuclide Imaging

Pat B. Zanzonico

6.1 INTRODUCTION

Radionuclide imaging, including *single-photon emission computed tomography (SPECT)* and *positron emission tomography (PET)*, utilizes small amounts of radioactivity administered, almost always systemically and usually intravenously, in the form of radiotracers (also known as radiopharmaceuticals). Diagnostic radionuclide imaging of patients is part of the clinical specialty known as *nuclear medicine*.* In recent years, the term *molecular imaging*, has become firmly entrenched in the lexicon of both clinical practice and preclinical research; it is defined as "... the visualization, characterization, and measurement of biological processes at the molecular and cellular levels in humans and other living systems" [1]. While it is not modality specific, and includes many of the other modalities discussed in this book, the term is often closely associated with radionuclide imaging and, in particular, SPECT and PET.

The ionizing radiations that accompany the decay of administered radioactivity can be detected, measured, and imaged noninvasively with instruments such as gamma cameras and SPECT and PET scanners. Radionuclide imaging in general, and SPECT and PET in particular, offers a number of important advantages in the context of clinical practice as well as clinical and preclinical research. First, the *specific activity* (i.e., activity per unit mass) of radiopharmaceuticals and the detection sensitivity of radionuclide imaging instruments are sufficiently high that administered activities needed for imaging correspond to nonpharmacologic,

* Although nuclear medicine remains primarily a diagnostic specialty, radioactive sources are also used therapeutically, that is, in sufficiently large amounts to deliver a high-enough radiation dose to destroy a diseased tissue such as a tumor. Therapeutic applications of nuclear medicine are beyond the scope of this chapter, however.

nonperturbing mass doses (typically in the sub-nmol range). This is in contrast to computed tomography (CT) and magnetic resonance imaging (MRI), for example, where the mass doses of contrast agents are usually far higher—typically in the μmol to mmol range—and thus may affect the system being studied. Second, radionuclide images are quantitative or at least semiquantitative, meaning that image "intensity" (i.e., counts) reflects the radiotracer-derived activity concentration. For PET, images can be absolutely quantitative and may be parameterized, for example, in terms of activity concentration. For some other imaging modalities, the relationship between the contrast-agent concentration and image intensity is typically not as direct. Third, a large number and variety of molecularly targeted and/or pathway-targeted radiotracers (such as metabolites and metabolite analogs, neurotransmitters, drugs, receptor-binding ligands, antibodies and other immune constructs, etc.) have been and continue to be developed for increasingly specific characterization of *in situ* biology.

Radionuclide imaging is not without its drawbacks, however. These include the relatively coarse spatial resolution ranging from ~5 mm for clinical PET to ~15 mm for clinical SPECT (although considerably better resolution can be achieved in preclinical systems). This is about an order of magnitude poorer than the spatial resolution of CT and MRI. Further, radionuclide imaging is an ionizing radiation-based modality and thus delivers low but nonnegligible radiation doses to patients or experimental animals. Typical injected activities are on the order of a few hundred MBq for human and large animal subjects (where 1 Bq = 1 nuclear disintegration per second) and ~4–40 MBq for small animal subjects. This leads to effective doses typically on the order of 10 millisieverts (mSv) and maximal organ absorbed doses up to several milligrays (mGy) [2] in clinical studies. Lastly, radionuclide images generally include only limited anatomic information, which may complicate their analysis and interpretation. With the increasingly widespread availability of multimodality (i.e., PET-CT and SPECT-CT) devices, radionuclide images reflecting *in vivo* function may be registered and fused with anatomic images, largely overcoming this limitation.

This chapter reviews the underlying physical principles and design and operation as well as the capabilities and limitations of SPECT and PET scanners used clinically and preclinically for *in vivo* radionuclide imaging. SPECT and PET radiotracers are also briefly reviewed and illustrative applications presented.

6.2 GENERAL CONSIDERATIONS IN RADIONUCLIDE IMAGING

6.2.1 Interactions of Radiation with Matter

Radiations emitted as a result of radioactive decay, such as x- and γ-rays and β-particles, are "ionizing" radiations. Such radiations ionize the atoms or molecules of a medium and produce free negative electrons and positive ions. X- and γ-rays are far more penetrating than β-rays. In soft tissue, x- and γ-rays with energies of several hundred keV will typically travel 5–10 cm before interacting, while β-particles with similar energies will travel no further than approximately 1 mm. "Diagnostic" x- and γ-rays interact with matter by the photoelectric effect or by Compton scatter (Figure 6.1a and b) [3,4]. In the *photoelectric effect*, an x- or γ-ray's energy is completely transferred to an orbital electron in an atom of the medium, ejecting the electron from the atom as a so-called photoelectron. The x- or γ-ray photon thus disappears in the process. In *Compton scatter*, only a portion of the incident x- or γ-ray's energy is transferred to an orbital electron, which is ejected from the atom as a so-called recoil electron.

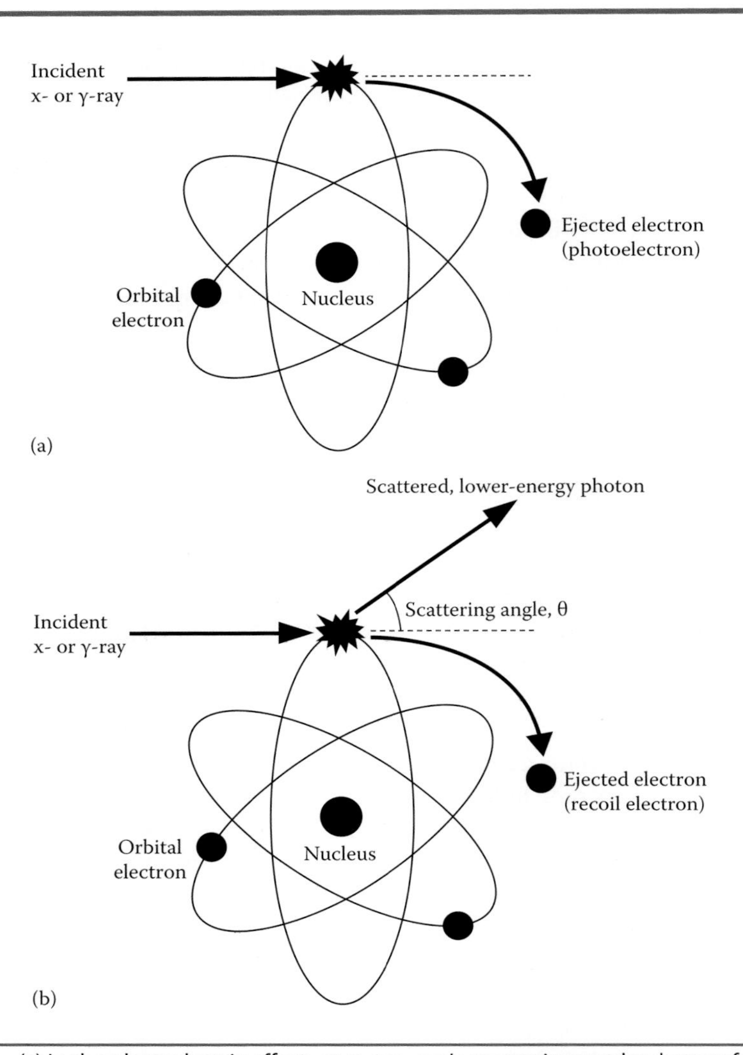

FIGURE 6.1 (a) In the photoelectric effect, an x- or γ-ray's energy is completely transferred to an orbital electron in an atom of the stopping medium, ejecting the electron from the atom as a "photoelectron." The x- or γ-ray thus disappears in the process. (b) In Compton scatter, only a portion of the incident x- or γ-ray's energy is transferred to an orbital electron, which is ejected from the atom as a "recoil electron." The scattered x- or γ-ray's energy is therefore less than that of the incident x- or γ-ray, and it travels in a different direction at a scattering angle θ relative to the original direction of travel. (From Zanzonico, P., *Radiat. Res.*, 177, 2012. With permission. With kind permission from Springer Science+Business Media: *Clinical Nuclear Medicine*, Physics, Instrumentation, and Radiation Protection, 2007, 1–33, Zanzonico, P., and S. Heller.)

The scattered x- or γ-ray's energy is therefore less than that of the incident x- or γ-ray, and it travels in a different direction. Importantly, because of their change in direction, x- or γ-rays that are Compton-scattered in the patient's body and detected with a gamma camera or other imaging device may erroneously appear to originate from a direction different from that of the original x- or γ-ray. Compton scatter, which is the predominant mode of interaction of "diagnostic" x- and γ-rays in tissues, thus represents one of the major impediments to accurate spatial localization and high-contrast detection of radionuclides. However, if the detection system has the ability to distinguish radiations of different energies, many of these

Compton-scattered x- or γ-rays can be effectively removed based on the fact that their energy is lower than that of the original (i.e., unscattered) x- or γ-rays.

Many medical radionuclides decay by emitting positrons. Positrons travel only ~1–2 mm or less in tissue. At the end of their range, positrons then undergo mutual annihilation with an "ordinary" electron in the medium, since positrons and electrons are antiparticles of one another. As a result of the positron-electron annihilation, their rest mass energies (the energy equivalent of their mass, 511 keV each) are converted to two 511 keV γ-rays emitted in opposite directions, or 180° apart. These back-to-back annihilation photons are utilized in coincidence detection and PET (see Section 6.5.3). The range of the positron prior to annihilation depends on the kinetic energy of the emitted positrons, which for many radionuclides is on the order of 1 mm or less. Some radionuclides emit more energetic positrons that can have a range up to ~5 mm.

6.2.2 Planar versus Tomographic Imaging

In *planar*, or 2-D, radionuclide imaging, radiations emanating from activity at all depths of the subject are projected onto an imaging detector. Ignoring effects such as attenuation and scatter, therefore, the image counts in a given pixel represent the ray sum, or integral, of detected radiations emitted over the full-depth volume of tissue corresponding to (i.e., underlying) that pixel. For a structure of interest such as a tumor, therefore, the tumor as well as background-tissue portions of the image include counts from activity outside the section, or slice, of tissue in which the tumor lies. This degrades image contrast (expressed, for example, as the tumor-to-background counts-per-pixel ratio) and compromises the visualization and quantitation of activity in the tumor—in some instances, to the point that tumors or other tissues of interest may be completely obscured.

A "*tomogram*," in contrast, is literally a picture of a slice through the subject. Tomographic imaging thus eliminates or at least minimizes the counts in the image arising from activity outside the tissue section of interest and thus improves image contrast and overall visualization of tumors and organs. *Tomography* may be characterized as either transmission or emission tomography, depending on the origin of the radiation. In transmission tomography (i.e., CT; see Chapter 2), x-rays are transmitted through the subject; in emission tomography, x- or γ-rays are emitted from within the subject. Emission tomography can be further characterized on the basis of the nature of the emitted radiation. Single photons, such as γ-rays associated with isomeric transition and x-rays associated with electron capture or internal conversion, form the basis of SPECT. The two 511 keV annihilation photons simultaneously emitted following positron-electron annihilation and associated with positron emission form the basis of PET.

The basic paradigm of tomographic imaging includes acquisition of images from multiple angles around a patient (multiple projections), correction of the acquired data for non-uniform response of the imaging system, and mathematical reconstruction of transverse tissue-section images. In SPECT and PET, the transverse images are essentially contiguous, with no intersection gaps. Therefore, the reconstructed 3-D array of volume elements, or voxels, may be rearranged at any angle relative to the longitudinal axis of the patient and thus yield coronal, sagittal, or oblique as well as transverse images. As noted, the principal advantage of tomography lies in its improved image contrast and greater quantitative accuracy. Another important advantage of emission tomography is the ability to visualize the 3-D distribution of activity *in situ*, that is, to ascertain the depths of foci of activity.

6.2.3 Static, Dynamic, and Whole-Body Imaging

Several types of radionuclide imaging—static, dynamic, and whole-body—may be performed in either planar or tomographic formats. *Static imaging* involves acquisition of an image over a time period where the distribution of the radiotracer is relatively stable.

Dynamic imaging involves the acquisition of a temporally varying distribution of activity as a series of images (or frames). Often, frames of different durations are used in a single study, with shorter-duration frames comprising the early times (when the *in vivo* distribution of a radiotracer often is changing rapidly) and progressively longer-duration frames comprising the later times (when the radiotracer distribution changes only slowly or stabilizes). The time-varying data provided by such studies may be used to derive information on dynamic processes. In conjunction with compartmental or other types of kinetic models, such data may yield estimates of functional parameters in absolute terms (see Chapter 7). For example, oxygen-15 (^{15}O)-labeled water has been used to measure regional perfusion in milliliters per minute per gram of tissue (mL/min/g).

Gated imaging is a type of dynamic imaging particularly important in cardiology, for example, for estimating left ventricular (LV) function (ejection fraction) and assessing LV wall motion. In gated imaging, a physiological "event" (such as the *R* wave of the electrocardiogram) in a repeating physiological process (such as the cardiac cycle) triggers the start of acquisition of a series of frames over each repetition of the process. The corresponding frames in each repetition are summed to yield a statistically reliable sequence of images over the cyclical process.

Dynamic acquisition can be performed in *frame mode*, with the number(s) and duration(s) of frames preset prior to acquisition. Alternatively, dynamic studies may be performed in so-called list mode, with the acquisition of a list of individual counts and with each count identified by its acquisition time and position coordinates. Importantly, the time binning is completely flexible and can be done and redone as often as necessary in post-processing to optimize the dynamic framing. This provides great flexibility for dynamic imaging in instances where the temporal resolution required may not be known prior to the study or, in the case of gated studies, the duration of the (cardiac) cycle may vary somewhat (e.g., due to arrhythmia). In contrast to frame-mode studies, there is really no distinction between the collection of static and dynamic images in list mode: If the acquired data are binned (summed) into a single frame, a static study results; binning of the same data into multiple frames yields a dynamic study.

Whole-body imaging is actually a type of static imaging, with either the detector(s) slowly translated over the stationary patient or the patient table slowly translated between the stationary detectors. Scan speeds are typically 5–10 cm/min, with scans including the entire length of the body. Whole-body scans thus disclose the distribution of a radiopharmaceutical throughout the entire body in a single image. Despite the motion of the detector(s) relative to patient, there is little to no perceptible degradation in image quality with whole-body scanning.

6.2.4 Statistical Considerations

Radioactive decay is a random process, described mathematically by the Poisson distribution, and therefore, random fluctuations will occur in the measured counts or count rates arising from decay of radioactivity. Such random fluctuations complicate the accurate detection, measurement, and imaging of radioactivity. If a detector were used to repeatedly measure the counts from a given radioactive sample, slightly different values would be obtained among the measurements. If an average of N counts were obtained, the standard deviation, σ, of the number of counts would be

$$\sigma = \sqrt{N}, \tag{6.1}$$

and the percent standard deviation (or "noise"), $\%\sigma$,

$$\%\sigma = \frac{100\%}{\sqrt{N}}. \tag{6.2}$$

Counting for a longer interval of time and acquiring a greater number of counts produces less random variation in the measurement (i.e., the $\%\sigma$ would be reduced) and the measured number of counts would be, on average, closer to the "true" value. Importantly, Equations 6.1 and 6.2 apply to planar radionuclide imaging (i.e., to the counts per unit picture element [or pixel] in a planar image) as well as to counting of radioactive samples. The more counts in an image and the greater the number of counts per pixel, the better the statistical accuracy of the image and the less mottled (or grainy) it appears. However, these equations do not directly apply to tomographic (PET and SPECT) images because the counts in individual volume elements (voxels) in such image sets are *not* independent of one another (i.e., do not follow Poisson statistics). As a result of the propagation of error in the image reconstruction process required for tomographic imaging, for the same number of counts per voxel in a tomographic image set and per pixel in a planar image, the statistical variation (i.e., noise) is considerably *greater* in the former than in the latter. Nonetheless, the same trend, where the noise level is inversely proportional to the square root of the number of acquired counts, still holds in many situations.

6.3 BASIC PRINCIPLES OF RADIATION DETECTION

Radiation detectors are generally characterized as either scintillation or ionization detectors [3,4]. In *scintillation detectors*, visible light is produced as radiation excites atoms of a scintillation crystal and is converted to an electronic signal, or pulse, and amplified by a *photomultiplier tube* (PMT) or other photodetector. In *ionization detectors*, free electrons produced when radiation ionizes a stopping material are collected to produce a small electronic signal. For radionuclide imaging, scintillation detectors are preferred because of their high sensitivity and lower cost. However, clinical devices based on solid-state ionization detectors are available as well for single-photon (i.e., non-PET) imaging but remain far less common than scintillation detector–based devices.

6.3.1 Scintillation Detectors

In scintillation detectors (Figure 6.2) [3,4], radiation interacts with and deposits energy in a scintillator, most commonly, a crystalline solid such as thallium-doped sodium iodide [NaI(Tl)]. The radiation energy thus deposited is converted to visible light, with the amount of light produced being proportional to the radiation energy deposited. Because the light is emitted isotropically, the inner surface of the light-tight crystal housing is coated with a reflective material so that light emitted toward the sides and front of the crystal are

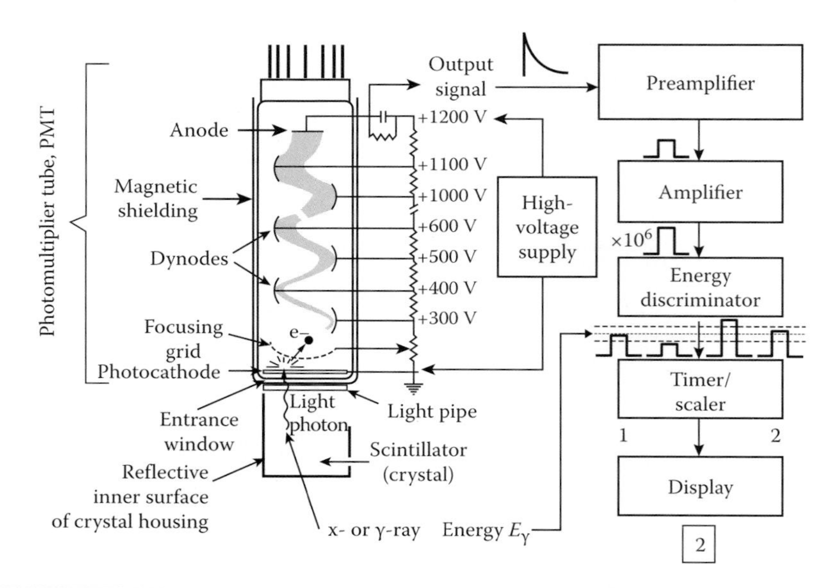

FIGURE 6.2 Basic design and operation of a scintillation detector. Note that only two of the four pulses have heights lying within the preset pulse height range (indicated by the two dashed horizontal lines below the energy discriminators, that is, corresponding to photon energies within the preset photopeak energy window; see Figure 6.3). Thus, only those two photons are counted (in the case of a radiation counter) or included in the image (in the case of a radiation imager). The other two photons, with pulse heights (and therefore energies) outside the photopeak energy window, are not counted or included in the image. (Adapted from Zanzonico, P., *Radiat. Res.*, 177, 2012. With permission. With kind permission from Springer Science+Business Media: *Clinical Nuclear Medicine*, Physics, Instrumentation, and Radiation Protection, 2007, 1–33, Zanzonico, P., and S. Heller.)

reflected back toward the PMT; this maximizes the amount of light collected and, therefore, the overall sensitivity of the detector and ensures that the amount of light detected is proportional to the energy of the absorbed photon. Interposed between the back of the crystal and the entrance window of the PMT is the light guide, sometimes simply a thin layer of transparent optical gel. The light guide optically couples the crystal to the PMT and thus maximizes the transmission of the light signal from the crystal into the PMT.

The PMT consists of an evacuated glass enclosure, containing a series of electrodes maintained at different voltages. Coated on the inner surface of the PMT's entrance window is the photocathode. When struck by the light from the crystal, the photocathode ejects electrons. The probability that each visible light photon will eject an electron from the photocathode is 15%–25%. Immediately beyond the photocathode is the focusing grid, maintained at a relatively low positive voltage relative to the photocathode. Once the "focused" electrons pass through the focusing grid, they are attracted by a relatively large positive voltage relative to the photocathode, ~300 V, on the first of a series of small metallic elements called dynodes. The resulting high-speed impact of each electron results in the ejection from the dynode surface of an average of three electrons. The electrons thus ejected are then attracted by the even-larger positive voltage, ~400 V, on the second dynode. The impact of these electrons on the second dynode surface ejects an additional three electrons on average for each incident electron. Typically, a PMT has 10–12 such dynodes (or stages), each ~100 V more positive than the preceding dynode, resulting in an overall electron amplification factor of 3^{10}–3^{12} for the entire PMT. At the collection anode, an output signal is generated.

The PMT output signal may undergo further processing by a preamplifier, or shaping amplifier, and is then digitized. The amplitudes of the digitized electrical pulses are proportional to the number of electrons produced at the PMT photocathode and thus the energy of the incident radiation. These pulses can then be sorted according to their respective energies, and those pulses with an amplitude within a preset energy range ("window") are counted or included in the image.

In recent years, new PMT configurations have been developed that allow enhanced approaches to position determination in gamma cameras and SPECT and PET scanners. These include the position-sensitive PMT (PSPMT), which provides 2-D position information via two sets of wire anodes perpendicular to one another. The electrons from the last dynode stage are read out on each set of anode wires. The position of the incident light on the photocathode can then be estimated from the electron distribution on each anode wire. Readout of the wire anodes is typically accomplished via separate internal resistors for x and y position information.

The silicon photodiode is an alternative to the PMT for conversion of scintillation light into electronic signals. Photodiodes typically have a gain of only 1 (compared to the ~10^6-fold gain of PMTs) and thus require low-noise electronics. So-called avalanche photodiodes (or APDs), in which the number of electrons produced by the visible light is amplified, have considerably higher gains, on the order of 100 to 1000, but still require low-noise readout electronics. An alternative to traditional vacuum-tube PMTs, *silicon photomultipliers* (SiPMs) are single-photon-sensitive devices built from an APD array on a silicon substrate [5]. The dimension of each single APD can vary from 20 to 100 μm, and their density can be up to 1000/mm^2. The supply voltage typically varies between 25 and 70 V, 30- to 50-fold lower than that required for traditional PMTs. Performance parameters of SiPMs are comparable to those of traditional PMTs but with a much more compact form factor.

The scintillation detection materials most widely used in nuclear medicine—all inorganic scintillators—are bismuth germanate (BGO, $Bi_4Ge_3O_{12}$), cerium-doped lutetium oxyorthosilicate [LSO(Ce) or LSO, Lu_2SiO_5:Ce], and cerium-doped lutetium-yttrium oxyorthosilicate [LYSO(Ce) or LYSO, Lu_2YSiO_5:Ce] as well as NaI(Tl) (Table 6.1) [3,4,6,7]. PET scanners primarily use BGO, LSO, or LYSO, and gamma cameras and SPECT systems use NaI(Tl). The most important practical features of scintillation detectors include the following:

- High mass density (ρ) and effective atomic number (Z_{eff})—to maximize the photon stopping power (i.e., intrinsic efficiency ε) of the detector;
- High light (scintillation) output—to maximize the signal and thus minimize statistical uncertainty in the energy of the detected signal;
- For PET, speed of the output light pulse—to get the best possible timing information to shorten the coincidence timing window (τ) and thus minimize the number of random events (see Section 6.5.3).

As noted, materials with higher-ρ and -Z_{eff} such as BGO, LSO, and LYSO have emerged as the detectors of choice for PET because of their greater stopping power for 511 keV annihilation γ-rays. The mean free path (MFP) for 511 keV γ-rays is at least twice as long in NaI(Tl) as in BGO, LSO, or LYSO. LSO and LYSO have almost eightfold faster decay

TABLE 6.1
Physical Properties of Scintillation Detector (Crystal) Materials

Material	Composition	Density, ρ (g/cm³)	Effective Atomic Number, Z_{eff}	Linear Attenuation Coefficient, μ, for 511 keV γ-rays (/cm)	Relative Probability of Photoelectric Interaction (%)	Light Output (photons per MeV)	Scintillation Decay Time (ns)	Scintillation Wavelength, λ (nm)	Energy Resolution at 511 keV (%FWHM)	Hygroscopic?
Bismuth germanate (BGO)	$Bi_4Ge_3O_{12}$	7.13	75	0.95 $\varepsilon(2\text{ cm}) = 0.72$[a]	40	9000	300	480	16	No
Cerium-doped lutetium oxyorthosilicate (LSO)	Lu_2SiO_5:Ce	7.4	66	0.88 $\varepsilon(2\text{ cm}) = 0.69$[a]	32	~30,000	~40	420	12	No
Cerium-doped lutetium-yttrium oxyorthosilicate (LYSO)	$LuYSiO_5$:Ce	~7.0	66	0.87 $\varepsilon(2\text{ cm}) = 0.68$[a]	30	~30,000	~42	430	12	No
Thallium-doped sodium iodide [NaI(Tl)]	NaI:Tl	3.67	51	0.34 $\varepsilon(2\text{ cm}) = 0.24$[a]	17	38,000	230	415	8	Yes

Source: Zanzonico, P., *Radiat. Res.*, 177, 2012; Humm, J.L. et al., *Eur. J. Nucl. Med. Mol. Imaging*, 30, 2003; Zanzonico, P., *Semin. Nucl. Med.*, 34, 2004; With kind permission from Springer Science+Business Media: *Clinical Nuclear Medicine*, Physics, Instrumentation, and Radiation Protection, 2007, 1–33, Zanzonico, P., and S. Heller.

[a] The intrinsic efficiency of 2-cm-thick coincidence detectors for 511 keV annihilation γ-rays.

times and a threefold greater light output than BGO. However, lutetium-based scintillators are relatively expensive. NaI(Tl) is the material of choice for most gamma cameras and SPECT scanners, as it is relatively inexpensive, can be fabricated with large areas, and has adequate stopping power for the 80–300 keV gamma rays typically imaged with these systems.

6.3.2 Semiconductor-Based Ionization Detectors

Semiconductor radiation detectors represent the main alternative to scintillator detector–based imaging systems. Such detectors are so-called direct-conversion devices, a major advantage of which is that they avoid the random effects associated with scintillation production and propagation and conversion of the optical signal to an electronic signal. When an x- or γ-ray interacts in a semiconductor detector, one or more energetic electrons are created and subsequently lose energy through ionization, among other processes. The ionization creates electron-hole (e-h) pairs, where a hole is the positively charged electron vacancy in the valence band left when the electron has been raised into the conduction (i.e., mobile-electron) band. Application of a bias voltage creates an electric field that causes the two types of charge carriers to migrate in opposite directions. These moving charges induce transient currents in the detector electrodes, thereby allowing measurement of the detector's response to an incident x- or γ-ray.

Semiconductor detectors offer several potential advantages over scintillator detectors [8]. By eliminating the need for bulky PMTs, semiconductor imaging systems can be made much more compact, simplifying their mechanics in comparison to those of a conventional PMT-based imaging system (see Sections 6.5.1 and 6.5.2). More importantly, the direct conversion of energy deposited by x- or γ-rays into e-h pairs eliminates the light-to-electrical signal transduction step and the associated loss of signal. Further, since the energy required to create an e-h pair in most semiconductors employed as radiation detectors is small (typically 3–5 eV), each incident photon generates a large number of charge carriers. In principle, therefore, Poisson (statistical) noise is considerably less and energy resolution (see Section 6.3.3) considerably better in semiconductor detectors than in scintillation detectors. Defects (i.e., inherent irregularities in the crystal lattice) can trap electrons produced by radiation and thus reduce the total charge collected. As a result of such incomplete charge collection due to "charge trapping," statistical uncertainty (i.e., noise) is increased and energy spectra broadened, that is, the otherwise excellent energy resolution of semiconductors is degraded. Thin semiconductors have fewer traps overall than thick detectors but also have a lower intrinsic efficiency. Trapping can be reduced by increasing the bias voltage across the detector but at the expense of increasing its leakage current. This results in greater electronic noise and, therefore, poorer overall performance. Practical, reasonably economical crystal-growing techniques have been developed for cadmium telluride (CdTe), mercuric iodide (HgI_2), and cadmium zinc telluride (CdZnTe, also known as CZT), which have been incorporated into commercial intraoperative gamma probes and gamma cameras [8,9].

6.3.3 Radiation Detector Performance

Radiation detectors may be quantitatively characterized by many different performance parameters. Among the most important of these, however, are sensitivity (or efficiency), energy resolution, count-rate performance, and, for devices that localize (image) as well as count radiation, spatial resolution and uniformity [10]. This section discusses these parameters in general terms; rigorous definitions and protocols for their measurement are modality specific and described in the references [11–14].

Sensitivity. Sensitivity (or efficiency) is the detected count rate per unit activity. Because the count rate detected from a given activity is highly dependent on the source-detector geometry and intervening absorbing media, characterization of sensitivity can be ambiguous. There are generally two major components of overall detector sensitivity, geometric and intrinsic sensitivities. Geometric sensitivity is the fraction of emitted radiations that strike the detector, that is, the fraction of the total solid angle subtended by the detector. It is therefore directly proportional to the detector area and, for a point source of radioactivity, inversely proportional to the square of the source-detector distance. Intrinsic sensitivity is the fraction of radiations striking the detector that are stopped within the detector. Intrinsic sensitivity is directly related to the detector thickness and density (and therefore effective atomic number Z_{eff} and mass density ρ) and generally decreases with increasing photon energy.

Energy resolution. Characteristic x- and γ-rays are emitted from radioactively decaying atoms with well-defined discrete energies. Even in the absence of Compton scatter, however, photon counts will be detected over a range of energies due to the finite ability of the detector to determine the energy deposited. With scatter, as in a patient, this becomes more pronounced due to the change in energy of photons undergoing Compton scatter. Therefore, many radiation detectors employ some sort of energy-selective counting: an energy range, or window, is selected such that radiations are counted only if their detected energies lie within that range. For example, a 20% photopeak energy window, $E_\gamma \pm 10\%$ (e.g., 126–154 keV for the 140 keV γ-ray from ^{99m}Tc), might be employed, where E_γ is the primary (or photopeak) energy of the radiation (Figure 6.3). For such energy-selective counting, overall sensitivity appears to increase as the photopeak energy window is widened. However, this results in acceptance of more scattered as well as primary (i.e., unscattered) radiations. Energy resolution quantifies the ability of a detector to separate, or discriminate, radiations of different energies and is generally specified as the full-width at half-maximum height (FWHM = ΔE), expressed as a percentage of the photopeak energy (E_γ) of the bell-shaped photopeak, $\text{FWHM}~(\%) = \frac{\Delta E}{E_\gamma} \times 100\%$ (Figure 6.3).

As illustrated in Figure 6.3, the primary energy spectrum (dotted curve) includes relatively few low-energy counts and is typical of a source in air while the total spectrum (solid curve) includes many lower-energy counts. These lower-energy counts result from Compton scatter (dot-dash curve) and are typical of a source in tissue or other medium. The importance of energy resolution thus lies in scatter rejection, particularly for imaging detectors. Radiation loses energy when scattered, and the lower-energy scattered radiations may therefore be discriminated from the primary radiations on the basis of their lower energies. However, the finite energy resolution of radiation detectors means that there will be some overlap of scattered and primary radiations, as illustrated in Figure 6.3. For systems with superior energy resolution (i.e., with a lower FWHM [%] and a narrower photopeak), scattered radiation is more effectively excluded while primary radiation is still efficiently counted. For routine gamma camera imaging of ^{99m}Tc with a 20% photopeak energy window, the scatter fraction (i.e., the fraction of the total counts in the photopeak energy window and therefore in the image) can easily be 0.2 or greater, with larger scatter fraction associated with larger sources (i.e., patients).

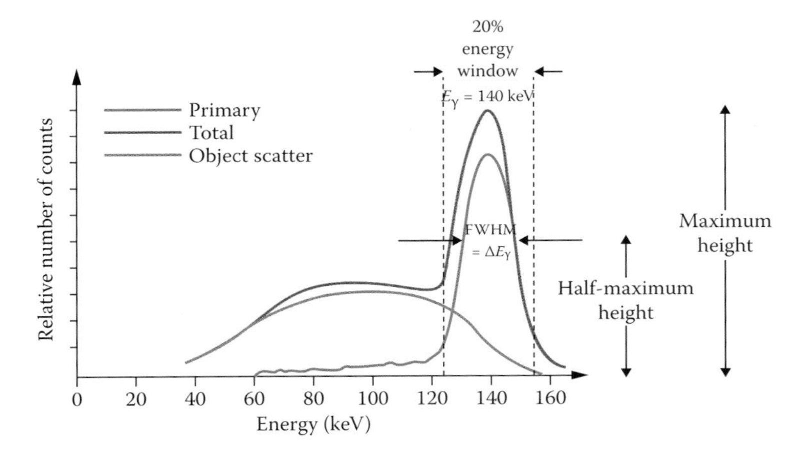

FIGURE 6.3 Energy spectrum (i.e., the graph of the number of x- or γ-rays at a given detected energy versus the detected energy) for a thallium-doped sodium iodide [NaI(Tl)] scintillation detector for the 140 keV γ-rays emitted by ^{99m}Tc, illustrating its 20% energy window and the contributions of primary (unscattered) and Compton-scattered radiation counts. This spectrum, with the large number of lower-energy Compton-scattered counts (the dot-dash curve), is typical of a source of ^{99m}Tc at some depth in a water-equivalent medium. The dotted curve, corresponding to primary radiation counts, would be typical of a source in air (i.e., in the absence of a scattering medium). (From Cherry, S.R. et al., *Physics in Nuclear Medicine*, Saunders, Philadelphia, PA, 2012. With permission.)

Spatial resolution. For imaging detectors, spatial resolution is one of the most important performance parameters. It reflects the ability of the detector to accurately determine the size and location of a source or, similarly, the ability to visually discriminate two neighboring sources. This is usually measured by imaging a point or line source and examining the spread (or "blurring") of its image. The detector's spatial resolution can then be expressed as the FWHM of this point (or line) spread function.

Uniformity. Uniformity is likewise a critical parameter of imaging devices. In principle, a uniform source of radioactivity should yield a uniform image (i.e., an image in which the counts per pixel or voxel are constant over the entire image). In practice, this is never achieved—even if one discounts the effects of fluctuations in counting statistics (noise)—because of inevitable point-to-point variations in sensitivity of an imaging detector. Corrections for the nonuniform response of a SPECT or PET scanner (often referred to as a "flood correction" or "normalization," respectively) can be measured by imaging of a uniform source of an appropriate radionuclide. Such corrections may need to be periodically updated due to changes in detector performance over time.

6.4 RADIONUCLIDES FOR SPECT AND PET IMAGING

Implicit in the suitability of a radionuclide for *in vivo* imaging is that it emits in sufficient abundance radiations penetrating enough to escape from the body and interact with external detectors. These emissions include γ-rays and characteristic x-rays ("single photons") used for SPECT and planar gamma camera imaging generally or the 511 keV annihilation γ-rays associated with positron (β^+) decay and used for PET. In addition, imaging radionuclides ideally should emit few or no nonpenetrating radiations, that is, particulate radiations such as β-particles and electrons (excluding, of course, positrons necessary for signal generation in the case of PET). Such particulate radiations typically have ranges in tissue on the order of 1 mm or less and thus cannot escape from the body and be detected externally. These contribute to the radiation dose to tissues and organs without providing any imageable signal. Further, an imaging radionucide ideally should have a physical *half-life* comparable to the time required for the administered radiotracer to localize in the tissue of interest. This will provide sufficient time for it to localize in that tissue while still retaining a near-maximal imaging signal, followed by elimination of the radioactivity by a combination of physical decay and biological clearance. The radiation dose to patients and individuals around them and the potential problem of radioactive contamination are thereby minimized. The pertinent physical properties of single-photon- and positron-emitting radionuclides used for imaging are summarized in Tables 6.2 and 6.3 [3,7,15,16], respectively.

For planar imaging and SPECT, a radionuclide emitting x- and γ-rays with energies of 100–200 keV, with an abundance of 100% (i.e., one x- or γ-ray emitted per decay), and with minimal emission of particulate radiations and of higher-energy x- and γ-rays is ideal. The abundance of imageable x- or γ-rays relative to particulate radiations is expressed by the x- and γ-ray–to–β-particle, conversion electron, and Auger electron energy-per-decay ratio (Table 6.2); the higher the value of this parameter, the better a radionuclide will be for imaging. With an MFP in soft tissue on the order of 10 cm and in NaI(Tl) of less than 0.5 cm, 100 to 200 keV photons provide adequate penetrability through tissue but are low enough in energy to be efficiently collimated (see Section 6.5.1) and stopped in relatively thin scintillation detectors, yielding optimum-quality images with reasonably low radiation doses. The

TABLE 6.2
Physical Properties of Single-Photon-Emitting Radionuclides Used in Gamma Camera Imaging (Including SPECT)

Radionuclide	Physical Half-Life	Decay Mode	Energies of Imageable x- and γ-rays (MeV)	Abundance of Imageable x- and γ-rays (%)	x- and γ-ray–to–β-Particle, Conversion Electron, and Auger Electron Energy-per-Decay Ratio[a]	Production
Gallium-67	3.26 days	Electron capture	0.0933	40	4.5	Cyclotron
			0.185	24		
			0.300	16		
Technetium-99m	6.01 h	Isomeric transition	0.140	89	7.8	Generator (molybdenum-99)
Indium-111	2.83 days	Electron capture	0.172	90	12	Cyclotron
			0.247	94		
Iodine-123	13.2 h	Electron capture	0.159	84	6.1	Cyclotron
Iodine-131	8.04 days	β^- decay	0.364	82	2.0	Reactor
Thallium-201	3.05 days	Electron capture	0.068–0.080	95	2.1	Cyclotron
			0.167	10		

Source: Zanzonico, P., *Radiat. Res.*, 177, 2012; Zanzonico, P., *Semin. Nucl. Med.*, 34, 2004; Firestone, R.B. and V.S. Shirley: *Table of Isotopes, 8th ed.* 1996. Copyright Wiley-VCH Verlag GmbH & Co. KGaA. Reproduced with permission; Weber, D. et al., *MIRD: Radionuclide Data and Decay Schemes*, Society of Nuclear Medicine, New York, 1989.

[a] As discussed, the ratio of the energy emitted per decay in the form of β-particles, conversion electrons, and Auger electrons (i.e., nonpenetrating particulate radiations) to that emitted in the form of x- and γ-rays should be high for a radionuclide for gamma camera imaging.

TABLE 6.3
Physical Properties of Positron-Emitting Radionuclides Used in PET

Radionuclide	Physical Half-Life $T_{1/2}$	β^+ Branching Ratio Abundance (%)	Maximum β^+ Energy (MeV) Production	β^+ Range in Water[a] (mm) R_e	R_{rms}	x- and γ-rays >0.25 MeV Energy (MeV)	Abundance (%)	Production
Carbon-11	20.4 min	99	0.96	3.9	0.4	N/A	0	Cyclotron
Nitrogen-13	9.96 min	100	1.2	5.1	0.6	N/A	0	Cyclotron
Oxygen-15	2.05 min	100	1.7	8.0	0.5	N/A	0	Cyclotron
Fluorine-18	1.83 h	97	0.64	2.3	0.1	N/A	0	Cyclotron
Copper-62	9.74 min	98	2.9	15	1.6	0.876–1.17	0.5	Generator (Zinc-62)
Copper-64	12.7 h	19	0.58	2.0	0.2	N/A	0	Cyclotron
Gallium-66	9.49 h	56	3.8	20	3.3	0.834–4.81	73	Cyclotron
Gallium-68	1.14 h	88	1.9	9.0	1.2	1.08 1.88	3.1	Generator (Germanium-68)
Bromine-76	16.1 h	54	3.7	19	3.2	0.473–3.60	146	Cyclotron
Rubidium-82	1.3 min	95	3.4	18	2.6	0.777	13	Generator (Strontium-82)
Yttrium-86	14.7 h	32	1.4	6.0	0.7	0.440–1.920	240	Cyclotron
Zirconium-89	78.4 h	23	0.90	3.8	0.4	0.909 1.62–1.75	99 10	Cyclotron
Iodine-124	4.18 days	22	1.5	7.0	0.8	0.603–1.69	23	Cyclotron

Source: Zanzonico, P., *Radiat. Res.*, 177, 2012; Zanzonico, P., *Semin. Nucl. Med.*, 34, 2004; Firestone, R.B. and V.S. Shirley, *Table of Isotopes, 8th ed.* 1996. Copyright Wiley-VCH Verlag GmbH & Co. KGaA. Reproduced with permission; Weber, D. et al., *MIRD: Radionuclide Data and Decay Schemes*, Society of Nuclear Medicine, New York, 1989.

[a] R_e is the maximum extrapolated range and R_{rms} the root mean square (i.e., mean) range of the β^+ particle in water and other soft tissue–equivalent media.

absence of higher-energy x- and γ-rays (i.e., with energies in excess of several hundred keV) is important because such radiations cannot be efficiently collimated and detected. Yet, they may undergo scatter in the patient and/or detector hardware and contribute, even with energy discrimination, mispositioned and otherwise spurious counts to the image. Based on the foregoing criteria (Table 6.2), ^{99m}Tc is a near-ideal radionuclide for gamma camera imaging, emitting only a 140 keV γ-ray and little particulate radiations. Iodine-131 (^{131}I), on the other hand, emits a relatively high-energy, difficult-to-collimate 364 keV γ-ray and abundant β-particles. Other gamma camera imaging radionuclides and their physical properties are also included in Table 6.2.

For PET, a radionuclide emitting low-energy, short-range positrons with a 100% abundance (i.e., a 100% positron branching ratio) and no high-energy prompt γ-rays is ideal; fluorine-18, for example, is such a PET radionuclide. The positron range places a lower limit on spatial resolution, so the lower the energy and the shorter the range of the positron, the better the spatial resolution that is ultimately achievable (see Section 6.5.3). Many positron-emitting radioisotopes also emit significant numbers of high-energy prompt γ-rays, and such γ-rays may be in cascade with each other or with the positron (Table 6.3). These can result in spurious events that are spatially uncorrelated but nonetheless counted as true events [17,18]. Although such coincidences degrade overall quality and quantitative accuracy, isotopes such as copper-62 (^{62}Cu), gallium-66 (^{66}Ga), gallium-68 (^{68}Ga), bromine-75 (^{75}Br), rubidium-82 (^{82}Rb), yttrium-86 (^{86}Y), zirconium-89 (^{89}Zr), and iodine-124 (^{124}I) are used effectively in PET [17,18]. Table 6.3 also includes, for selected positron emitters, the energy and abundance of photons with sufficient energy (i.e., greater than 250 keV) to fall within the energy windows typically used to count the 511 keV annihilation γ-rays in PET as well as other pertinent properties of positron emitters.

6.5 GAMMA CAMERAS, SPECT, AND PET SCANNERS

6.5.1 Gamma Camera

Developed in the late 1950s by Hal Anger, the *gamma camera* (Figure 6.4), also known as the scintillation or Anger camera, has long been the predominant imaging device in nuclear medicine [3,4]. Its large detector area allows simultaneous and, therefore, rapid data acquisition over a large area of the body. Gamma camera crystals, almost universally, are composed of a plate of NaI(Tl), and a 9.53-mm-thick (3/8") crystal provides a reasonable balance between sensitivity and resolution and is the most widely used for general gamma camera imaging. About 95% of the 140 keV photons from ^{99m}Tc are absorbed in a 9.54-mm-thick crystal. For clinical applications, gamma camera crystals are most commonly rectangular in shape and ~50 × 60 cm in area for general-purpose imaging. Crystals smaller in area are used on dedicated cardiac systems.

Once the incident radiation passes through the collimator (see Figure 6.4), it strikes and may produce scintillation light within the crystal. The resulting light signal is distributed among a 2-D array of PMTs backing the crystal, the light intensity varying inversely with the distance between the scintillation and the respective PMT: the farther the PMT is from the scintillation, the less light it receives and the smaller is its output pulse. This inverse relationship is the basis of the Anger position logic circuitry for determining the precise position of a scintillation within the crystal. In the older gamma cameras, the x and y coordinates were calculated by analog circuitry, that is, using matrices of resistors. In current models, this is done by digitizing the output signal from each PMT and using digital electronics.

The gamma camera *collimator*, almost always comprised of lead, "directionalizes" the incoming radiation (Figure 6.4). Any radiation traveling at an oblique angle to the axes of the holes (apertures) will strike the lead walls (septa) between the holes and not reach the crystal. As a result, only radiations traveling perpendicular to the crystal surface pass through the apertures and contribute counts to the resulting image. Otherwise, without a collimator, radiations would strike the crystal at positions unrelated to the locations of the radiation emission within the subject, and a usable image could not be formed. A certain fraction of γ-rays striking the septa will nonetheless pass through them and reach the crystal; this phenomenon, which degrades image quality, is known as septal penetration. Almost all collimators used clinically are parallel-hole collimators, with the apertures and septa parallel to one another. In addition, single-aperture pinhole collimators, most commonly used for thyroid imaging because of their pronounced magnifying effect, are available as well (Figure 6.5a). Pinhole collimators, however, suffer from low sensitivity, limited field of view (FOV), and geometric distortion, since image magnification varies with both source-aperture distance and lateral position in the field of view, and therefore are rarely used for clinical imaging other than for the thyroid (normally a small, relatively flat organ).

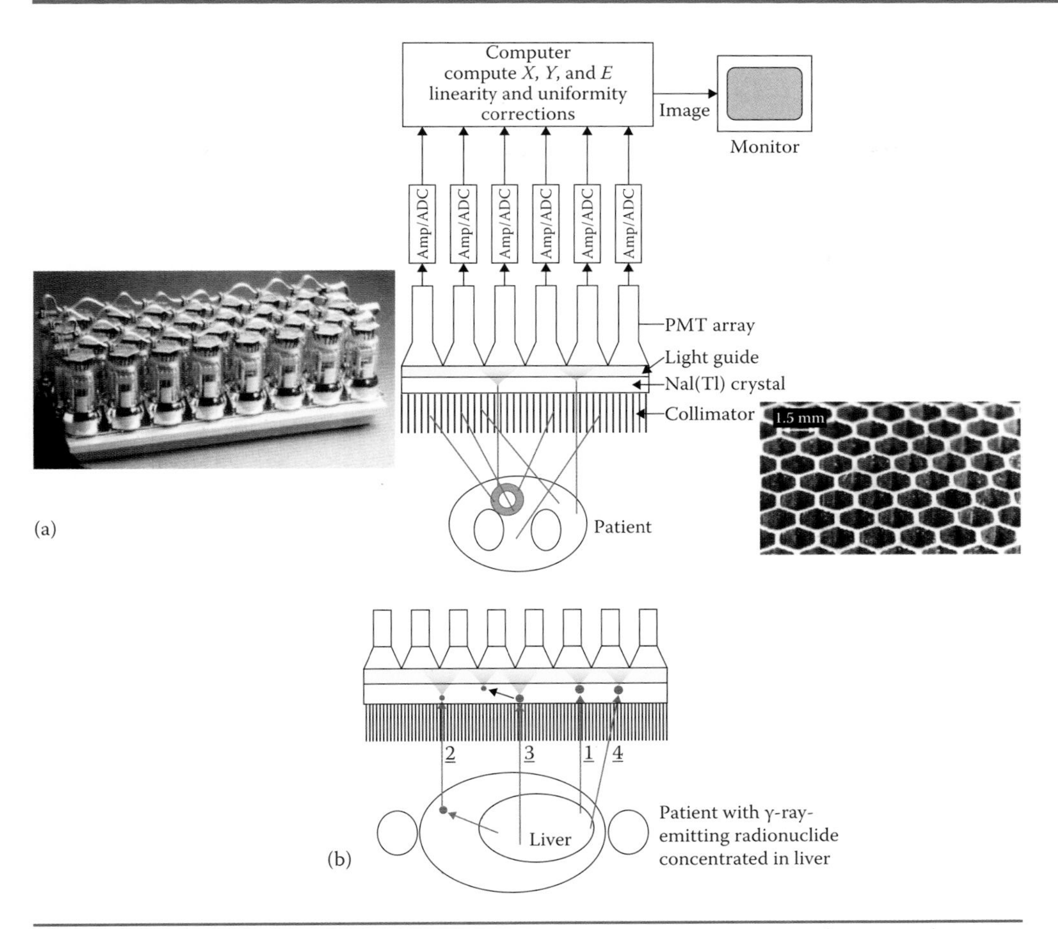

FIGURE 6.4 (a) Basic design of a gamma camera, consisting of a multihole collimator, a thin large-area NaI(Tl) crystal, a 2-D array of PMTs and associated electronics (high-voltage power supply, preamplifier, amplifier ["amp"], and analog-to-digital converter ["ADC"]), and image display. In current-day gamma cameras, the output signal from each PMT is digitized and the x and y position as well as the energy E of each event determined with computer software. The left inset shows a photograph of the 2-D PMT array backing the crystal in a typical rectangular field-of-view gamma camera. The right inset shows a drawing of a portion of a parallel-hole collimator showing typical dimensions of the apertures. (b) The "desirable" events (arrows labeled "1") are unscattered (i.e., photopeak) photons traveling in a direction parallel or nearly parallel to the axes of the collimator apertures and thus yielding correctly positioned counts in the gamma camera image. "Undesirable" events include photons scattered in the patient but traveling in a direction parallel or nearly parallel to the axes of the apertures ("2"). Most, though not all, of such scattered photons are eliminated by energy discrimination; any such photons not eliminated by energy discrimination will thus yield mispositioned counts. Other undesirable events include photons scattered in the detector ("3") and septal-penetration photons (i.e., photons traveling in a direction oblique to the axes of the apertures yet passing through the septa, "4"), yielding mispositioned counts. (Adapted from Cherry, S.R. et al., *Physics in Nuclear Medicine*, Saunders, Philadelphia, PA, 2012. With permission.)

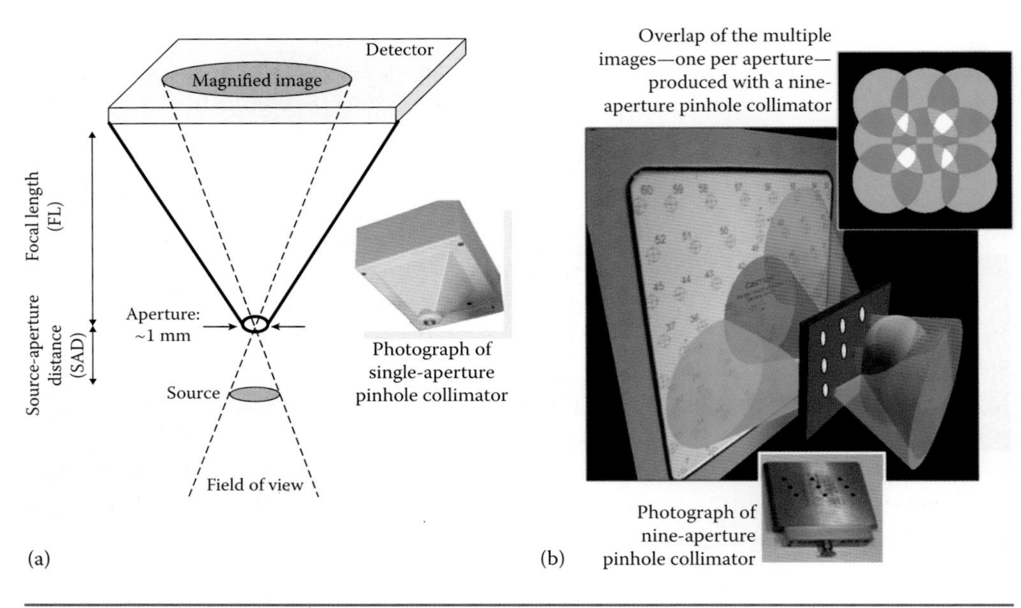

FIGURE 6.5 Pinhole collimation for gamma cameras. (a) A single-aperture pinhole collimator. Pinhole collimation provides a magnification effect and thereby improved "effective" resolution. For example, if a source were placed at a distance of 3 cm from the aperture (i.e., source-to-aperture distance [SAD] = 3 cm) of a pinhole collimator having a focal length of 9 cm (i.e., focal length [FL] = 9 cm) fitted to a detector with an intrinsic spatial resolution of 3 mm (i.e., FWHM = 3 mm), the magnification is FL/SAD = 9 cm/3 cm = 3, and the effective resolution therefore improves to FWHM/(FL/SAD) = 3 mm/(9 cm/3cm) = 3 mm/3 = 1 mm. (b) A multiaperture pinhole collimator, now the standard approach to preclinical (i.e., mouse and rat) SPECT imaging. As shown, however, each of the apertures produces an image, and the resulting multiple images—one per aperture—overlap. The apertures are angled with respect to one another to minimize this overlap. If larger-area detectors are used, this angulation can be increased and, therefore, the image overlap further minimized and image quality improved. ([a] Photograph courtesy of Gamma Medica, Northridge, California. [b] Courtesy of Bioscan, Washington, District of Columbia.)

For preclinical (i.e., mouse and rat) imaging, multiaperture pinhole collimation is now the standard approach, since it combines the magnification effect (and improved resolution) of pinhole collimation with the greater sensitivity afforded by multiple apertures (Figure 6.5b).

Gamma camera collimators are "rated" with respect to photon energy and resolution/sensitivity. Low-energy collimators, including "low-energy all-purpose (LEAP)" (or "general all-purpose [GAP]"), "low-energy high-resolution" (LEHR)," and "low-energy high-sensitivity" (LEHS)" collimators, are designed to image radionuclides emitting x- and γ-rays less than 200 keV in energy. These include ^{99m}Tc (photopeak energy; 140 keV) as well as thallium-201 (^{201}Tl; 68–80 keV) and iodine-123 (^{123}I; 159 keV). Medium-energy collimators are designed for radionuclides emitting x- and γ-rays 200–300 keV in energy, including gallium-67 (^{67}Ga; 93, 185, and 300 keV) as well as indium-111 (^{111}In; 172 and 247 keV). High-energy collimators are designed to image radionuclides emitting x- and γ-rays greater than 300 keV in energy, including ^{131}I (364 keV). In progressing from low- to medium- to high-energy collimation, the collimators are made longer and the septa thicker in order to interpose more lead between the subject and the crystal. This is done in order to maintain septal penetration (i.e., expressed as the percentage of counts in an image attributable to photons penetrating the septa) at or below an acceptably low level, typically set at 5%. This, in turn, reduces the overall fraction of emitted x- and γ-rays reaching the crystal. To compensate, at least in part, for the resulting lower sensitivity, the apertures are made wider in progressing from low- to medium- to high-energy collimators. This, however, degrades spatial resolution by dispersing the counts passing through each aperture over a large area of the crystal (Figure 6.6). Overall, therefore, gamma camera images are progressively poorer in quality for radionuclides emitting low- versus medium- versus high-energy x- and γ-rays. For each energy rating (and as indicated above for low-energy collimators), collimators may also be further rated as "general purpose" (or "all purpose"), "high resolution," or "high sensitivity." High-resolution collimators have narrower apertures (and therefore lower sensitivity), and high-sensitivity collimators have wider apertures (and therefore coarser resolution), than general-purpose collimators.

The FWHM spatial resolution, $FWHM_{system}$, of gamma cameras is determined by a combination of physical and instrumentation factors. Intrinsic resolution, $FWHM_{intrinsic}$, is the component of spatial resolution contributed by the crystal and associated electronics and is related to statistical fluctuations in pulse formation; typical values are on the order of 3–5 mm. These statistical fluctuations include variations in the production of light photons resulting from x- or γ-ray interactions in the crystal and in the number of electrons emitted by the photocathode and the series of dynodes in the PMTs. Collimator, or geometric, resolution, $FWHM_{collimator}$, represents the major contribution to system resolution and is determined by the collimator design. The spatial resolution of a parallel-hole collimator, by far the most common type, is determined by the geometric radius of acceptance of each aperture, as illustrated in Figure 6.6:

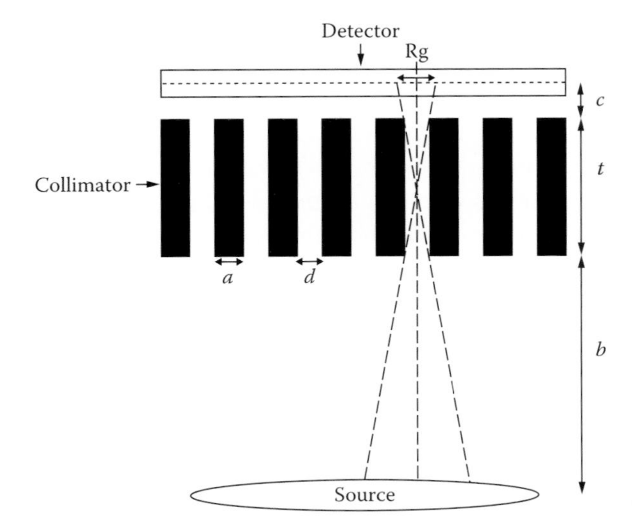

FIGURE 6.6 Collimator spatial resolution, $FWHM_{collimator}$, for a parallel-hole collimator with thickness *t*, aperture diameter *d*, septal thickness *a*, and source-to collimator face distance *b*. The collimator is fitted to a detector whose midplane is at a distance *c* from the back surface of the collimator. (From Saha, G.S., *Physics and Radiobiology of Nuclear Medicine*, Springer-Verlag, New York, 1993. With permission.)

$$\mathrm{FWHM}_{\mathrm{collimator}} = \frac{d(t_e + b + c)}{t_e}, \quad (6.3)$$

where d is the diameter of the collimator aperture, b the source-to-collimator face distance, c the collimator back–to–midcrystal distance, and $t_e = t - 2/\mu$ the effective thickness of the collimator where t is the actual collimator thickness and μ the radiation energy–dependent linear attenuation coefficient of the collimator material [19]. As indicated by Equation 6.3, collimator resolution is improved (i.e., lowered) by reducing the diameter d of the collimator aperture, the source-to-collimator face distance b, and the collimator thickness t. System resolution is further degraded by the contributions of septal-penetration resolution, $\mathrm{FWHM}_{\mathrm{penetration}}$, and by scatter resolution, $\mathrm{FWHM}_{\mathrm{scatter}}$. The spatial resolution of a gamma camera system, $\mathrm{FWHM}_{\mathrm{system}}$, can be obtained by combining the resolution of the respective components of the system [19]:

$$\mathrm{FWHM}_{\mathrm{system}} = \sqrt{\mathrm{FWHM}_{\mathrm{system}}{}^2 + \mathrm{FWHM}_{\mathrm{collimator}}{}^2 + \mathrm{FWHM}_{\mathrm{penetration}}{}^2 + \mathrm{FWHM}_{\mathrm{scatter}}{}^2}. \quad (6.4)$$

In clinical practice, the collimator resolution generally dominates the system resolution.

6.5.2 SPECT Systems

A SPECT system consists of one or more gamma cameras detector heads that rotate around the subject to provide the necessary information to reconstruct tomographic images. Although there are many possible combinations of detector number, geometry, and motion that can acquire the necessary projection data, rotating gamma camera–based SPECT systems are by far the most common [20]. Nowadays, two-detector-head systems predominate clinically and two- to four-detector-head systems preclinically. The basic SPECT imaging paradigm includes acquisition of planar projection images from multiple angles around the subject, correction of the acquired data for nonuniform scanner response and possibly other signal-degrading effects, and mathematical reconstruction of thin (several-millimeter-thick) transverse tissue-section images [20]. The raw data are acquired as a series of discrete planar images at multiple angles about the longitudinal axis of the subject (Figure 6.7). The number of counts recorded in each projection image pixel represents the ray sum, or line integral, of the sampling line perpendicular to and extending from the detector through the subject. The following are typical SPECT acquisition parameters: 20–30 min for data acquisition; ~60 to ~120 projection images at ~6° to ~3° angular increments, respectively; and a 180° or 360° rotation for cardiac or noncardiac studies, respectively. An angular increment in excess of 6° between successive projection images will result in prohibitive undersampling artifacts in the reconstructed images. Because of the length of time (20–30 min) required for an acquisition, dynamic SPECT and whole-body SPECT remain largely impractical at the current time. It should be emphasized that SPECT images can, in principle, be quantitative in absolute terms, with voxel values representing the local activity concentration [21–23]. However, in contrast to PET (see Section 6.5.3), this is often not the case in routine practice, because of the confounding effects of scatter and attenuation. Accurate correction for these effects remains more challenging in SPECT than in PET [21,22].

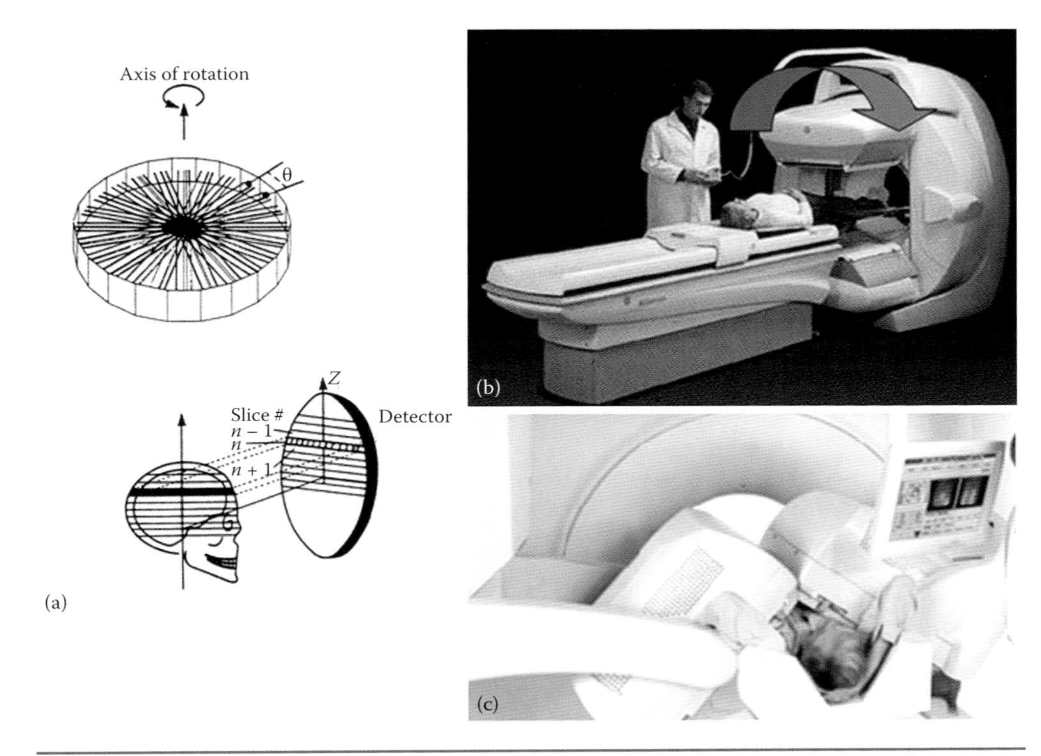

FIGURE 6.7 (a) The basic data acquisition paradigm in rotating gamma camera SPECT. Photographs of a dual-detector gamma camera, with (b) the two detectors in opposed positions, as routinely used for a 360° rotation and general (noncardiac) SPECT and (c) the two detectors perpendicular to each other, as routinely used for a 180° rotation and cardiac SPECT (with projection images acquired from approximately right anterior oblique to left posterior oblique). The advantage of such two-detector systems is that two projection images can be acquired simultaneously and the acquisition time therefore halved. (From Zanzonico, P.B., Technical Requirements for SPECT: Equipment and Quality Control, in *Clinical Applications* in SPECT, E.L. Kramer and J.J. Sanger, editors, Raven Press, New York, 1995. With permission.)

Although PET offers important advantages over SPECT (i.e., generally better spatial resolution, higher sensitivity, and more accurate activity quantitation), SPECT offers the capability of multi-isotope imaging. Because different SPECT radionuclides emit x- and γ-rays of different energies, multiple isotopes and, therefore, multiple radiotracers can be imaged simultaneously using distinct, isotope-specific photopeak energy windows. In contrast, all PET radionuclides emit positrons and, consequently, annihilation photons of the same energy, 511 keV. Therefore, PET radiotracers cannot be distinguished on the basis of energy discrimination, and multiple PET radiotracers cannot be imaged simultaneously.

6.5.3 PET Scanners

PET is based on the *annihilation coincidence detection* (ACD) of the two colinear (approximately 180° apart) 511 keV γ-rays resulting from the mutual annihilation of a positron and an electron (Figure 6.8a) [3,4,7]; a typical PET scanner and the detector configurations used in modern scanners are shown in Figure 6.8b. Each individual annihilation photon is referred to as a "single" event, and the total count rate of individual

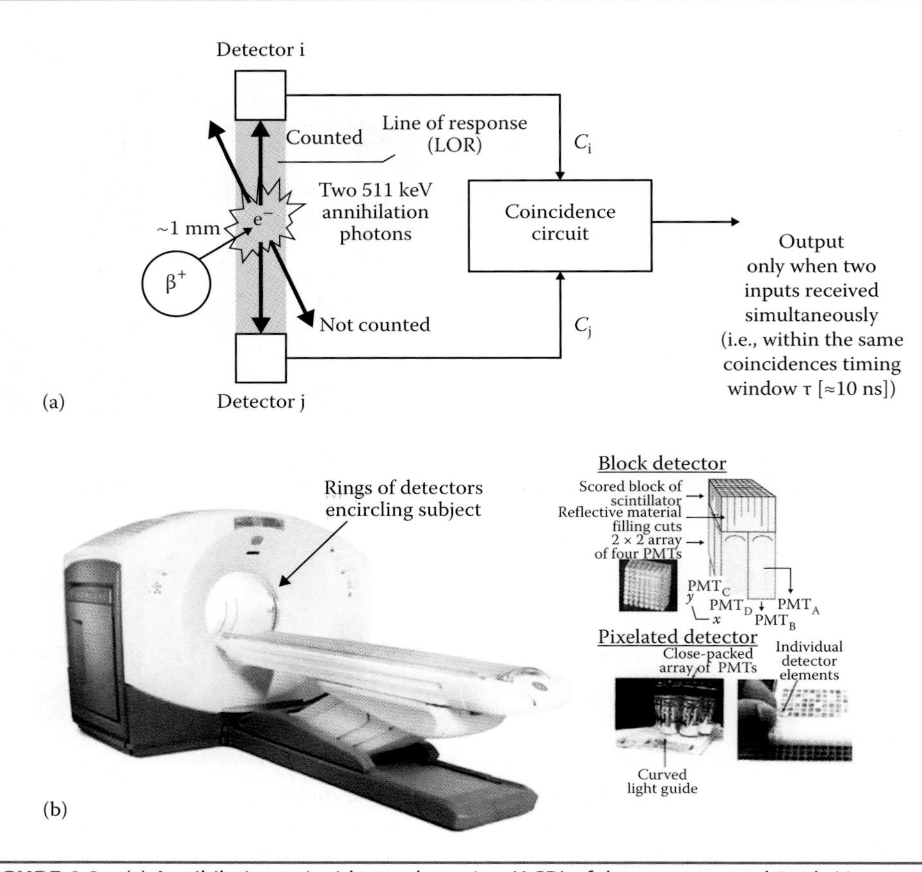

FIGURE 6.8 (a) Annihilation coincidence detection (ACD) of the two opposed 511 keV γ-rays resulting from positron decay and positron-electron annihilation. Note that the true coincidence (or "trues") count rate for detectors i and j is much less than their respective individual (or "singles") count rates C_i and C_j. The short coincidence timing window, τ ($\approx$ 10 ns), minimizes the number of random coincidence events. (b) A photograph of a PET scanner that consists of rings of discrete detectors. In the insert are shown a block detector (top) and pixelated detectors (bottom) used in PET scanners. The block detector consists of a cubic piece of scintillator scored to variable depths into a 2-D array of detector elements, typically backed by a 2 × 2 array of PMTs. Pixelated detectors consist of individual scintillator detector elements backed by a continuous light guide and a close-packed array of PMTs. For both the block and the pixelated detectors, the individual detectors elements are typically ~4–6 mm in width and 20–30 cm thick. (Adapted from Zanzonico, P., *Semin. Nucl. Med.*, 34, 2004. With permission.)

annihilation photons is called the "singles count rate." When both photons from an annihilation are detected simultaneously (in coincidence), this triggers the coincidence circuit, and a "true coincidence event" ("true") is generated. The various events associated with ACD of positron-emitting radionuclides, including trues, randoms, scatter, and spurious coincidences, are illustrated in Figure 6.9 [7]. The singles count rate in PET is typically much higher than the trues count rate. The volume between the opposed coincidence detectors absorbing the two annihilation photons (the shaded area in Figure 6.8a) is referred to as a *line of response* (LOR) even though it is actually a volume of response. Figure 6.10 illustrates the substantial difference in lines (volumes) of response

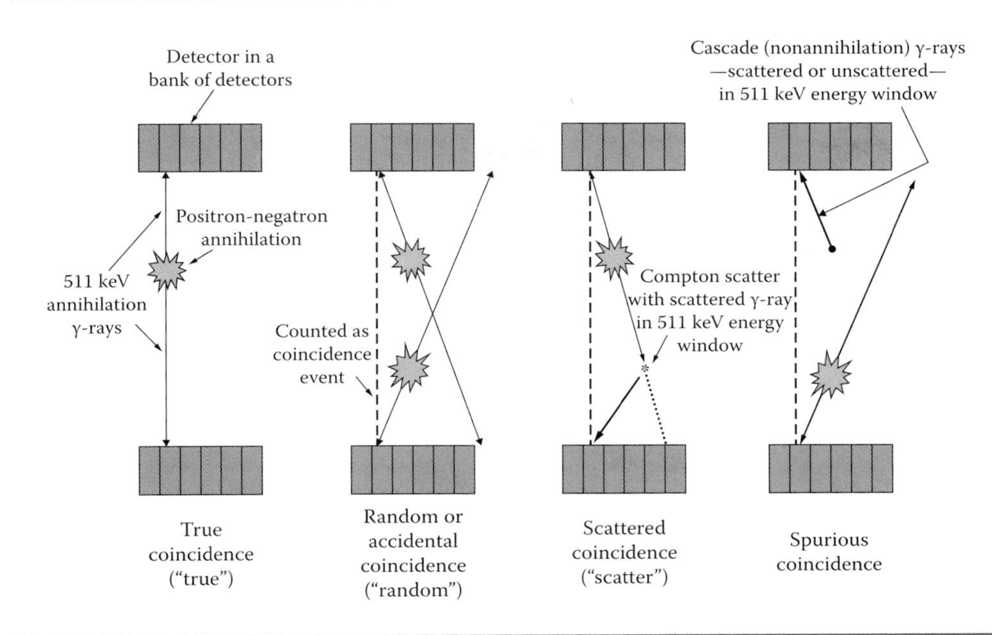

FIGURE 6.9 The various events associated with ACD of positron-emitting radionuclides, illustrated for two opposed banks of coincidence detectors and assuming only one opposed pair of detectors are in coincidence. A true coincidence ("true") is counted only when each of the two 511 keV annihilation γ-rays for a single positron-electron annihilation are not scattered and are detected within the timing window τ of the two coincidence detectors. A random or accidental coincidence ("random") is an inappropriately detected and positioned coincidence (dashed line) that arises from two separate annihilations, with one γ-ray from each of the two annihilations detected within the timing window τ of the coincidence-detector pair. A scattered coincidence ("scatter") is a mispositioned coincidence (dashed line) resulting from a single annihilation, with one of the γ-rays undergoing a small-angle Compton scatter but retaining sufficient energy to fall within the 511 keV energy window. A spurious coincidence is an inappropriately detected and positioned coincidence (dashed line) that arises from an annihilation and a cascade γ-ray, scattered or unscattered but having sufficient energy to fall within the 511 keV energy window. Spurious coincidences occur only for radionuclides that emit both positrons and high-energy prompt cascade γ-rays, that is, γ-rays with energies (either scattered or unscattered) lying within the 511 keV energy window. (From Zanzonico, P., *Semin. Nucl. Med.*, 34, 2004. With permission.)

between gamma camera imaging and PET and the effect on spatial resolution. In PET, LORs are defined electronically, and an important advantage of ACD is that absorptive collimation (as is used in gamma cameras) is not required. As a result, the sensitivity of PET is two to three orders of magnitude higher than that of gamma camera or SPECT imaging. Modern PET scanners generally employ a series of rings of discrete, small-area detectors (i.e., scored block detectors or pixelated detectors) encircling the subject (Figure 6.8b) and, in clinical PET scanners, typically spanning a distance of 15–20 cm in the patient's longitudinal direction. Thus, a whole-body PET scan will typically require data acquisition at six to seven discrete bed positions and subsequently merge, or "knit," the discrete images into a single whole-body image.

PET ring scanners originally employed lead or tungsten walls, or septa, positioned between and extending radially inward from the detector blocks [7] (Figure 6.11). In this

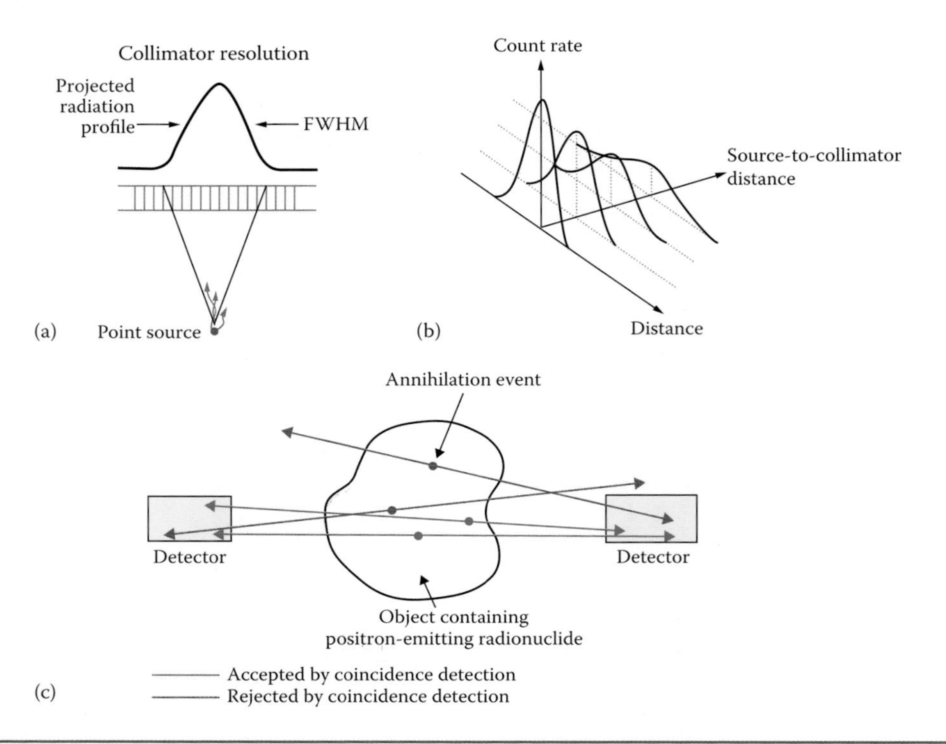

FIGURE 6.10 (a and b) In gamma camera imaging with parallel-hole collimation, the 3-D volume subtended by a point source of radiation is a cone, with the area of the detector over which radiations from the source are dispersed increasing, and therefore, the spatial resolution worsening, with increasing source-to-collimator distance. (c) In contrast, for PET and annihilation coincidence detection, the 3-D "line" (actually, volume) of response corresponds to the elongated cube whose cross-sectional area is defined by that of the opposed detector elements. As a result, in first-order spatial resolution is independent of the position of the source between the detector elements. (Adapted from Cherry, S.R. et al., *Physics in Nuclear Medicine*, Saunders, Philadelphia, PA, 2012. With permission.)

approach, known as 2-D PET, these inter-ring annular septa define plane-by-plane LORs and largely eliminate out-of-plane annihilation γ-rays. By eliminating most of the contribution from out-of-plane randoms and scatter, image quality is improved, especially for large-volume sources (i.e., as in whole-body PET). However, 2-D PET also eliminates most of the trues as well and thus reduces sensitivity. Removing the septa altogether and including coincidence events from all of the LORs among all the detectors significantly increases PET detector sensitivity. This is known as 3-D PET, and is the prevailing design among state-of-the-art PET scanners [7,24]. Sensitivity is increased up to approximately fivefold in 3-D relative to 2-D PET—but with a considerable increase in the randoms and scatter count rates. Clinically, the scatter–to–true count rate ratios range from 0.2 (2-D) to 0.5 (3-D) in the brain and from 0.4 (2-D) to 2 (3-D) in the whole body [25].

A notable refinement in PET detectors has been the use of adjacent layers of two different materials with significantly different scintillation decay times (such as LSO and gadolinium oxyorthosilicate [GSO], with decay times of approximately 40 and 60 ns, respectively); these are known as phoswich detectors [25]. Based on the pulse shape of the scintillation signal, the depth of the interaction of the annihilation γ-ray in the detector

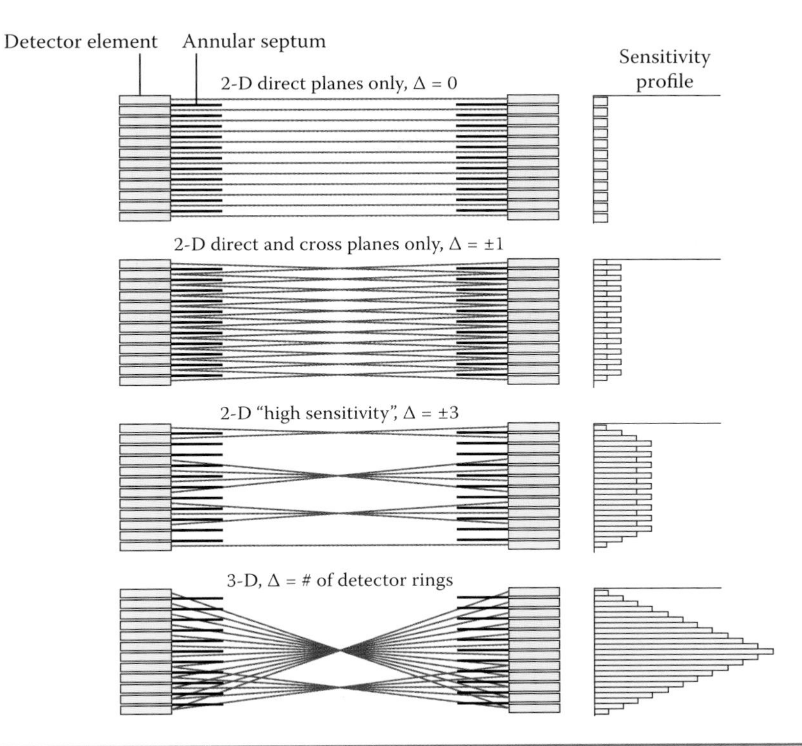

FIGURE 6.11 2-D and 3-D PET data acquisition schemes (axial cross-sectional views of a multi-ring scanner) and the corresponding axial sensitivity profiles. 2-D data acquisition schemes with a maximum ring difference Δ of 0, 1, and 3 are shown. The ring difference refers to the number of different detector element rings between which coincidence events are allowed. A ring difference of Δ of 0, for example, indicates that only intradetector ring (i.e., direct planes) coincidences are allowed. 3-D (septa-less) data acquisition can accept coincidences between all detector rings. For clarity, only the allowed ring differences for selected axial locations are shown in the lower two schematics. The sensitivity profiles show the nonuniformity of response as a function of position along the axial field of view (FOV). (Adapted from Cherry, S.R. et al., *Physics in Nuclear Medicine*, Saunders, Philadelphia, PA, 2012. With permission.)

can therefore be localized in the upper or lower layer [26]. As a result, the resolution-degrading depth-of-interaction (DOI) effect (see below) is reduced. However, the fabrication of phoswich detectors is more complex than that of single-component detectors, and to date, it has seen only limited use in commercial PET scanners, specifically, preclinical scanners.

Increasingly important, time-of-flight (TOF) PET scanners utilize the measured difference between the detection times of the two annihilation photons arising from the decay of a positron. This allows at least approximate spatial localization (12–18 cm) of the annihilation event along the LOR with current values, 400–600 ps, of the coincidence time resolution [27,28]. This does not improve the spatial resolution of state-of-the-art PET scanners (~5 mm) but reduces the random coincidence rate and improves the signal-to-noise ratio (S/R), especially for large subjects [29]. This is important, as conventional (i.e., non-TOF) PET image quality is degraded with increasing patient size due to more pronounced attenuation and scatter and fewer trues.

As with gamma cameras, the overall spatial resolution of PET scanners results from a combination of physical and instrumentation factors. There are several important limitations imposed on resolution by the basic physics of positron-electron annihilation. First, for a given radionuclide, positrons are emitted over a spectrum of initial kinetic energies ranging from 0 to a characteristic maximum, or end point, energy, E_{max}; the associated average positron energy, $\bar{E}$, is approximately one-third of its end-point energy, $\bar{E} \approx \frac{1}{3} E_{max}$. As a result, emitted positrons will travel a finite distance from the decaying nucleus prior to annihilating and producing 511 keV photons. This distance ranges from 0 to a maximum called the extrapolated range, R_e, corresponding to its highest-energy positrons (Figure 6.12a) [30].

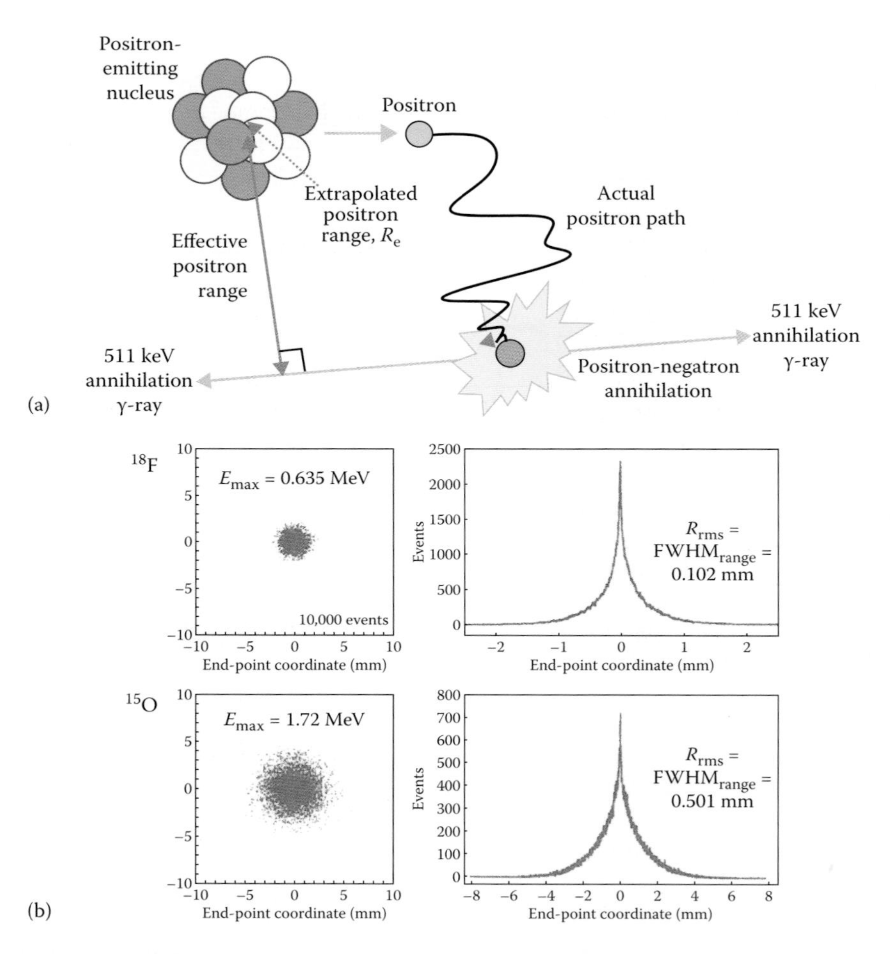

FIGURE 6.12 Effect of positron range on PET spatial resolution. (a) Positrons travel a finite distance before undergoing annihilation, resulting in blurring of PET images. (b) The spatial dispersion of positron-electron annihilations for ^{18}F (E_{max} = 0.640 MeV) and ^{15}O (E_{max} = 1.72 MeV) as determined by Monte Carlo simulation. The annihilations for the higher-energy ^{15}O positrons are clearly more widely dispersed than those for the lower-energy ^{18}F positrons (left panel). A graphical representation of the resulting range-related blurring in PET, $FWHM_{range} = R_{rms}$: 0.102 and 0.501 mm for ^{18}F and ^{15}O, respectively. ([a] Adapted from Cherry, S.R. et al., *Physics in Nuclear Medicine*, Saunders, Philadelphia, PA, 2012. With permission. [b] Reproduced from Levin, C.S. and E.J. Hoffman, *Phys. Med. Biol.*, 44, 1999. With permission.)

For positron emitters used in PET, the maximum energies (E_{max}) vary from 0.58 to 3.7 MeV, the extrapolated ranges (R_e) from 2 to 20 mm, and the root mean square (rms) ranges (R_{rms}) from 0.2 to 3.3 mm (Table 6.3). Although the finite positron range acts to blur PET images (i.e., degrade spatial resolution), the range-related blurring is mitigated by the spectral distribution of positron energies for a given radioisotope as well as the characteristically tortuous path positrons travel [30,31]; these effects are reflected by the fact the rms positron ranges are much shorter than the extrapolated positron ranges (Table 6.3). The radial distance the positron travels is thus considerably shorter than its actual path length. The overall effect of positron range on PET spatial resolution, $FWHM_{range}$, is illustrated quantitatively in Figure 6.12b [30]. The positron range degrades spatial resolution by only ~0.1 mm for ^{18}F (E_{max} = 0.640 MeV) and ~0.5 mm for ^{15}O (E_{max} = 1.72 MeV) [30]; these values are much shorter than the respective extrapolated positron ranges.

The second physics-related limitation on PET performance is the noncolinearity of the two annihilation photons: Because a positron typically has some small residual (nonzero) momentum and kinetic energy at the end of its range, the two annihilation photons are not always emitted exactly back to back (i.e., 180° apart) but deviate from colinearity by an average of 0.25° [32]. The noncolinearity-related blurring, $FWHM_{180°}$, varies from ~2 mm for an 90-cm-diameter whole-body PET to ~0.7 mm for a 30-cm-diameter brain PET to ~0.3 mm for a 12-cm-diameter small animal PET (Figure 6.13) [25].

Among instrumentation-related determinants of overall spatial resolution are the intrinsic detector resolution and the DOI effect. For discrete detector elements, the intrinsic resolution, $FWHM_{intrinsic}$, is determined by the detector element width (d), increasing from $d/2$ midway between opposed coincidence detectors to d at the face of either detector [25].

For PET systems employing rings of discrete, small-area detectors, the depth (x) of the detector elements (2–3 cm) results in a degradation of spatial resolution termed the DOI, or parallax, effect [25]. As illustrated in Figure 6.14, with increasing radial offset of a source

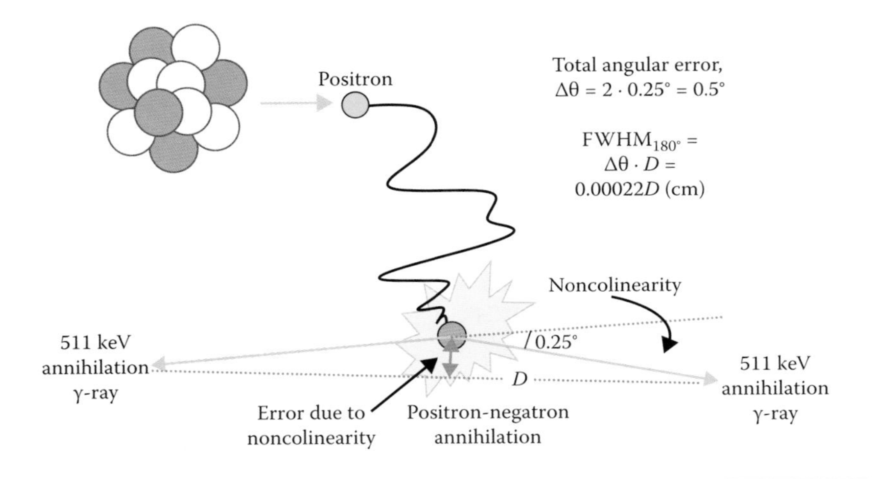

FIGURE 6.13 Effect of the noncolinearity of the annihilation γ-rays on PET spatial resolution. The 511 keV annihilation γ-rays resulting from positron-electron annihilation are not always exactly colinear but may be emitted 180° ± 0.25° apart [32]. The noncolinearity-related blurring, $FWHM_{180°}$, may be calculated geometrically and depends on the separation, D, of the coincidence detectors. (Reproduced from Cherry, S.R. et al., *Physics in Nuclear Medicine*, Saunders, Philadelphia, PA, 2012. With permission.)

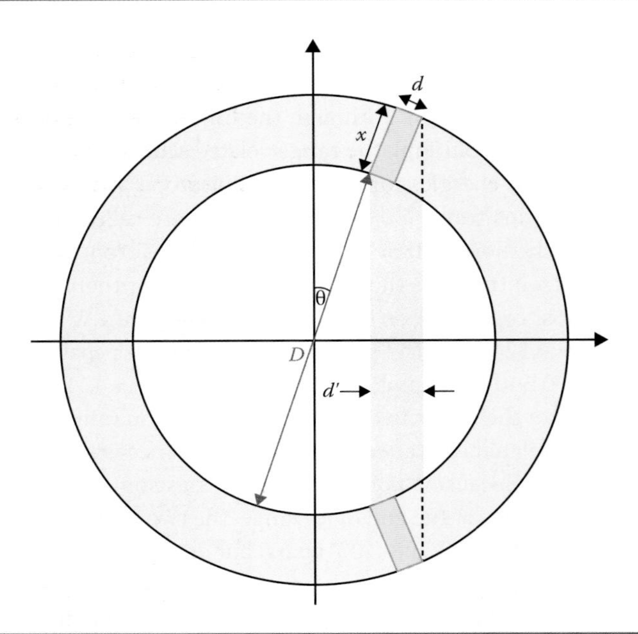

FIGURE 6.14 The depth-of-interaction (DOI) effect. In PET scanners, the apparent width of a detector element, d', increases with increasing radial offset from the center of the scanner. Because the depths with the detector elements at which the annihilation γ-rays interact are unknown, the apparent detector element width and, therefore, the effective line of response (i.e., the shaded volume) are wider than those corresponding to the actual detector element width d. The angle θ increases, and therefore, the magnitude of this resolution-degrading effect increases for sources offset further from the center of the rotation and for scanners with a smaller diameter D and a larger detector element depth x. (Reproduced from Cherry, S.R. et al., *Physics in Nuclear Medicine*, Saunders, Philadelphia, PA, 2012. With permission.)

from the center of a detector ring, the effective detector width (d') and, with it, the intrinsic resolution, $\text{FWHM}_{\text{intrinsic}}$, increases [25]:

$$d' = d\cos\theta + x\sin\theta \tag{6.5}$$

and, along chords of the ring,

$$\text{FWHM}_{\text{intrinsic}} = \frac{d'}{2} \tag{6.6a}$$

$$= \frac{d}{2}\left(\cos\theta + \frac{x}{d}\sin\theta\right), \tag{6.6b}$$

where θ is the angular position of the coincident detector pair, d is the detector element width, and x is the detector element depth. In whole-body PET scanners, the detector depth (x) is typically 2–3 cm (20–30 mm), the detector width (d) about 4 mm, and the detector ring diameter about 80 cm (800 mm), and the DOI effect thus degrades spatial resolution by up to 50% at 10 cm from the center of the detector ring. Because the DOI effect decreases as the detector ring diameter increases (Figure 6.14), clinical PET systems have detector rings substantially larger in diameter than that needed to accommodate patients. A variety of approaches have been developed to correct for the DOI effect in small-diameter, preclinical PET scanners, where the DOI effect is pronounced [25,26,33].

In a manner analogous to Equation 6.4, the spatial resolution at the center of the field of view (where the DOI effect is negligible) of a PET system, $FWHM_{system}$, can be obtained by combining the resolution of the respective components of the system:

$$FWHM_{system} = \sqrt{{FWHM_{intrinsic}}^2 + {FWHM_{range}}^2 + {FWHM_{180°}}^2} \quad (6.7)$$

6.6 DATA PROCESSING AND IMAGE RECONSTRUCTION IN SPECT AND PET*

6.6.1 Data Corrections

Normalization (uniformity correction). Even optimally performing SPECT or PET scanners exhibit some nonuniformity of response [7,10,20]. Among the thousands of pixels in a SPECT projection image and the thousands of detector elements in a PET scanner, slight variations in detector thickness, light emission or coupling properties, electronics performance, and so forth result in slightly different measured count rates for the same activity. In principle, such nonuniform response can be corrected by acquiring data from a uniform flux of γ-rays and normalizing to the *mean* count rate from all the pixels in SPECT or LORs in PET. This "normalization" table or "uniformity map" corrects for the nonuniform count rate of the individual pixels or LORs to thereby yield a pixel-by-pixel or LOR-by-LOR uniformity correction. The effects of nonuniformity of scanner response and of the correction for nonuniform response are illustrated in Figure 6.15 [10].

For gamma camera imaging, a correction table may be acquired using either a uniform flood source placed on the detector or a point source placed sufficiently far (typically ~2 m) from the uncollimated detector to deliver a uniform photon flux. For PET, it may be acquired using a positron-emitting rod source (e.g., germanium-68 [^{68}Ge]) spanning the entire axial FOV and rotating around the periphery of the FOV, exposing the detector pairs to a uniform photon flux per revolution. Alternatively, a uniform cylinder of a positron-emitting radionuclide can be scanned and the data thus acquired analytically corrected for attenuation; for a well-defined geometry such as a uniform cylindrical source, this correction is straightforward. However, for 3-D PET, the contribution of and correction for scatter with such a large-volume source are nontrivial. For both planar imaging as well as SPECT and PET, acquisition of the data required for uniformity correction is somewhat problematic in practice because of statistical considerations: tens of millions (SPECT) to hundreds of millions (PET) of counts must be acquired to avoid possible "noise"-related artifacts in the uniformity correction table.

Dead-time correction. Radiation detectors are characterized by a finite "dead time" and associated count losses [25]. The dead time, typically on the order of 1–10 μs, is the interval of time required for a counting system to record and process an event, during which

* Other aspects of data processing and image analysis are discussed in Chapter 7 of this volume.

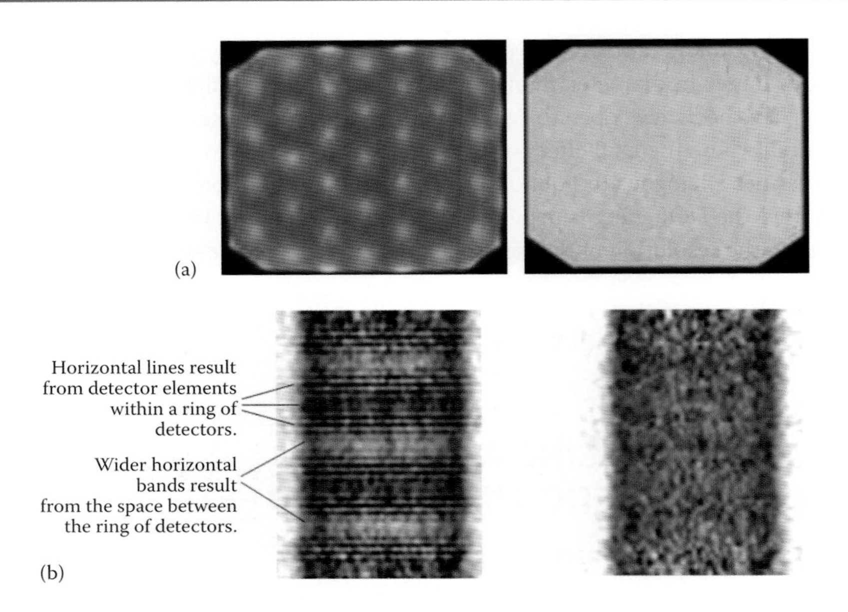

FIGURE 6.15 Normalization (uniformity) corrections and their effects. (a) The left panel shows a gamma camera image of a uniform source of radioactivity without the uniformity correction applied. The pattern of PMTs is apparent in this grossly nonuniform, uncorrected image. In contrast, the corrected image in the right panel is uniform. (b) Reconstructed PET images (coronal view) of a germanium-68 uniform-cylinder phantom without (left panel) and with (right panel) the normalization applied. The unnormalized (i.e., uncorrected) image has obvious artifacts attributable to the differences in sensitivities between direct and cross planes (Figure 6.11) and the presence of separate rings of detectors. Appropriate normalization virtually eliminates these and other artifacts related to nonuniformity of the scanner response. (From Zanzonico, P., *J. Nucl. Med.*, 49, 2008. With permission.)

additional events cannot be recorded. As a result, the measured count rate is systematically lower than the actual count rate. Such count losses are significant, however, only at "high" count rates (i.e., greater than ~100,000 counts per second [cps] per detector, which is on the order of the inverse of the dead time, for modern detectors). Dead-time count losses are generally minimal at diagnostic administered activities. Nonetheless, a real-time correction for dead-time count losses is routinely applied in PET (though not in SPECT) to the measured count rates, most commonly by scaling up the measured count rate based on an empirically derived mathematical relationship between measured and true count rates. As noted, count rates encountered in PET are much higher than in SPECT—in part because of the used of electronic rather than absorptive collimation—and therefore, accurate dead-time correction is more critical in PET.

Center-of-rotation (COR) correction (SPECT). In rotating gamma camera SPECT, the location of the projection of the COR on the projection image matrix must be constant [10,34,35]. If the mechanical and electronic CORs are aligned, the pixel location of the projection of the COR onto the projection image matrix will be the same for all projection images, and for all such images, the counts in each pixel will then be projected across the appropriate row of pixels in the tomographic image matrix. If, however, the mechanical and electronic CORs are not aligned, the pixel location of the COR will vary among the projection images, the counts in each projection image pixel will be projected across different locations in the tomographic image matrix, and blurred images will result (Figure 6.16a). In today's SPECT systems, COR misalignment may be easily measured and corrections created and automatically applied using the system's software (Figure 6.16b). In contrast, PET scanners typically utilize fixed rings of detectors and thus do not suffer from COR misalignment.

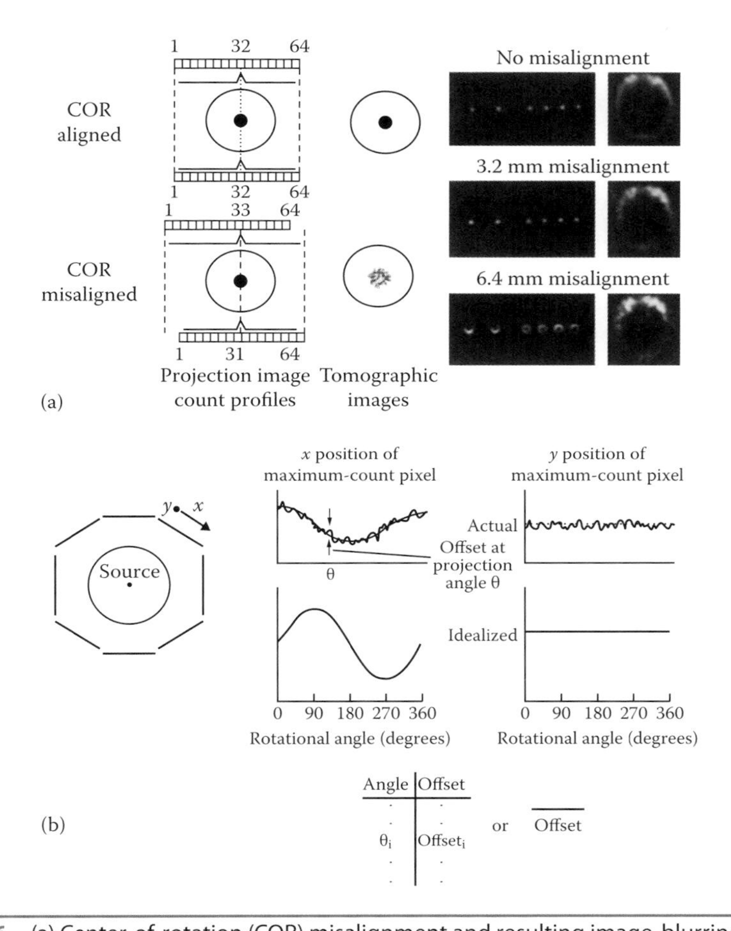

FIGURE 6.16 (a) Center-of-rotation (COR) misalignment and resulting image-blurring artifacts in rotating gamma camera SPECT. The degree of blurring is related to the magnitude of the spatial misalignment of the mechanical and electronic CORs. Note that for a cross-sectional image of a line source, COR misalignment blurs the expected point into a full or partial circle depending on the position of the source in the FOV: if it is at or near the center of the FOV, the line source appears as a full circle in cross section; if it is near the periphery of the FOV, it appears as a partial circle. (b) COR misalignment can be measured and corrected based on acquiring a 360° circular SPECT study of a ^{99m}Tc point source and constructing graphs of the *x* and *y* positions (perpendicular and parallel to the axis of rotation, respectively) of the position of the maximum-count pixel in each projection image versus angular position. The *x* and *y* position–versus–angle graphs should be a sinusoidal curve and a straight line, respectively. The angle-by-angle deviation between the *x* position on the best-fit sine curve and the *x* position of the actual maximum-count pixel thus yields a correction table indicating the offset by which each projection image must be shifted at each angular position to align the CORs. ([a] Adapted from Greer, K. et al., *J. Nucl. Med. Technol.*, 13, 1985. With permission. [b] Adapted from Zanzonico, P., *J. Nucl. Med.*, 49, 2008 and Greer, K. et al., *J. Nucl. Med. Technol.*, 13, 1985. With permission.)

Randoms correction (PET). In PET, randoms (accidental coincidences) increase the detected coincidence count rate by introducing mispositioned events and thus reduce image contrast and distort the relationship between image intensity and activity concentration [7]. One approach to randoms correction is the so-called delayed-window method, based on the fact that the random coincidence γ-rays are temporally uncorrelated (i.e., not simultaneously emitted) [36]. Briefly, once events in the coincidence timing window (typically ~10 ns or less) are detected, the number of events in a timing window equal in duration to, but much later (>50 ns later) than, the coincidence timing window is determined. More specifically, the coincidence timing window accepts events whose time difference is $\pm\tau$, while the delay window accepts events whose time difference is the delay $\pm\tau$. The number of events in the delayed timing window thus provides an estimate of the number of randoms in the coincidence timing window. Real-time subtraction of the delayed-window counts from the coincidence-window counts for each LOR thus corrects for randoms.

Another randoms-correction method, the "singles" method, uses the measured singles count rate for each detector and the coincidence timing window τ to calculate the randoms rate for each pair of coincidence detectors [37]. This approach does not require a separate measurement in a delayed-coincidence window and thus lessens the data-processing requirements on the coincidence electronics. Further, because the singles count rate is typically at least an order of magnitude higher than the coincidence-event count rate, the statistical uncertainty (noise) in the estimate of the number of randoms is much smaller than that in the number of prompt coincidences. The singles method is thus the method of choice for randoms correction in modern PET scanners.

Scatter correction. Scatter results in generally diffuse background counts in reconstructed images, like randoms, reducing contrast and distorting the relationship between image intensity and activity concentration [21,38,39]. Scatter as a portion of the total PET events is far more abundant in 3-D than in 2-D PET (especially for body imaging of larger [i.e., adult] patients), and its correction is more challenging in 3-D than in 2-D PET. Perhaps the simplest scatter correction method is the "Gaussian-fit" technique [40,41] (Figure 6.17).

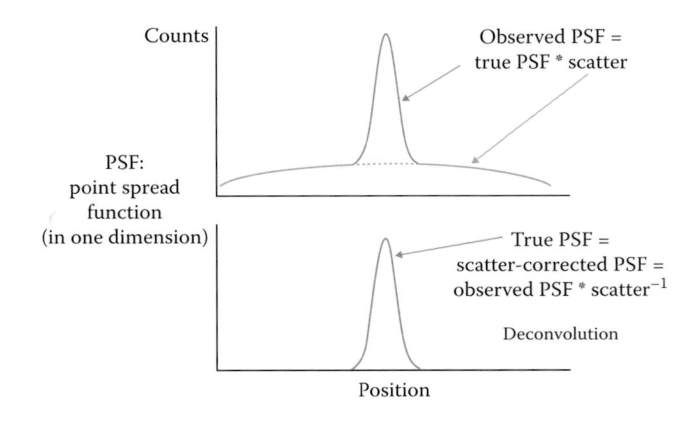

FIGURE 6.17 The "Gaussian-fit" scatter correction method. In practice, the point spread function (PSF) of a PET scanner includes both primary (unscattered) and scattered counts, with the former corresponding to the Gaussian, or bell-shaped, curve and the latter corresponding to the peripheral tails extending from the Gaussian curve. The tails can be subtracted (deconvolved) from the total PSF to yield the scatter-corrected PSF. In a similar manner, this scatter correction can be applied to projection images prior to reconstruction to yield scatter-corrected reconstructed images.

Once the randoms correction has been applied, the peripheral "tails" in the projection image count profiles, presumably due to scatter, are fit to a mathematical function and then subtracted (or deconvolved) from the measured profile to yield scatter-corrected profiles for tomographic image reconstruction. While this approach works reasonably well for 2-D PET and small source volumes (e.g., the brain) in 3-D PET, it is not adequate for 3-D PET generally. Scatter corrections for 3-D PET include dual-energy window–based approaches, convolution/deconvolution-based approaches (analogous to the correction in 2-D PET), direct estimation of scatter distribution (by computer modeling of the imaging system), and iterative reconstruction-based scatter compensation approaches (also employing computer modeling) [38,39].

Several methods, characterized as energy distribution– or spatial distribution–based methods, have been proposed for SPECT scatter correction [23]. However, only the dual-energy window–based or triple-energy window (TEW)–based estimation and the effective scatter source estimation (ESSE) method have been implemented clinically. In multienergy window–based methods [42] (Figure 6.18), the number of scattered photons in the photopeak energy window is estimated based on counts in one or more scatter windows. In the TEW method, for example, the scatter is estimated as the area of the trapezoid beneath the line joining the two adjacent narrow scatter windows; for radionuclides emitting a single x- or γ-ray, the upper scatter window can be omitted. For each projection image pixel i,j, the number of scatter counts in the photopeak window, $(C_{i,j})_{scatter}$, is calculated as

$$(C_{i,j})_{scatter} = \left(\frac{(C_{i,j})_{lower}}{W_{lower}} + \frac{(C_{i,j})_{upper}}{W_{upper}} \right) \times \frac{W_{photopeak}}{2}, \tag{6.8}$$

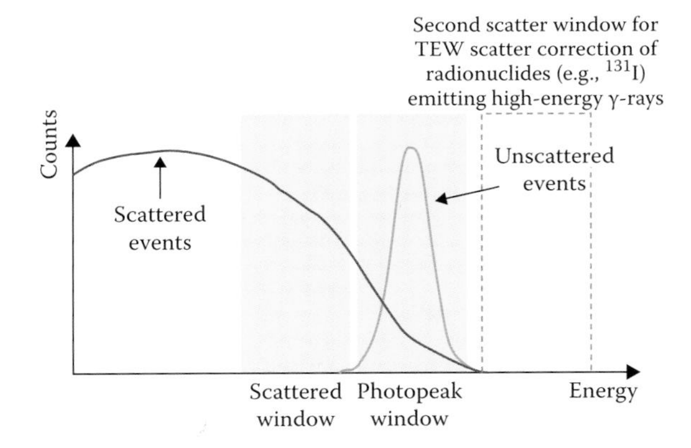

FIGURE 6.18 Dual-energy windows (shaded boxes) superimposed on the energy spectrum of unscattered and scattered events for a patient-sized phantom filled with ^{99m}Tc. In contrast to ^{99m}Tc, a radionuclide such as ^{131}I emits a higher-energy γ-ray, which will produce scattered events in the photopeak (unscattered-event) energy window in addition to those from scatter of the primary "imaging" γ-ray. For such radionuclides, the dual-energy window method extended to the triple-energy window (TEW) method by using a second scatter energy window (indicated by the dashed-line box) immediately above the photopeak window. (Adapted from Cherry, S.R. et al., *Physics in Nuclear Medicine*, Saunders, Philadelphia, PA, 2012. With permission.)

where $(C_{i,j})_{lower}$ and $(C_{i,j})_{upper}$ are the counts in the lower and upper scatter energy windows, respectively, for projection pixel i,j, and $W_{photopeak}$, W_{lower}, and W_{upper} are the widths (in keV) of the photopeak, lower scatter, and upper scatter windows, respectively. Typically, 20% photopeak and ~5% scatter energy windows are used. Because of the use of such narrow scatter energy windows, the TEW estimate of scatter counts tends to be noisy. In the ESSE method [43], a model-derived scatter function is used to calculate and correct for the scatter in the projection data based on the estimate of the activity distribution. An effective scatter source is calculated, and the attenuated projection of that source yields the scatter component of the SPECT projections. Other, more sophisticated and presumably more accurate approaches to scatter modeling include Monte Carlo simulation–based methods [44–46] and analytic methods using the Klein-Nishina formula for Compton scattering [47]. These methods are computationally intensive, however, and have been limited for use in research settings but likely will become clinically practical as computing power continues to increase.

Attenuation correction. Correction for the attenuation of the γ-rays as they pass through tissue is generally the largest correction in SPECT and PET. The correction factors can range from ~2 for a ^{99m}Tc SPECT scan of the brain up to ~15–40 for a PET scan of the abdomen. The magnitude of the correction depends on the energy of the γ-rays (variable for SPECT studies and 511 keV for PET studies), the thickness of tissue(s) that the γ-rays must travel through, and the attenuation characteristics of the tissue(s). One of the attractive features of PET is the relative ease of applying accurate and precise corrections for attenuation, based on the fact that attenuation depends only on the total thickness of the attenuation medium (at least for a uniformly attenuating medium) (Figure 6.19). For a positron-emitting source and a volume of thickness L, the attenuation factor is $e^{-\mu L}$ and the attenuation correction factor (ACF) $e^{\mu L}$ regardless of the position (i.e., depth) of the source. In fact, the source could even be outside the subject, and for a given LOR joining two detectors i and j, the detected counts will suffer the same degree of attenuation as a source inside the subject located at any depth along that same line. Accordingly, if an external rod source of a positron emitter such as ^{68}Ge is extended along the axial FOV and rotated around the periphery of the FOV first with and then without the subject in the imaging position—the transmission and the blank scans, respectively—the *ACF* can be derived from the ratio of the counts in these respective scans:

$$\mathrm{ACF}_{ij} = e^{\mu L}_{ij} = \frac{(C_{\mathrm{Blank}})_{ij}}{(C_{\mathrm{Trans}})_{ij}}, \tag{6.9}$$

where ACF_{ij} is the ACF for coincidence events between detectors i and j, L_{ij} is the thickness of the volume between coincident detectors i and j, and $(C_{Blank})_{ij}$ and $(C_{Trans})_{ij}$ are the external-source counts between detectors i and j in the blank and transmission scans, respectively. This expression also holds when the attenuation characteristics of the tissue are nonuniform. In practice, a blank scan is acquired only once a day. The transmission scan can be acquired before the subject has been injected with the radiotracer, after the subject has been injected with the radiotracer but before or after the emission scan, or after the patient has been injected with the radiotracer and at the same time as the emission scan. Preinjection transmission scanning avoids any interference between the emission and transmission data but requires that the subject remain on the imaging table before, during,

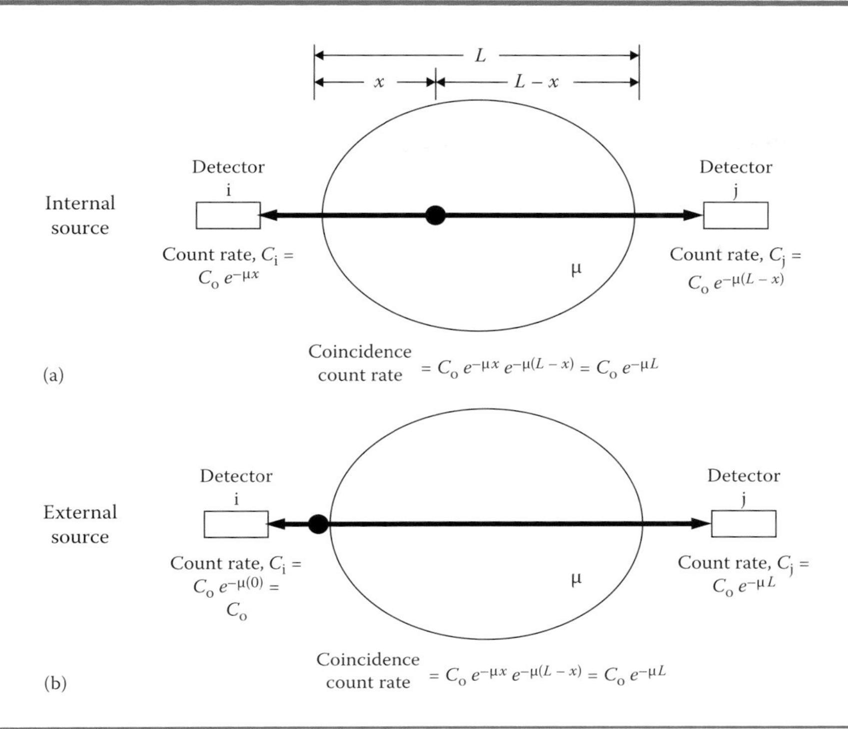

FIGURE 6.19 In PET, attenuation of the annihilation γ-rays for a uniformly attenuating medium depends only on the total thickness (*L*) of the absorber, that is, it is independent of the position (depth *x*) of the source in the absorber (a). An external positron emitter can therefore be used for attenuation correction (b). C_o is the actual source count rate, and C_i and C_j are the count rates measured by detectors i and j, respectively; μ is the linear attenuation coefficient of the absorber for 511 keV photons. (Adapted from Zanzonico, P., *Semin. Nucl. Med.*, 34, 2004. With permission.)

and after injection of the radiotracer. It is the least efficient operationally and is rarely used in practice. Postinjection transmission scanning minimizes the effects of motion, relying on the much higher external-source count rates for reliable subtraction of the emission counts from the transmission counts. It was the most commonly used approach in "PET-only" scanners. Simultaneous emission/transmission scanning is obviously the most efficient (fastest) approach but may result in excessively high randoms and scatter counter rates in the emission data.

With the introduction of hybrid PET-CT scanners (see Section 6.7), attenuation correction is now performed using CT rather than transmission sources. A CT image is basically a 2-D map of the attenuation coefficients (μ in Equation 6.9) at the CT x-ray energy (~80 keV).* For attenuation correction of the PET emission data, however, these values must be appropriately scaled to the 511 keV energy of the annihilation γ-rays. The mass-attenuation coefficients (μ_m) for CT x-rays (~80 keV) and for 511 keV annihilation γ-rays are 0.182 and 0.096 cm²/gm, 0.209 and 0.093 cm²/gm, and 0.167 and 0.087 cm²/gm in soft tissue, bone, and lung, respectively. The corresponding μ_m ratios are therefore 1.90, 2.26, and 1.92, respectively. Thus, ACFs derived from CT images cannot be scaled to those for

* X-ray (i.e., transmission x-ray) imaging in general and CT scanning in particular are discussed in Chapter 2 of this volume.

511 keV annihilation γ-rays simply using a global factor. Accordingly, CT-based attenuation correction in PET has been implemented using a combination of segmentation (to delineate soft tissue, bone, and lung compartments) and variable scaling (to account for the different μ_m ratios in these respective tissues).

Like scatter corrections, attenuation corrections in SPECT are not yet as well developed or as reliable as those in PET—because, for single photons, the ACF also depends on the depth of the source [20]. For many years, if attempted at all, SPECT ACFs were calculated (as in Chang's first-order correction and the Sorenson method [48,49]) based on the assumptions—neither of which is generally accurate—that the body is a uniform medium with a single, well-defined value of μ and that the body's contour is known. More recently, manufacturers have incorporated long-lived radioactive sources (such as gadolinium-153 [^{153}Gd]) into SPECT scanners to perform attenuation correction. As part of the SPECT procedure, a shutter opens at each projection image angle to expose a highly collimated line source, and a transmission image is acquired. The transmission images thus acquired are reconstructed into an ACF map for correction of the SPECT study. The introduction of hybrid SPECT-CT scanners (see Section 6.7) is resulting in more practical and more accurate CT-based attenuation correction in SPECT [50].

6.6.2 Image Reconstruction

In emission tomography, the emission data correspond to the projected sum of counts (or line integrals) at various angles about the axis of the scanner (Figure 6.20a and b). The full set of 2-D projection data is usually represented and stored as a 2-D matrix in polar coordinates (distance r, angle ϕ) known as a *sinogram* (Figure 6.20c and d) [51]. In 3-D PET, the projections are 2-D (x_r,y_r) parallel line integrals with azimuthal angle ϕ and oblique, or polar, angle θ (Figure 6.21) [51]. Correction of the emission data (normalization, scatter correction, and then attenuation correction) is typically performed in sinogram space prior to reconstruction.

There are two basic classes of image reconstruction methods, analytic and iterative. One of the most widely used analytic algorithms for reconstruction of tomographic images from 2-D data (or 3-D data rebinned into 2-D projections)—in SPECT as well as PET—has been *filtered backprojection* (FBP) [20,51]. The basic procedure is as follows: Each angular projection is Fourier transformed from real to frequency (or "*k*") space; the projection is filtered in frequency space using a ramp filter (to enhance high and suppress low spatial frequencies and thereby minimize blurring artifacts) (Figure 6.22a through c); the filtered projection is inverse Fourier transformed from frequency back to real space; and the filtered projection data in real space are uniformly distributed, or back-projected, over the reconstructed image matrix [20,51]. The methods are completely analogous to those used in the reconstruction of CT images (see Chapter 2). To compensate for the high noise levels resulting from the relatively low counts in most nuclear imaging studies, low-pass, or apodizing, filters (known as Shepp-Logan, Hann, etc.) are used in place of the ramp filter to eliminate those spatial frequencies above a specified cutoff frequency (Figure 6.22d through g). In this way, the high spatial frequencies characteristic of statistical noise are eliminated (or at least minimized) in the reconstructed images. The 3-D reprojection (3DRP) algorithm is an extension of the standard 2-D FBP algorithm and has been implemented on commercial 3-D scanners [52,53]. In 3DRP, unsampled data (caused by the fact the subject is not completely enclosed by detectors and, therefore, some lines of response are not measured) are first estimated by forward-projection through an initial stack of images obtained by standard 2-D reconstruction [51].

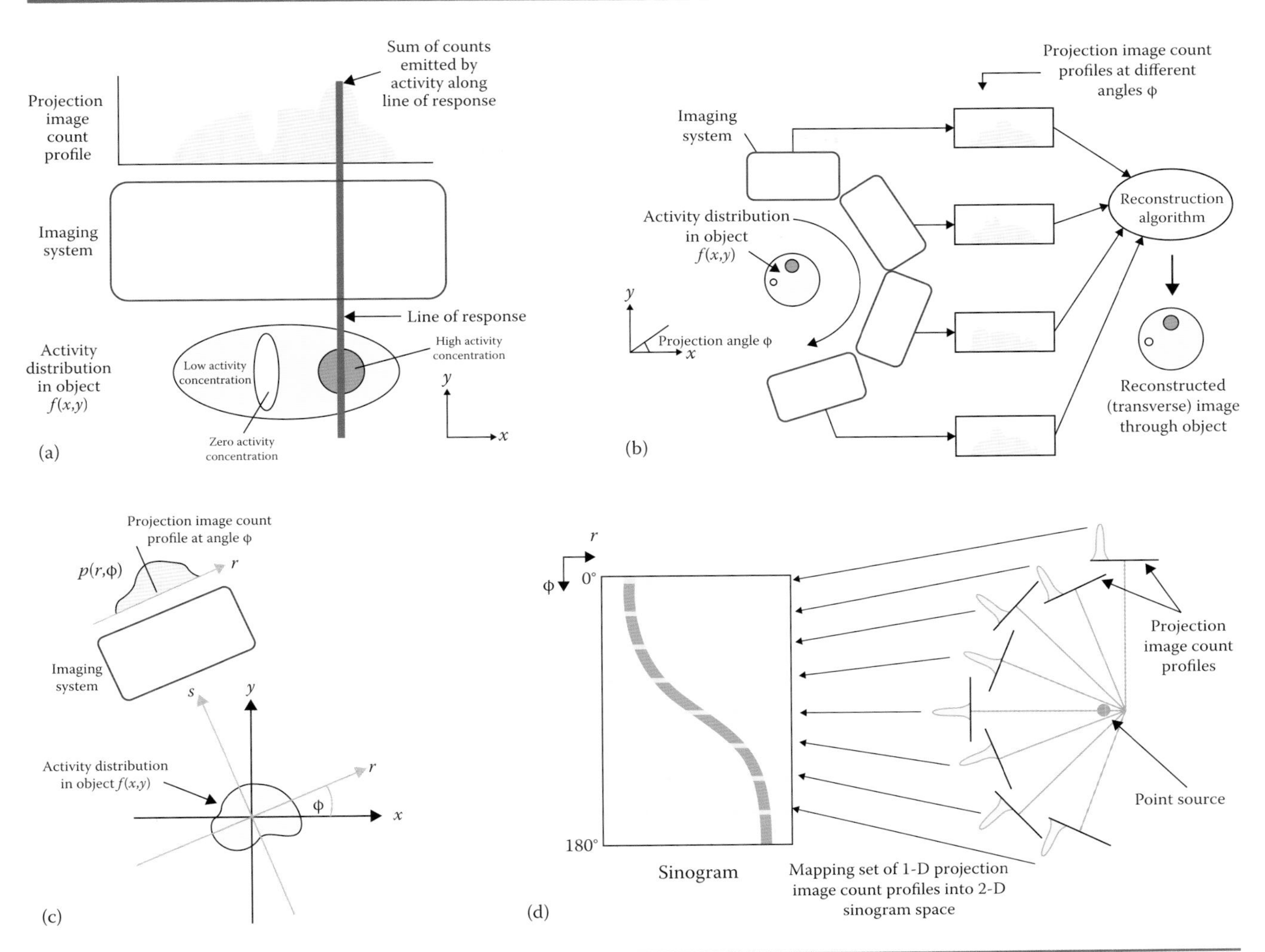

FIGURE 6.20 Emission tomography data acquisition and sinogram display. (a) Projection image count profile under idealized conditions (no attenuation or scatter). (b) Acquisition of multiple projection images and their corresponding count profiles at various projection angles ϕ about the longitudinal axis of the object being imaged. (c) The (r,s) coordinate system is rotated by projection angle ϕ with respect to the (x,y) coordinate system of the cross section of the object being imaged. The (r,s) coordinate system is fixed with respect to the imaging system, and the (x,y) coordinate system is fixed with respect to the object. (d) 2-D count display, or sinogram, of a set of projection image count profiles. Each row in the sinogram corresponds to an individual projection image count profile, displayed for increasing projection image angle ϕ from top to bottom. As shown, a point source of radioactivity traces out a sinusoidal path in the sinogram. (Adapted from Cherry, S.R. et al., *Physics in Nuclear Medicine*, Saunders, Philadelphia, PA, 2012. With permission.)

Iterative algorithms [54] attempt to progressively refine estimates of the activity distribution, rather than directly calculating the distribution, by maximizing or minimizing some target function. The solution is said to *converge* when the difference of the target function between successive estimates (iterations) of the activity distribution is less than some prespecified value. Importantly, iterative reconstruction algorithms allow incorporation of realistic modeling of the data acquisition process (including effects of attenuation and of scatter), modeling of statistical noise, and inclusion of pertinent a priori information (e.g., only nonnegative count values). The maximum-likelihood expectation maximization

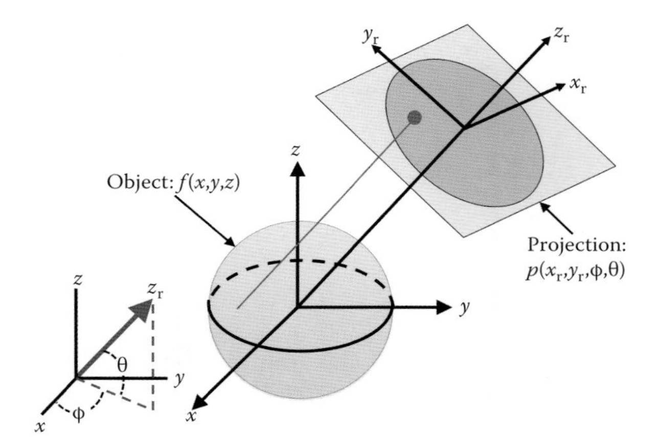

FIGURE 6.21 In 3-D PET, the projections are 2-D (x_r,y_r) parallel line integrals with azimuthal angle ϕ and oblique angle θ. The 3-D projection data are represented as a set of sinograms, with one sinogram per polar angle θ, each row representing the projected intensity across a single polar angle θ and each column the projected intensity at the same position x_r across the projection at successive azimuthal angles ϕ. (From Defrise, M. and P. Kinahan, Data Acquisition and Image Reconstruction for 3D PET, in *The Theory and Practice of 3D PET*, B. Bendriem and D.W. Townsend, editors, Kluwer Academic Publishers, Dordrecht, Netherlands, 1998. With permission.)

(MLEM) algorithm is based on maximizing the logarithm of a Poisson-likelihood target function [55,56]. The MLEM algorithm suppresses statistical noise, but large numbers of iterations typically are required for convergence, and therefore, processing times are long. To accelerate this slow convergence, the ordered-subset expectation maximization (OSEM) algorithm [57] groups the projection data into subsets comprised of projections uniformly distributed around the source volume. The OSEM algorithm, which is a modified version of the MLEM algorithm in that the target is still maximization of the log-likelihood function, converges more rapidly than MLEM and is now the most widely used iterative reconstruction method in PET as well as SPECT [58]. The row-action maximization-likelihood (RAMLA) algorithm, related to the OSEM algorithm, has been implemented for direct reconstruction of 3-D PET data. The so-called 3D-RAMLA algorithm, which eliminates 2-D rebinning of the 3-D data, employs partially overlapping, spherically symmetric volume elements called "blobs" in place of voxels [58–60]. Reconstruction times are fairly long by clinical standards, but the results have been excellent [61].

3-D reconstruction algorithms remain computer intensive and rather slow by clinical standards. In addition, 3-D PET emission data files are very large—typically more than two orders of magnitude larger than 2-D data sets. It may be preferable, therefore, to reduce 3-D data sets to a more manageable size for image reconstruction—by rebinning of the 3-D set of oblique sinograms into a smaller number of direct 2-D sinograms. The simplest method is "single-slice rebinning (SSRB)," wherein true oblique LORs are assigned to the direct plane midway between the two detector elements actually in coincidence [62]. SSRB distorts off-axis activity and thus is accurate only for activity distributions close to the detector axis, as in brain or small animal imaging. A second method is multislice rebinning (MSRB) [63], which is fast but is susceptible to noise-related artifacts. The current method of choice is Fourier rebinning (FORE), based on the 2-D Fourier transform of the oblique

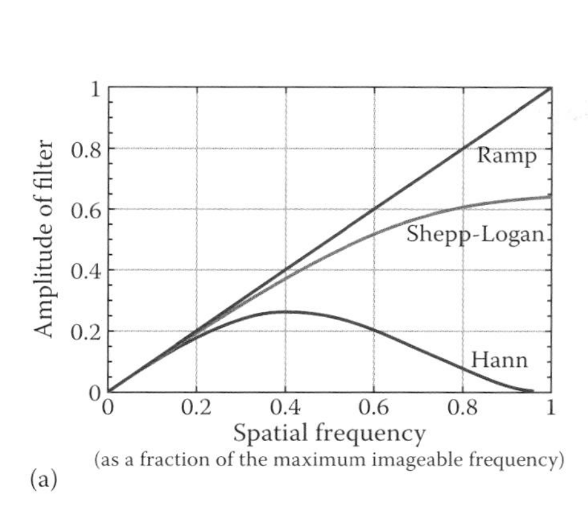

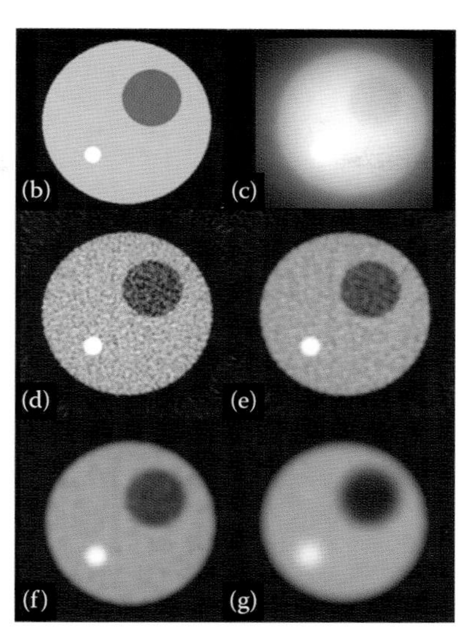

FIGURE 6.22 (a) Graphical representation of the ramp and two other mathematical filters used in reconstruction of tomographic images. Note that the cutoff spatial frequencies of all three filters are set to the maximum frequency that can be imaged with the system being used to reduce noise amplification artifacts in the reconstructed images. Note further that the Shepp-Logan and Hann filters "roll off" gradually with increasing spatial frequency to further reduce noise amplification and other artifacts. (b) Cross section through a phantom containing high (white), intermediate (light gray), and low (dark gray) activity concentrations. (c) Tomographic image of phantom reconstructed by "simple" backprojection, that is, with no mathematical filtering of the projection images. Note the resulting blurring and other artifacts in the reconstructed image. (d through g) Tomographic images of phantom reconstructed by filtered backprojection using a Shepp-Logan filter with cutoff frequencies equal to 1, 0.8, 0.6, and 0.2 times the maximum spatial frequency that can be imaged with the system. Note the reduction in noise (i.e., graininess or mottle) but coarser delineation of edges (i.e., poorer resolution) as the cutoff frequency is reduced. For clinical SPECT studies, which generally have limited numbers of counts, cutoff frequencies lower than the maximum imageable frequency are used routinely. (Adapted from Cherry, S.R. et al., *Physics in Nuclear Medicine*, Saunders, Philadelphia, PA, 2012. With permission.)

sinograms (Figure 6.23) [54,64]. In contrast to SSRB and MSRB, however, FORE cannot be performed in real time and thus requires the full 3-D data set.

6.6.3 Quantitation

Once the SPECT or PET emission data have been corrected for dead time, randoms, system response (by normalization), scatter, and attenuation, the count rate per voxel in the reconstructed tomographic images is proportional to the local activity concentration [21,38,39]. In routine practice, however, scatter and attenuation corrections are generally less accurate for SPECT than for PET (as previously noted) and reconstructed SPECT images less quantitatively accurate than PET images. Further, acquired SPECT data generally include far fewer counts than do PET data because of the use of absorptive collimation for SPECT and electronic collimation for PET, increasing the statistical uncertainties in SPECT.

Another consideration that adversely effects the quantitative accuracy of both SPECT and PET is partial-volume averaging: For a source that is "small" relative to the spatial

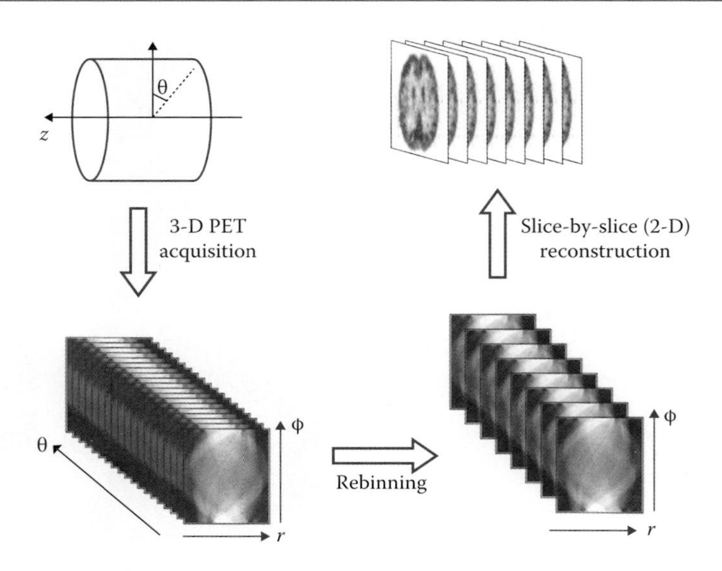

FIGURE 6.23 The Fourier rebinning (FORE) algorithm for rebinning 3-D PET data into 2-D data. The basic steps of the FORE algorithm are creating a stack of oblique sinograms, calculating the 2-D Fourier transform for each oblique sinogram, normalizing the resulting 2-D sinogram for the number of oblique slices contributing to the oblique sinogram, and calculating the 2-D inverse Fourier transform of the resulting normalized sinograms to yield the 2-D rebinned data for image reconstruction. (With kind permission from Springer Science+Business Media: *Positron Emission Tomography: Basic Sciences*, Image Reconstruction Algorithms in PET, 2005, 63–91, Defrise, M. et al.)

resolution of the imaging system, the image-derived activity or activity concentration in such a source underestimates the actual value, and the smaller the source, the greater the degree of the underestimation [21,65]. The dimensions of a source must be at least two times the FWHM spatial resolution of the imaging system to avoid such underestimation of the activity or activity concentration [65]. Since the source-size dependence of the partial-volume effect can be measured (e.g., by phantom studies), the underestimated activity or activity concentration can be corrected if the source dimensions can be independently determined (e.g., by CT) [66].

Another key factor that limits the quantitative accuracy of SPECT and PET is subject motion. As in photography, subject motion blurs SPECT and PET images, dispersing the counts in a source region over an apparently larger volume than the actual volume of that region. This results in not only an overestimate of the source region volume but also an underestimate of the source region activity concentration. For sources (e.g., tumors) in the lung and the dome of the liver, a particularly important source of such inaccuracies is respiratory motion [67–70]. Practical, reasonably accurate methods for respiratory gating of PET studies have been developed and result in more reliable estimates of the volumes and activity concentrations in pulmonary lesions [67–70].

To make the images absolutely quantitative, the count rate per voxel (cps), $\dot{C}_{ijk}$, in voxel i,j,k must be divided by a measured system calibration factor (CF) [(cps/voxel)/(μCi/cm^3)] to yield the activity concentration [7]:

$$[A]_{\text{ijk}} = \frac{\dot{C}_{\text{ijk}}}{\text{CF}}, \tag{6.10}$$

where $[A]_{\text{ijk}}$ is the activity concentration (e.g., μCi/cm^3) in voxel i,j,k. The CF can be derived by scanning a calibrated standard, that is, a water-filled or water (tissue)-equivalent volume source with all linear dimensions at least twice that of the system spatial resolution (FWHM) and with a uniform, well-defined activity concentration at the time of the scan. The requirement for water equivalence is to ensure that effects such as scatter and attenuation are comparable in both the subject and the standard. And the requirement for linear dimensions at least twice that of the system spatial resolution is to ensure that the effect of partial-volume averaging and associated underestimation of local count rates are negligible. Activity concentration is often expressed in terms of the decay-corrected fraction or percentage of the administered activity per cubic centimeter or per gram of tissue. In clinical PET, activity concentrations are often expressed in terms of the *standardized uptake value* (SUV):

$$\text{SUV} = \frac{[A](\mu\text{Ci/cc})}{\text{injected activity } (\mu\text{Ci}) \times \text{body mass (g)}}. \tag{6.11}$$

An advantage of using SUV to express activity concentration is that it is independent of the body mass of the subject.

The practical quantitative accuracy of SPECT and PET (expressed, for example, as the percent difference between the image-derived and actual activity or activity concentration in a source region) is difficult to characterize unambiguously, as it depends on such factors as the count statistics, the size of the source region, the source-to-background activity concentration ratio, subject motion, and the accuracy of all applicable corrections. Under favorable circumstances (i.e., for a high-count study of a large, stationery source region with a high source-to-background ratio), the accuracy of the PET is on the order of 10% [65]. A comparable accuracy, ~10%, is achievable for SPECT with state-of-the-art attenuation and scatter corrections [71]. With routinely available corrections, however, the quantitative accuracy of SPECT generally exceeds 20% [72].

Quantitative nuclear medicine studies performed dynamically yield regional radiotracer concentrations as a function of time postadministration of the tracer. Based on an understanding of the biologic fate of the radiotracer *in vivo*, mathematical models (also known as tracer kinetic models) may be constructed and fit to the observed time-activity curves. More specifically, one or more parameters of the equations representing the model may be adjusted to yield an optimum match (or "best fit") of the model-derived time-activity curves to the observed time-activity curves. In some cases, the model parameters can be related directly to physiologic or biologic quantities such as tissue perfusion (measured in mL/min/g) and the rate of glucose utilization (in mol/min/g). Further, such a physiologic parameter is sometimes derived on a voxel-by-voxel basis to yield a so-called parametric map of the parameter [73]. See Chapter 7 for further details on kinetic modeling and parametric imaging.

6.7 MULTIMODALITY DEVICES: SPECT-CT AND PET-CT SCANNERS

Image registration and fusion have become increasingly important components of both clinical and laboratory imaging and have led to the development of a variety of pertinent software and hardware tools, including multimodality (e.g., SPECT-CT and PET-CT) devices that "automatically" provide registered and fused 3-D image sets [50,74–77]. Since information derived from multiple images is often complementary (e.g., localizing the site of an apparently abnormal metabolic process to a pathologic structure such as a tumor), integration of image data in this way may be helpful and even critical. In addition to anatomic localization of signal foci, image registration and fusion provide intramodality as well as intermodality corroboration of diverse images and more accurate and more certain diagnostic and treatment-monitoring information. The problem, however, is that differences in image size and dynamic range, voxel dimensions and depth, image orientation, subject position and posture, and information quality and quantity make it difficult to unambiguously colocate areas of interest in multiple image sets. The objective of image *registration* and *fusion*, therefore, is (1) to appropriately modify the format, size, position, and even shape of one or both image sets to provide a point-to-point correspondence between images and (2) to provide a practical integrated display of the images thus aligned.

In both clinical and laboratory settings, there are two practical approaches to image registration and fusion, "software" and "hardware" approaches. In the software approach, images are acquired on separate devices, imported into a common image-processing computer platform, and registered and fused using the appropriate software. This is discussed in Chapter 7. In the hardware approach, images are acquired on a single, multimodality device and transparently registered and fused with the manufacturer's integrated software. Both approaches are dependent on software sufficiently robust to recognize and import diverse image formats. The availability of industry-wide standard formats, such as the American College of Radiology and National Electrical Manufacturers Association for Digital Imaging and Communications in Medicine (ACR-NEMA DICOM) standard [78,79], is therefore critical.

All manufacturers of SPECT, PET, and CT scanners market multimodality scanners, combining PET or SPECT with CT scanners in a single device. These instruments provide near-perfect registration of images of *in vivo* function (SPECT or PET) and anatomy (CT) using a measured, and presumably fixed, rigid transformation (i.e., translation and rotation in three dimensions) between the image sets. These devices have already had a major impact on clinical practice, particularly in oncology, and PET-CT devices, for example, have eliminated PET-only scanners from the clinic [77]. Although generally encased in a single seamless housing, the PET and CT gantries in such multimodality devices are separate; the respective fields of view are separated by a distance on the order of 1 m, and the PET and CT scans are performed sequentially. Most recently, hybrid PET-MRI scanners (some even permit simultaneous PET and MRI acquisition) have also become commercially available.

In addition to PET-CT scanners, SPECT-CT scanners are now commercially available [50]. The design of SPECT-CT scanners is similar to that of PET-CT scanners in that the SPECT and CT gantries are separate and the SPECT and CT scans are acquired sequentially, not simultaneously. In such devices, the separation of the SPECT and CT scanners is more apparent because the rotational and other motions of the SPECT detectors effectively preclude encasing them in a single housing with the CT scanner. Multimodality imaging devices for small animals (i.e., rodents)—PET-CT, SPECT-CT, PET-MRI, and even SPECT-PET-CT devices—are now commercially available as well.

Multimodality devices simplify image registration and fusion—conceptually as well as logistically—by taking advantage of the fixed geometric arrangement between the PET and CT scanners or the SPECT and CT scanners in such devices. Further, because the time interval between the sequential scans is short (i.e., a matter of minutes) and the subject remains in place, it unlikely that subject geometry will change significantly between the PET or SPECT scan and the CT scan. Accordingly, a predefined rigid transformation matrix can be used to align the PET or SPECT and the CT image sets.

Image fusion may be as simple as simultaneous display of images in a juxtaposed format. A more common, and more useful, format is an overlay of the registered images, where one image is displayed in one color table and the second image in a different color table. Typically, the intensities of the respective color tables as well as the "mixture" of the two overlaid images can be adjusted. Adjustment (e.g., with a slider on the computer screen) of the mixture allows the operator to interactively vary the overlay so that the designated screen area displays only the first image, only the second image, or some weighted combination of the two images, each in its respective color table.

6.8 PRECLINICAL SPECT AND PET IMAGING

Imaging-based experimentation in small animal (i.e., mice and rats) models is now an established and widely used approach in basic and translational biomedical research and will no doubt remain an important component of such research [80]. Small animal imaging provides a noninvasive means of assaying biological structure and function *in vivo*, yielding quantitative, spatially and temporally indexed information on normal and diseased tissues such as tumors. Importantly, because of its noninvasive nature, imaging allows serial (i.e., longitudinal) assay of rodent models of human cancer and cardiovascular, neurological, and other diseases over the entire natural history of the disease process, from inception to progression, and monitoring of the effectiveness of treatment or other interventions (with each animal serving as its own control and thereby reducing biological variability). This also serves to minimize the number of experimental animals required for a particular study. With the ongoing development of genetically engineered (i.e., transgenic and knockout) rodent models of cancer and other diseases, such models are increasingly more realistic in recapitulating the natural history and clinical sequelae of the corresponding human disease, and the ability to track these disease models long term is therefore critical. Importantly, in contrast to cell culture-or tissue culture–based experiments, studies in intact animals incorporate all of the interacting physiological factors—neuronal, hormonal, nutritional, immunological, and so forth—present in the complex *in vivo* milieu. Intact whole-animal models also facilitate investigation of systemic aspects of disease such as cancer metastasis, which are difficult or impossible to replicate in ex vivo systems. Further, because many of the same imaging modalities—PET, SPECT, CT, MRI, and ultrasound—used in the clinic are also used in the laboratory setting, the findings of small animal imaging are readily translatable to patients. It should be noted, however, that radiation doses to experimental animals in PET, SPECT, and CT studies are considerably—one to two orders of magnitude—higher than those encountered in the corresponding clinical studies. Indeed, at absorbed doses on the order of 100 cGy, they approach single-fraction doses used in external-beam radiation therapy. Investigators should be aware of the magnitude of absorbed doses encountered in small animal PET, SPECT, and CT studies and potential radiogenic perturbation of their experimental system.

6.9 COMMERCIAL DEVICES

6.9.1 Clinical

All of the major manufacturers of medical imaging equipment market clinical PET and SPECT devices. Since 2005, PET scanners have been marketed exclusively as multimodality (i.e., PET-CT) scanners; as noted, clinical PET-only scanners are no longer marketed. The PET subsystems of commercial PET-CT scanners utilize BGO, LSO, or LYSO as the detector material, with most now operating exclusively in 3-D mode and all except the BGO-based systems incorporating TOF capability. The dimensions of the detector elements are typically $4 \times 4 \times 20$ mm^3, though slightly larger-area and thicker detector elements are found in some units. The axial fields of view are typically 20 cm in length, and whole-body studies therefore require six or seven bed positions. PET system spatial resolution (FWHM) is on the order of 5 mm among clinical systems. "SPECT-only" scanners remain popular and continue to be marketed. SPECT scanners overwhelmingly continue to utilize dual NaI(Tl) detectors (most commonly ~10 mm in thickness) in an opposed configuration; many commercial systems allow rotation of the detectors to an angle of 90° with respect to one another for cardiac studies. SPECT system spatial resolution (FWHM) with LEHR collimation is typically 7–8 mm for ^{99m}Tc at a source-to-collimator distance of 10 cm. In contrast to PET-CT scanners, in which the CT subsystems are uniformly "diagnostic grade," different models of SPECT-CT devices are equipped with distinctly different types of CT scanners: either a diagnostic-grade scanner (as in PET-CT devices) or a coarser-resolution cone-beam scanner equipped with a flat-panel detector. The latter type of CT scanner is suitable for attenuation correction of SPECT images and for anatomic orientation but is inadequate for radiological diagnosis. On the other hand, such devices are less expensive and occupy less space than diagnostic-grade CT scanners.

Several "special-purpose" clinical scanners are also currently marketed. These include positron emission mammography systems and other breast imaging devices [81,82] and dedicated SPECT scanners for cardiac (i.e., myocardial perfusion) imaging [83].

6.9.2 Preclinical

A number of preclinical PET and SPECT devices are commercially available. Preclinical PET scanners, which operate exclusively in 3-D mode, are diverse in design, utilizing different scintillation detectors in combination with either PSPMTs or APDs. The superior spatial resolution of preclinical versus clinical PET scanners, 1–2 mm versus 4–6 mm, is due in part to the much smaller gantry diameter and, therefore, a less pronounced resolution-degrading effect of the noncolinearity of the annihilation γ-rays. Preclinical SPECT scanners, on the other hand, are rather similar in design, generally utilizing multiple NaI(Tl) scintillation detectors fitted with multiaperture pinhole collimators. The superior spatial resolution of preclinical versus clinical SPECT scanners, ~1 versus ~10 mm, is due to the magnification effect afforded by the pinhole collimation. Of course, this is achieved at the cost of lower sensitivity, though the use of multiaperture collimators, and multiple (up to four) detectors as least partially mitigates the reduction in sensitivity. Preclinical devices are currently marketed as PET-only or SPECT-only scanners or as multimodality devices with integrated CT scanners, typically cone-beam devices with flat-panel detectors, or more recently with MRI scanners. There also are trimodality PET-SPECT-CT systems available.

6.10 RADIOTRACERS

Over the years, a large number and variety of radiotracers (also known as radiopharmaceuticals) for SPECT and PET have been developed. And, as noted in the introduction, a large number and variety of molecularly targeted and/or pathway-targeted radiotracers have been and continue to be developed for increasingly specific characterization of *in situ* biology. Although a detailed discussion of radiochemistry and radiopharmacology is beyond the scope of this chapter, a tabulation of some available SPECT and PET radiotracers is presented in Table 6.4 to illustrate the rich variety of such agents that have been developed [84–87]. Of course, any such tabulation is inevitably incomplete and almost immediately outdated, given the breadth of the field and its rapid pace of advancement. Notable properties and several illustrative examples of SPECT and PET radiotracers, respectively, are discussed briefly in this section.

SPECT radiotracers are generally labeled with radiometals (e.g., ^{99m}Tc or ^{111}In) or radioiodines (i.e., ^{123}I or ^{131}I). An important SPECT radiotracer is ^{99m}Tc-labeled methoxyisobutylisonitrile (also known as MIBI or sestamibi and marketed as Cardiolite), a lipophilic cationic complex used in place of, or in addition to, ^{201}Tl-thallous chloride for evaluation of myocardial perfusion and perfusion abnormalities resulting from coronary artery disease. Another example of a SPECT radiotracer is ^{111}In-labeled octreotide (Octreoscan), an analog of the peptide hormone somatostatin used for detection and localization of primary and metastatic neuroendocrine tumors (such as carcinoid tumors) expressing or overexpressing somatostatin receptors. Radiolabeled octreotide and analogs have also been used for targeted radiation therapy of such tumors.

PET radiotracers more commonly utilize radionuclides, ^{11}C, ^{13}N, and ^{15}O, of the "physiologic" elements, carbon, nitrogen, and oxygen, respectively. However, their half-lives are generally too short for routine clinical use. Instead, therefore, ^{18}F, which has a 110 min half-life, is often covalently incorporated into organic compounds with minimal perturbation of their structure and biologic behavior. A notable feature of PET is the number and variety of biologically important molecules (such as metabolites and metabolite analogs, drugs, receptor-binding ligands, neurotransmitters, antibodies and other immune constructs, etc.) that have been developed as radiotracers. ^{18}F-fluoro-2-deoxyglucose (FDG) is by far the most widely used PET radiotracer and has dramatically impacted the patient management in oncology. It is a partially metabolized, metabolically trapped structural analog of glucose whose uptake is related to the levels of expression of glucose transporters and of glucose metabolism (i.e., glycolysis). ^{18}F-fluoro-thymidine (FLT) is an analog of the deoxynucleotide thymidine and has been widely investigated as a tracer of cellular proliferation.

6.11 SELECTED EXAMPLES OF APPLICATIONS FOR SPECT AND PET

For many years, SPECT and PET have been important, widely used imaging modalities in clinical practice, clinical investigation, and preclinical research, with the number of

TABLE 6.4
SPECT and PET Radiotracers and Their Applications

SPECT Radiotracers

Blood Flow (Perfusion) and Blood Volume

^{99m}Tc-ethyl cysteinate dimer (ECD)—brain

^{99m}Tc-furifosmin—heart

^{99m}Tc-hexamethylpropyleneamine oxime (HMPAO)—brain

^{99m}Tc-macroaggregated albumin (MAA)—lung

^{99m}Tc-sestamibi—heart

^{99m}Tc-red blood cells (rbcs)

^{99m}Tc-teboroxine—heart

^{99m}Tc-tetrofosmin—heart

^{201}Tl-thallous chloride—heart

Bone and Bone Tumor Imaging

^{99m}Tc-phosphates and phosphonates such as hydroxyethylidene diphosphonate (HEDP) and methylene diphosphonate (MDP)

^{177}Lu-ethylenediamine-N,N,N′,N′-tetrakis(methylene phosphonic acid) (EDTMP)

^{177}Lu-1,4,7,10-tetraazacyclododecane-1,4,7,10-tetraaminomethylenephosphonate (DOTMP)

Endocrine Imaging (Nonthyroid)

^{67}Ga-/^{111}In-/^{177}Lu-1,4,7,10-tetraazacyclododecane-1,4,7,10-tetraacetic acid (DOTA)-octreotide (DOTATOC)—somatostatin receptor

^{111}In-pentetreotide—somatostatin receptor

^{111}In-diethylene triamine penta-acetate (DTPA)-octreotide (DTPATOC)—somatostatin receptor

^{123}I-Tyr-3-octreotide (TOC)—somatostatin receptor

^{123}I-/^{131}I-meta-iodobenzylguanidine (MIBG)—epinephrine transporter

^{131}I-6β-19-norepinephrine-cholesterol

Kidney Imaging

^{99m}Tc-diethylene triamine penta-acetate (DTPA)—glomerular filtration

^{99m}Tc-dimercaptosuccinic acid (DMSA)—cortical retention

^{99m}Tc-glucoheptonate (GH)—cortical retention and glomerular filtration

^{99m}Tc-L,L and D,D-ethylene dicysteine (EC)—glomerular filtration

^{99m}Tc-mercaptoacetyltriglycine (MAG3)—glomerular filtration

^{123}I-/^{131}I-ortho-iodohippurate (OIH)—tubular secretion

Infection and Inflammation Imaging

^{67}Ga-gallium citrate (also used for tumor imaging)

^{99m}Tc-/^{111}In—leukocytes

^{99m}Tc—antibodies and antibody fragments

Liver, Spleen, and Hepatobiliary Imaging

^{99m}Tc-iminodiacetic acid (IDA) derivatives such as 2,6-di-isopropylacetanilido-iminodiacetic acid (DISIDA) and bromo-2,4,6-trimethylacetanilido-iminodiacetic acid (BRIDA)—hepatobiliary tree

^{99m}Tc-sulfur colloid—reticuloendothelial (RE) system

Miscellaneous

^{99m}Tc-/^{111}In-^{123}I-/^{131}I—antibodies and antibody fragments

^{99m}Tc-annexin—apoptosis imaging

^{99m}Tc-glutarate—myocardial infarct imaging

^{99m}Tc-nitroimidazole derivative (HL91)—hypoxia

^{99m}Tc-pyrophosphate (PYP)—myocardial infarct imaging

^{123}I-bicyclic fluoropyrimidine deoxynucleoside analogs (BCNAs)—reporter gene imaging

^{123}I-β-carbomethoxy-3-β-(4-iodophenyltropane) (CIT)—dopaminergic system

TABLE 6.4 (CONTINUED)
SPECT and PET Radiotracers and Their Applications

^{123}I-N-ω-fluoropropyl-2β-carbomethoxy-3β-(4-iodophenyl)nortropan (FP-CIT)—dopaminergic system

^{123}I-(S)-2-hydroxy-3-iodo-6-methoxy-N-(1-ethyl-2-pyrrodinyl)-methyl)benzamide (IBZM)—dopaminergic system

^{123}I-/^{131}I-iodo-deoxyuridine (IUdR)—DNA synthesis/cell proliferation

^{123}I-/^{131}I-5-iodo-2′-fluoro-2′-deoxy-1-β-D-arabinofuranosy1-5-iodouracil (FIAU)—reporter gene imaging

^{191}Pt-/^{193m}Pt-/^{195m}-cisplatin—drug

Thyroid

^{123}I-/^{131}I-sodium iodide

^{99m}Tc-sodium pertechnetate

^{186}Re/^{188}Re-sodium perrhenate

Ventilation (Lung)

^{81m}Kr-krypton gas

^{99m}Tc-diethylene triamine penta-acetate (DTPA) aerosol

^{127}Xe-/^{133}Xe-xenon gas

PET Radiotracers[a]

β-Amyloid Plaques (Alzheimer's Disease)

^{11}C-2-(4-(methylamino)phyenyl-6-hydroxybenzothiazole (Pittsburgh compound B, PIB)

^{11}C-4-N-methylamino-4′-hydroxystilbene (SB-13)

^{18}F-(4S)-4-(3-fluoropropyl)-L-glutamate (BAY 94-9392)

^{18}F-1-(1-(6-((2-fluoroethyl)(methyl)amino)-2-naphthyl)ethylidene malononitrile (FDDNP)

^{18}F-2-(4-(methylamino)phyenyl-6-hydroxybenzothiazole (Pittsburgh compound B, PIB)

Blood Flow (Perfusion) and Blood Volume

^{11}C-carbon monoxide

^{11}C-epinephrine (EPI)—heart

^{11}C-hydroxyephedrine (HED)—heart

^{13}N-ammonia—heart

^{13}N-nitrogen gas

^{15}O-water

^{15}O-butanol

^{18}F-p-fluorobenzyl triphenyl phosphonium cation (FBnTP)—heart

^{38}K-potassium chloride

^{62}Cu-/^{64}Cu-pyruvaldehyde-bis-(N(sup 4)-methylthiosemicarbazide (PTSM)

^{82}Rb-rubidium chloride—heart

Amino Acids and Amino-Acid Analogs

^{11}C-α-amino-isobutyrate (AIB)

^{11}C-aminocyclobutane carboxylate (ACBC)

^{11}C-L-leucine

^{11}C-L-methionine

^{13}N-alanine

^{13}N-γ-amino-butyric acid

^{13}N-L-glutamate

^{13}N-L-glutamine

^{13}N-L-leucine

^{13}N-putrescine

^{13}N-L-valine

^{18}F-fluoro-aminocyclobutane carboxylate (FACBC, also known as FCCA)

(Continued)

TABLE 6.4 (CONTINUED)
SPECT and PET Radiotracers and Their Applications

^{18}F-*O*-2-fluoroethyl-L-tyrosine
^{18}F-L-3-fluoro-α-methyltyrosine (FMT)
Drugs and Drug Analogs
^{18}F-5-fluoro-uracil (5-FU)
^{18}F-fluoro-dasatinib
^{124}I-HU-P71
Hypoxia-Imaging Agents
^{18}F-1-(5-fluoro-5-deoxy-a-D-arabinofuranosyl)-2-nitroimidazole (FAZA)
^{18}F-fluoro-erythronitro-imidazole (FETNIM)
^{18}F-2-(2-nitro-1(H)-imidazol-1-yl)-*N*-(2,2,3,3,3-pentafluoropropyl)-acetamide (EF5)
^{18}F-fluoro-misonidazole (FMiso)
^{62}Cu-/^{64}Cu-copper (II)-(diacetyl-bis (N4-methylthiosemicarbazone)) (ATSM)
^{124}I-iodo-azomycin galactopyranoside (IAZGP)
Fatty Acids and Fatty-Acid Analogs and Related Compounds
^{11}C-choline
^{11}C-palmitate
^{18}F-fluoro-choline
^{18}F-fluoro-6-thia-heptadecanoic acid (FTHA)
Miscellaneous
^{11}C-4-(3-chloroanilino)-6,7-dimethoxyquinazoline (AG1478)—epidermal growth factor (EGFR) receptor (ligand)
^{11}C-nicotine—nicotinic receptor
^{15}O-oxygen gas
^{18}F-sodium fluoride—bone
^{64}Cu-/^{86}Y/^{89}Zr-/^{124}I—antibodies and antibody fragments
^{124}I-meta-iodobenzylguanidine (MIBG)—epinephrine transporter
^{124}I-sodium iodide
Neurotransmitters and Related Compounds
Adrenergic System
^{11}C-(2S)-4-(3-t-butylamino-2-hydroxypropoxy)-benzimidazol-2-one (CGP-12177)
^{11}C-hydroxyephedrine
^{18}F-fluorometaraminol
^{18}F-fluorodopamine
Benzodiazepine (BDZ) System
^{11}C-flumazenil
^{11}C-iomezanil
^{11}C-((*R*)-1-(2-chlorophenyl)-*N*-methyl-*N*-(1-methylpropyl)-3-isoquinoline carboxamide (PK11195)
Cholinergic System
^{11}C-dexetimide
^{18}F-fluoropropyl-thiadiazolyltetrahydropyridine (FP-TZTP)
Dopaminergic System
^{11}C-(−)-2-β-Carbomethoxy-3-β-(4-fluorophenyl)tropane (β-CFT, WIN 35,428)
^{11}C-chlorgyline
^{11}C-8-chloro-7-hydroxy-3-methyl-5-(7-benzofuranyl)-2,3,4,5-tetrahydro-IH-3-benzazepine (NNC-112)
^{11}C-7-chloro-3-methyl-1-phenyl-1,2,4,5-tetrahydro-3-benzazepin-8-ol (SCH23390)

TABLE 6.4 (CONTINUED)
SPECT and PET Radiotracers and Their Applications

^{11}C-cocaine
^{11}C-L-deprenyl
^{11}C-α-(+)-dihydro-tetra-benazine
^{11}C-N-methylspiperone (NMSP)
^{11}C-raclopride
^{18}F-2-β-carbomethoxy-3-β-(4-chlorophenyl)-8-(2-fluoroethyl)nortropane (FECNT)
^{18}F-desmethoxyfallypride
^{18}F-3,4-dihydroxy-6-fluoro-L-phenylalanine (FDOPA)
^{18}F-fallypride
^{18}F-N-ω-fluoropropyl-2β-carbomethoxy-3β-(4-iodophenyl)nortropan (FP-CIT)
^{18}F-haloperidol
^{18}F-3-*O*-methyl-3,4-dihydroxy-6-fluoro-L-phenylalanine (OMFD)
^{18}F-spiroperidol
^{18}F-M-tyrosine
Opiate System
^{11}C-carfentanil
^{11}C-diprenorphine
^{18}F-cyclofoxy
Serotonergic System
^{11}C-5-hydroxytryptophan (5-HTP)
^{11}C-N-(2-(4-(2-methoxyphenyl)-1-piperazinyl)ethyl)-N-(2-pyridyl)cyclohexanecarboxamide (WAY, 100635)
^{11}C-(+)-6-β-(4-Methylthiophenyl)-1,2,3,5,6-α,10-β-hexahydropyrrolo(2,1-a)isoquinoline (McN 5652)
^{11}C-(R)-(+)-4-(1-hydroxy-1-(2,3-dimethoxyphenyl)methy1)-N-2-(4-fluorophenylethyl)piperidine (MDL 100,907)
^{18}F-altanserin
^{18}F-4-(2;-methoxyphenyl)-1-(2;-(N-2″-pirydynyl)-p-fluorobenzamido)ethylpiperazine
^{18}F-setoperone
Nucleotides and Related Compounds
^{11}C-2′-deoxy-2′-fluoro-1-β-D-arabinofuranosyluracil (FMAU)
^{11}C-thymidine
^{18}F-2′-deoxy-2′-fluoro-1-β-D-arabinofuranosyluracil (FMAU)
^{18}F-3′-fluoro-3′-deoxythymidine (FLT)
^{124}I-iodo-deoxyuridine (IUdR)
Receptor Ligands
^{18}F-fluoro-dihydrotestosterone—androgen receptors
^{18}F-fluoro-16-α-fluoro-estradiol—estrogen receptors
^{66}Ga/^{68}Ga-1,4,7,10-tetraazacyclododecane-1,4,7,10-tetraacetic acid (DOTA)-octreotide (DOTATOC)—somatostatin receptor
Reporter Gene and Gene-Expression Substrates
^{11}C-bicyclic fluoropyrimidine deoxynucleoside analogs (BCNAs)
^{18}F-bicyclic fluoropyrimidine deoxynucleoside analogs (BCNAs)
^{18}F-2-fluoro-2-deoxyarabinofuranosyl-5-ethyluracil (FEAU)
^{18}F-9-(4-fluoro-3-(hydroxymethyl)butyl)guanine (FHBG)
^{18}F-ganciclovir
^{18}F-5-iodo-2′-fluoro-2′-deoxy-1-β-D-arabinofuranosyl-5-iodouracil (FIAU)

(Continued)

TABLE 6.4 (CONTINUED)
SPECT and PET Radiotracers and Their Applications

^{124}I-5-iodo-2′-fluoro-2′-deoxy-1-β-D-arabinofuranosyl-5-iodouracil (FIAU)
^{18}F-oligonucleotides
^{18}F-penciclovir
Sugars and Sugar Analogs
^{11}C-2-deoxyglucose (DG)
^{11}C-glucose
^{11}C-3-*O*-methyl-glucose
^{18}F-fluoro-2-deoxyglucose (FDG)

Source: Silinder, M. and A.Y. Ozer., *FABAD J. Pharm. Sci.*, 33, 2008; Vallabhajosula, S., Molecular Imaging: Radiopharmaceuticals for PET and SPECT, Springer-Verlag, Berlin, 2009; With kind permission from Springer Science+Business Media: *Clinical Nuclear Medicine*, Radiochemistry and Radiopharmacy, 2007, 34–76, Guhlke, S. et al.; With kind permission from Springer Science+Business Media: *Clinical Nuclear Medicine*, 2007, Biersack, H.-J. and L.M. Freeman, editors.

[a] Among PET radiotracers, only ^{13}N-ammonia, ^{82}Rb-rubidium chloride, ^{18}F-sodium fluoride, and ^{18}F-fluoro-2-deoxyglucose (FDG) are currently approved for routine clinical use by the United States Food and Drug Administration (FDA).

applications as innumerable as the number of SPECT and PET radiotracers. A comprehensive discussion of the applications of SPECT and PET is therefore beyond the scope of this chapter. Several applications—of clinical and preclinical SPECT, PET, and multimodality imaging—are nonetheless presented to illustrate the power and versatility of radionuclide-based molecular imaging.

6.11.1 PET Imaging of Alzheimer's Disease

Figure 6.24 compares PET images of ^{11}C-Pittsburgh compound B (PIB) [88] and ^{18}FDG PET scans of two older individuals, a healthy control subject and an Alzheimer's disease (AD) patient [89]. PIB binds to the pathologic β-amyloid plaques characteristic of AD, and the PIB images demonstrate a marked difference in uptake and retention in the brain between healthy control and AD individuals. In the healthy controls, there is a notable lack of PIB uptake in the entire gray matter (top left), consistent with absence of plaques, and normal cerebral glucose metabolism (bottom left). There is high PIB uptake in the frontal and temporoparietal cortices of the AD patient (top right) and a typical pattern of FDG hypometabolism present in the temporoparietal cortex (arrows, bottom right) along with preserved metabolism in the frontal cortex. The PIB and FDG scans were obtained within 3 days of each other. PET imaging techniques that label and thus allow noninvasive visualization of β-amyloid protein *in vivo* are promising approaches to the early diagnosis of AD, a devastating and increasingly widespread affliction of our aging population. Among the applications of PET will be earlier and more specific diagnosis of AD and assessment by serial (i.e., pretreatment versus posttreatment) imaging of new therapeutic interventions. A noteworthy feature of the study shown in the lower row in Figure 6.24 presents parametric maps of the [^{18}F]FDG image set in terms of the regional rate of cerebral glucose metabolism (rCMRglc in mol/min/100 g), illustrating the application of pharmacokinetic modeling to dynamic radionuclide imaging studies to derive physiological parameters in fully quantitative terms.

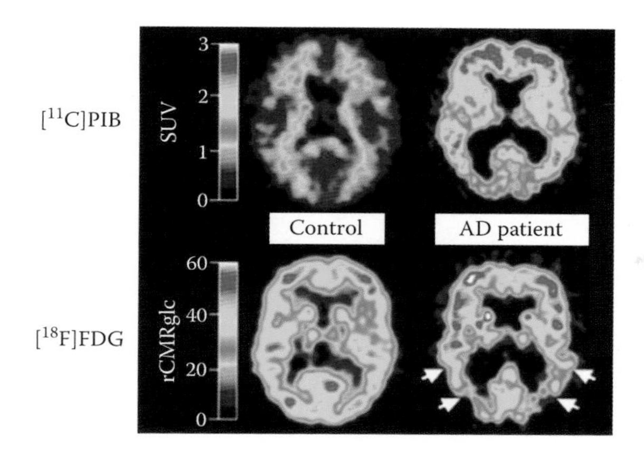

FIGURE 6.24 ^{11}C-Pittsburgh compound B (PIB) and ^{18}F-^{18}F-fluoro-2-deoxyglucose (FDG) PET scans (transverse-section images through brain) of a 67-year-old healthy control (HC) subject and a 79-year-old Alzheimer's dementia (AD) patient. Standardized uptake value (SUV) PIB images (top row) summed over 40–60 min postinjection and FDG images of the regional rate of cerebral glucose metabolism (rCMRglc in mol/min/100 g) (bottom row) are shown. (Adapted from Klunk, W.E. et al., *Ann. Neurol.*, 55, 2004. With permission.)

6.11.2 SPECT and PET Imaging of Myocardial Perfusion

The long-standing efficacy and cost-effectiveness of myocardial perfusion imaging by SPECT and its role in the management of coronary artery disease, as illustrated in Figure 6.25, are firmly established. PET may provide additional clinical and technical benefits for myocardial perfusion imaging, including better spatial resolution, higher sensitivity, use of more highly extracted radiotracers, and perhaps greater diagnostic accuracy overall [90]. PET myocardial perfusion imaging protocols are also shorter in duration and deliver a lower radiation dose [90]. Further, with more accurate and robust attenuation and scatter corrections, PET perfusion images may be parameterized in absolute quantitative terms (e.g., as blood flow in mL/min/g). Presently, SPECT remains the modality of choice for myocardial perfusion imaging in routine clinical practice. Figure 6.25 compares ^{99m}Tc-MIBI SPECT and ^{82}Rb-rubidium chloride PET images of stress and rest myocardial perfusion in the same patient [90]. The visualization and discrimination of ischemic and normally perfused myocardium are more apparent in the PET than in the SPECT images. Note, in particular, the suspicious area (white arrow) in the inferior wall, where there is apparently reduced uptake of ^{99m}Tc-MIBI; this patient weighs approximately 300 lb., and this may be a diaphragmatic attenuation artifact. This possible defect is not seen in the higher-quality, attenuation-corrected PET images, suggesting that this suspicious area is likely an attenuation artifact rather than ischemic myocardium. With the availability of SPECT-CT and PET-CT scanners and the ability to readily register and fuse images of cardiac perfusion and function (SPECT and PET) and anatomy (CT), imaging-based management of cardiac disease continues to advance. Among the applications of SPECT-CT and PET-CT are coronary calcium measurements combined with perfusion studies, clinically important because of the possibility of discrepant results, that is, normal myocardial perfusion with a high coronary calcium score or abnormal perfusion and little or no coronary calcium [90]. In addition to coronary calcium scoring, the multislice CT subsystem of a SPECT-CT or

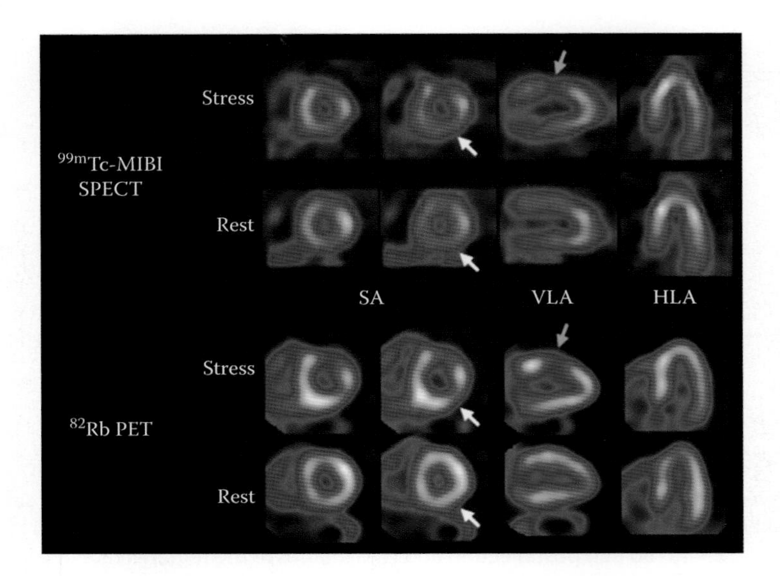

FIGURE 6.25 ^{99m}Tc-MIBI and ^{82}Rb-rubidium chloride PET images of stress and rest myocardial perfusion in the same patient. In each pair of left ventricle images, the two left columns are short axis (SA) views and the columns labeled "VLA" and "HLA" vertical and horizontal long axis views, respectively. An anterior wall perfusion defect is identified by the blue arrow and a suspicious, but likely normal, inferior wall area by the white arrow. (From Bengel, F.M., Integrated Assessment of Myocardial Perfusion and Coronary Anatomy by PET-CT, in *PET Imaging: Status Quo and Outlook*, Review articles based on a corporate Bayer Schering Pharma Symposium at EANM 2008, Bayer Healthcare, Bayer Schering Pharma, Berlin, 2009. With permission.)

PET-CT scanner can be used to perform contrast coronary angiography. Figure 6.26, for example, illustrates a potentially important diagnostic paradigm in cardiology [90]. In the case shown, all three coronary arteries exhibit stenoses (outlined in yellow) by CT coronary angiography. At the same time, the PET stress perfusion study identifies the myocardium perfused by the right coronary artery (RCA) as the functionally affected vascular territory. This suggests that a stress perfusion study, combined with CT angiography, might be sufficient for definitive diagnosis and localization of coronary artery disease, thereby eliminating the need for conventional coronary angiography.

6.11.3 [^{18}F]FDG PET Imaging in Oncology

[^{18}F]FDG, by far the most widely used PET radiotracer, has profoundly impacted the clinical management of cancer patients. An important application of [^{18}F]FDG PET-CT in oncology, determination of the extent of disease, and of its impact on patient management is illustrated in Figure 6.27 [91]. Reading PET and CT images separately or in the juxtaposed format shown (Figure 6.27b), it is difficult to definitively identify the anatomic site (i.e., tumor versus normal structure) of the focus of activity in the neck. The overlayed PET and CT images (Figure 6.27c), on the other hand, unambiguously demonstrate that the FDG activity is located within muscle, a physiological normal variant. The FDG activity in the neck thus was *not* previously undetected disease, a finding that would have significantly impacted the subsequent clinical management of the patient.

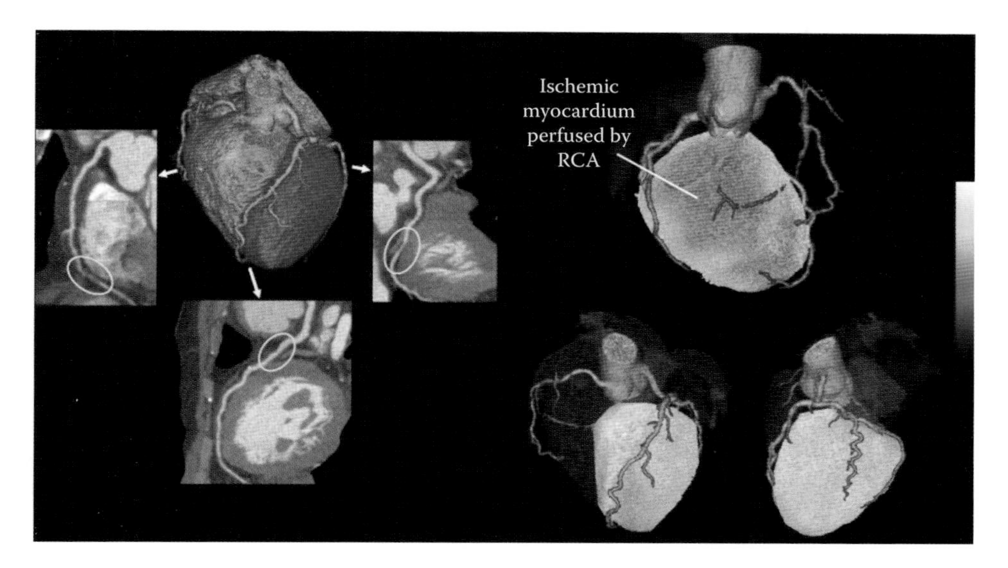

FIGURE 6.26 PET-CT study combining PET stress myocardial perfusion imaging (three thermal-scale images on right) and CT coronary angiography (three grayscale images on left). The registered contrast angiograms are superimposed on the PET perfusion images (right) and a 3-D volume rendering of the coronary anatomy (colorized image on the left) are also shown. Three coronary artery stenoses are outlined in yellow on the angiograms (left), and the area of myocardial ischemia, the vascular territory perfused by the right coronary artery (RCA), is also indicated (red-tone area on top image on right). (From Bengel, F.M., Integrated Assessment of Myocardial Perfusion and Coronary Anatomy by PET-CT, in *PET Imaging: Status Quo and Outlook*, Review articles based on a corporate Bayer Schering Pharma Symposium at EANM 2008, Bayer Healthcare, Bayer Schering Pharma, Berlin, 2009. With permission.)

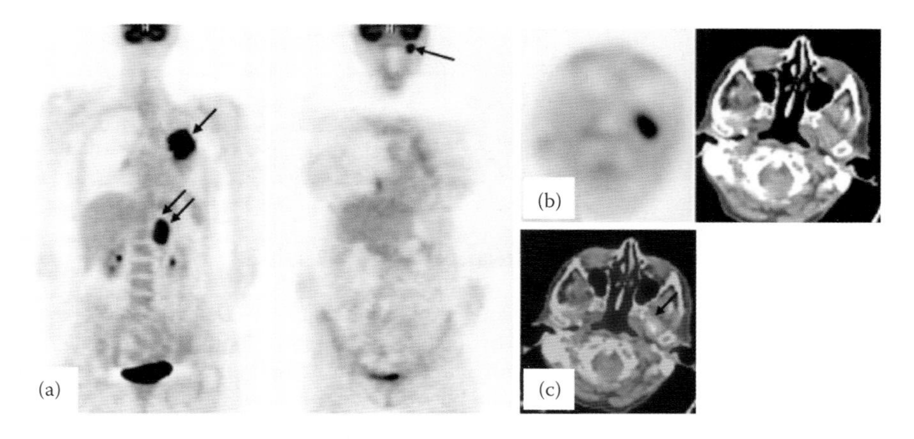

FIGURE 6.27 Registered and fused FDG PET and CT scans of a patient with lung cancer and an adrenal gland metastasis. (a) Coronal PET images show increased FDG uptake in the primary lung tumor (single arrow in left panel) and in the metastasis in the left adrenal gland (double arrow in left panel) but also in an area in the left side of the neck (arrow in right panel). (b) Transaxial PET and CT images through the focus of activity in the neck. (c) The registered PET-CT images, using the fused, or overlay, display. The arrow identifies the location in the neck of this unusual, but non-pathologic, focus of FDG activity on the fused images. (Adapted from Schoder, H. et al., *Eur. J. Nucl. Med. Mol. Imaging*, 30, 2003. With permission.)

6.11.4 Preclinical PET-Based Pharmacodynamic Imaging of Tumor Response to an Inhibitor of Heat Shock Protein 90

The development of inhibitors of key oncogenic signaling pathways as targeted therapies of cancer would be expedited by the ability to assess the effect of a drug on its target *in vivo*. Radionuclide-based molecular imaging, as illustrated by the preclinical PET study shown in Figure 6.28 [92], is a promising approach to such pharmacodynamic assessment of anticancer drugs. The heat shock protein 90 (HSP90) inhibitor 17-allylaminogeldanamycin (17-AAG) causes the degradation of human epidermal growth factor receptor 2 (HER2), overexpressed on certain aggressive breast cancers, and other oncogenic HSP90 client proteins, and has anti–breast cancer activity in preclinical models. Smith-Jones et al. [92] developed a PET-based method for quantitatively imaging the inhibition of HSP90 by 17-AAG, labeling the $F(ab')_2$ fragment of the anti-HER2 antibody Herceptin with the short-lived positron emitter ^{68}Ga. This method was used to noninvasively quantify the time-dependent 17-AAG-induced loss and recovery of HER2 in breast tumor xenografts in mice. The clinically translatable paradigm illustrated in Figure 6.28 allows noninvasive imaging of the pharmacodynamics of a molecularly targeted drug and should facilitate

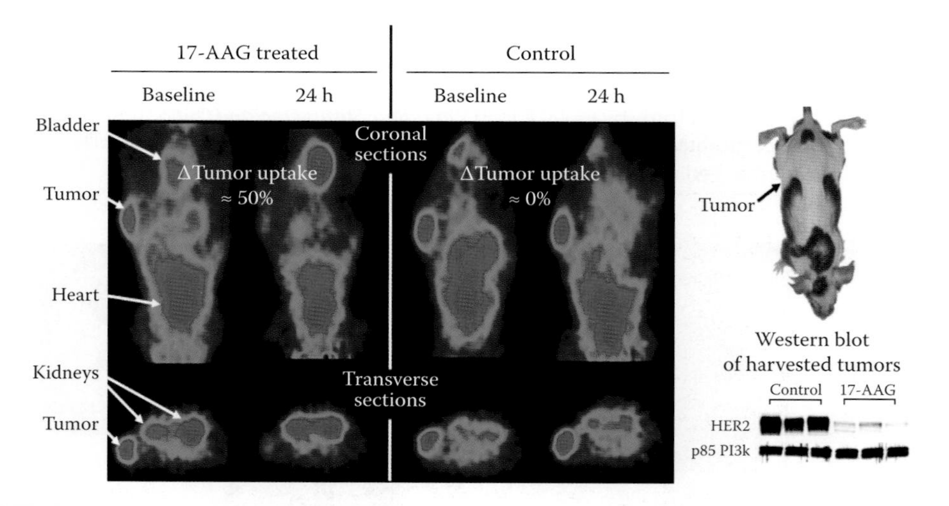

FIGURE 6.28 PET imaging of the pharmacodynamic effect of the HSP90 inhibitor 17-AAG (150 mg/kg) on expression of HER2 in BT474 breast tumor xenografts (right thigh) in mice. Imaging was performed 3 h post-intravenous injection of a ^{68}Ga-labeled $F(ab')_2$ fragment of Herceptin, an anti-HER2 antibody. For the 17-AAG-treated animals, the radiotracer was injected and imaging performed before (i.e., at baseline) and again 24 h after the 17-AAG administration; for the control (i.e., untreated) animals, repeat radiotracer injections and imaging were performed 24 h apart. The coronal PET images (upper row of images in the left panel) are oriented as shown in the mouse photograph to the right, and the transverse images (lower row of images in the left panel) are oriented with the dorsal surfaces of the animals to the top and the ventral surfaces to the bottom. The 17-AAG treatment induced a 50% reduction in the tumor uptake of the ^{68}Ga-labeled $F(ab')_2$ fragment of Herceptin at 24 h posttreatment, while there was no difference in repeat tumor uptakes measured in controls (animals). The Western blot (lower right) corroborates the 17-AAG-induced reduction in HER2 levels. (Adapted from Smith-Jones, P.M. et al., *J. Nucl. Med.*, 47, 2006.)

the rational design and dose and dose-schedule optimization of therapies based on target inhibition. Among other considerations, the study shown in Figure 6.28 illustrates the important advantage of quantitation provided by radionuclide imaging in general and PET in particular.

6.11.5 Preclinical SPECT Imaging of Progression of Bone Tumor Metastasis

A debilitating and painful consequence of advanced prostate and other cancers is the development of bone metastases. At the same time, it has been observed that many cancer cell lines as well as primary human tumors synthesize bombesin, which appears to act in an autocrine fashion to stimulate the growth of the tumor cells it originated from through membrane bombesin receptors. In a preclinical model in mice, Winkelman et al. [93] demonstrated that CT combined with SPECT of a bombesin receptor (BB2)-binding radioligand, ^{111}In-DOTA-8-Aoc-BBN(7-14)NH_2, can provide a combined structural and functional map—skeletal anatomy and bombesin receptor status—of metastatic bone lesions, as illustrated in Figure 6.29. By directly targeting metastatic tumor cells in bone using a specific (e.g., bombesin) receptor-binding radioligand, rather than observing by CT, the secondary effect of osteolysis, more sensitive and specific early diagnosis of skeletal metastases may be possible.

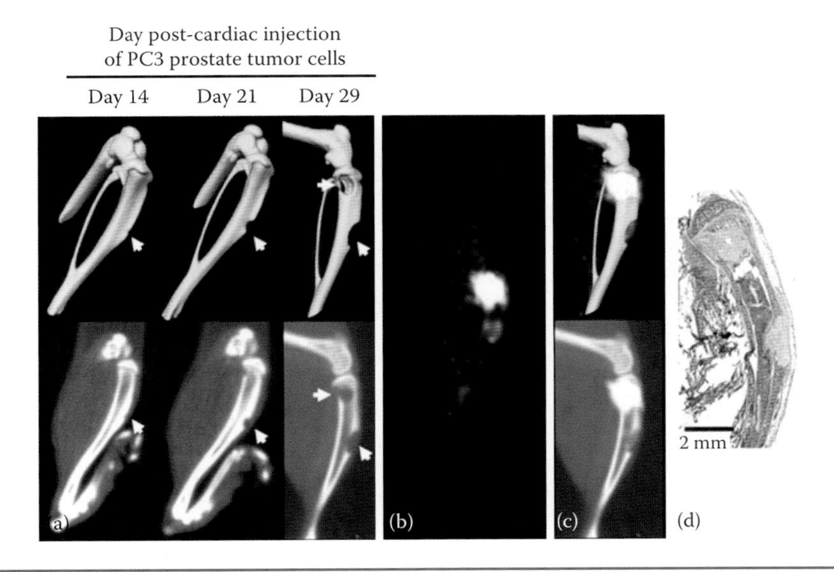

FIGURE 6.29 (a) Progression of two metastases to the tibia (arrows) visualized by CT surface renderings (top row) and sagittal images (bottom row) following intracardiac injection of PC3 prostate tumor cells. (b) Pinhole SPECT image obtained at 1 h postinjection of ^{111}In-DOTA-8-AOC-BBN(7-14) NH2. (c) Overlay of anatomic CT images with the radiotracer SPECT image, showing that the foci of radiotracer uptake correspond to the lytic bone metastases. (d) Photomicrograph of the histopathology of the two bone metastases. (From Winkelmann, C.T. et al., *Mol. Imaging Biol.*, 2012. With permission.)

6.12 CONCLUDING REMARKS

The use of nuclear medicine, particularly PET, has grown dramatically over the last several decades, with the annual number of nuclear medicine procedures increasing threefold (from 7 million to 20 million) between 1985 and 2005 [94]. The recent introduction of SPECT-MRI [95–97] and PET-MRI [98–101] multimodality devices will no doubt lead to new and important applications of radionuclide imaging [102], as SPECT or PET and MRI studies (including dynamic studies) can be performed simultaneously, at least in principle. This is in contrast to SPECT-CT and PET-CT, where there are temporal and spatial offsets between the SPECT or PET study and the CT study. It should be noted that in some of the PET-MRI scanners currently being marketed, the PET and MRI sub-systems are, in fact, offset from one another and the scans performed sequentially rather than simultaneously. It should be further noted that while clinical and preclinical PET-MRI devices are now commercially available, SPECT-MRI scanners are still very much in the developmental stage.

Although the spatial resolution of SPECT (~1 and ~10 mm for preclinical and clinical devices, respectively) and of PET (1–2 and ~5 mm for preclinical and clinical devices, respectively) is excellent by historical standards for these modalities, it remains relatively coarse compared to that of such high-resolution "anatomic" imaging modalities as CT and MRI (0.1–1 mm). Nonetheless, the distinctive and important advantages of radionuclide-based imaging—high detection sensitivity, "image-ability" of nonperturbing doses of radiotracers, quantitative accuracy (optimally ~10%), and a vast array of radiotracers—ensure that SPECT and PET (particularly in combination with CT or MRI) will remain invaluable molecular imaging modalities in clinical practice and in clinical and preclinical research.

REFERENCES

1. Mankoff, D. A. 2007. A definition of molecular imaging. *J Nucl Med* 48:18N, 21N.
2. Mettler, F. A., Jr., W. Huda, T. T. Yoshizumi, and M. Mahesh. 2008. Effective doses in radiology and diagnostic nuclear medicine: A catalog. *Radiology* 248:254–263.
3. Zanzonico, P. 2012. Principles of nuclear medicine imaging: Planar, SPECT, PET, multi-modality, and autoradiography systems. *Radiat Res* 177:349–364.
4. Zanzonico, P., and S. Heller. 2007. Physics, instrumentation, and radiation protection. In *Clinical Nuclear Medicine*, H.-J. Biersack, and L. M. Freeman, editors. Springer, Heidelberg, 1–33.
5. Otte, N. 2006. The silicon photomultiplier—A new device for high energy physics, astroparticle physics, Industrial and medical applications. In *SNIC Symposium*. Stanford University, Palo Alto, CA.
6. Humm, J. L., A. Rosenfeld, and A. Del Guerra. 2003. From PET detectors to PET scanners. *Eur J Nucl Med Mol Imaging* 30:1574–1597.
7. Zanzonico, P. 2004. Positron emission tomography: A review of basic principles, scanner design and performance, and current systems. *Semin Nucl Med* 34:87–111.
8. Zanzonico, P., and S. Heller. 2000. The intraoperative gamma probe: Basic principles and choices available. *Semin Nucl Med* 30:33–48.
9. Heller, S., and P. Zanzonico. 2011. Nuclear probes and intraoperative gamma cameras. *Semin Nucl Med* 41:166–181.
10. Zanzonico, P. 2008. Routine quality control of clinical nuclear medicine instrumentation: A brief review. *J Nucl Med* 49:1114–1131.
11. NEMA. 2001. *Performance Measurements of Scintillation Counters*. NEMA Standards Publication NU1-2001. National Electrical Manufacturers Association (NEMA), Rosslyn, VA.
12. NEMA. 2001. *Performance Measurements of Positron Emission Tomographs*. NEMA Standards Publication NU 2-2001. National Electrical Manufacturers Association (NEMA), Rosslyn, VA.

13. Hines, H., R. Kayayan, J. Colsher, D. Hashimoto, R. Schubert, J. Fernando, V. Simcic, P. Vernon, and R. L. Sinclair. 2000. National Electrical Manufacturers Association recommendations for implementing SPECT instrumentation quality control. *J Nucl Med* 41:383–389.
14. Hines, H., R. Kayayan, J. Colsher, D. Hashimoto, R. Schubert, J. Fernando, V. Simcic, P. Vernon, and R. L. Sinclair. 1999. Recommendations for implementing SPECT instrumentation quality control. Nuclear Medicine Section—National Electrical Manufacturers Association (NEMA). *Eur J Nucl Med* 26:527–532.
15. Firestone, R. B., and V. S. Shirley, editors. 1996. *Table of Isotopes*, 8th ed. John Wiley & Sons, New York.
16. Weber, D., K. Eckerman, L. Dillman, and J. Ryman. 1989. *MIRD: Radionuclide Data and Decay Schemes*. Society of Nuclear Medicine, New York.
17. Graham, M. C., K. S. Pentlow, O. Mawlawi, R. D. Finn, F. Daghighian, and S. M. Larson. 1997. An investigation of the physical characteristics of 66Ga as an isotope for PET imaging and quantification. *Med Phys* 24:317–326.
18. Pentlow, K. S., R. D. Finn, S. L. Larson, Y. E. Erdi, B. J. Beattie, and J. L. Humm. 2000. Quantitative imaging of yttrium-86 with PET: The occurrence and correction of anomalous apparent activity in high density regions. *Clin Positron Imaging* 3:85–90.
19. Saha, G. S. 1993. *Physics and Radiobiology of Nuclear Medicine*. Springer-Verlag, New York, 107–123.
20. Zanzonico, P. B. 1995. Technical requirements for SPECT: Equipment and quality control. In *Clinical Applications in SPECT*, E. L. Kramer, and J. J. Sanger, editors. Raven Press, New York, 7–41.
21. Frey, E. C., J. L. Humm, and M. Ljungberg. 2012. Accuracy and precision of radioactivity quantification in nuclear medicine images. *Semin Nucl Med* 42:208–218.
22. Tsui, B. M., X. Zhao, E. C. Frey, and W. H. McCartney. 1994. Quantitative single-photon emission computed tomography: Basics and clinical considerations. *Semin Nucl Med* 24:38–65.
23. Dewaraja, Y. K., E. C. Frey, G. Sgouros, A. B. Brill, P. Roberson, P. B. Zanzonico, and M. Ljungberg. 2012. MIRD pamphlet no. 23: Quantitative SPECT for patient-specific 3D dosimetry in internal radionuclide therapy. *J Nucl Med* 53:1310–1325.
24. Townsend, D. W., and B. Bendriem. 1998. Introduction to 3D PET. In *The Theory and Practice of 3D PET*, B. Bendriem, and D. W. Townsend, editors. Kluwer Academic Publishers, Dordrecht, The Netherlands, 1–10.
25. Cherry, S. R., J. A. Sorenson, and M. E. Phelps. 2003. *Physics in Nuclear Medicine*. Saunders, Philadelphia, PA.
26. Mosset, J. B., O. Devroede, M. Krieguer, M. Rey, J. M. Vieira, J. H. K. Jung, C. Kuntner, M. Z. Streun, K. Ziemons, E. Auffray, P. Sempere-Roldan, P. Lecoq, P. L. F. Bruyndonckx, S. Tavernier, and C. Morel. 2004. Development of an optimised LSO/LuYAP phoswich detector head for the ClearPET camera. In *Nuclear Science Symposium Conference Record, 2004 IEEE*, 2439–2443.
27. Lewellen, T. K. 1998. Time-of-flight PET. *Semin Nucl Med* 28:268–275.
28. Moses, W. W. 2007. Recent advances and future advances in time-of-flight PET. *Nucl Instrum Methods Phys Res A* 580:919–924.
29. Karp, J. S., S. Surti, M. E. Daube-Witherspoon, and G. Muehllehner. 2008. Benefit of time-of-flight in PET: Experimental and clinical results. *J Nucl Med* 49:462–470.
30. Levin, C. S., and E. J. Hoffman. 1999. Calculation of positron range and its effect on the fundamental limit of positron emission tomography system spatial resolution. *Phys Med Biol* 44:781–799.
31. Derenzo, S. E. 1986. Mathematical removal of positron range blurring in high-resolution tomography. *IEEE Trans Nucl Sci* NS-33:565–569.
32. Berko, S., and F. L. Hereford. 1956. Experimental studies of positron interactions in solids and liquids. *Rev Modern Phys* 28:299–307.
33. Yang, Y., Y. Wu, J. Qi, S. St James, H. Du, P. A. Dokhale, K. S. Shah, R. Farrell, and S. R. Cherry. 2008. A prototype PET scanner with DOI-encoding detectors. *J Nucl Med* 49:1132–1140.
34. Greer, K., R. J. Jaszczak, C. Harris, and R. E. Coleman. 1985. Quality control in SPECT. *J Nucl Med Technol* 13:76–85.
35. Harkness, B. A., W. L. Rogers, N. H. Clinthorne, and J. W. Keyes, Jr. 1983. SPECT: Quality control and artifact identification. *J Nucl Med Technol* 11:55–60.

36. Hoffman, E. J., S. C. Huang, M. E. Phelps, and D. E. Kuhl. 1981. Quantitation in positron emission computed tomography: 4. Effect of accidental coincidences. *J Comput Assist Tomogr* 5:391–400.
37. Rokitta, O., M. Casey, K. Wienhard, and U. Pictrzyk. 2000. Random correction for positron emission tomography using singles count rates. In *Nuclear Science Symposium Conference Record, 2000 IEEE*, D. Merelli, J. Surget, and M. Ulma, editors. Institute of Electrical and Electronics Engineers, Lyon, France, 17/37–17/40.
38. Meikle, S. R., and R. D. Badawi. 2005. Quantitative techniques in PET. In *Positron Emission Tomography: Basic Sciences*, D. L. Bailey, D. W. Townsend, P. E. Valk, and M. N. Maisey, editors. Springer-Verlag, London, 93–126.
39. Bailey, D. L. 1998. Quantitative procedures in 3D PET. In *The Theory and Practice of 3D PET*, B. Bendriem, and D. W. Townsend, editors. Kluwer Academic Publishers, Dordrecht, The Netherlands, 55–109.
40. Cherry, S. R., and S. C. Huang. 1995. Effects of scatter on model parameter estimates in 3D PET studies of the human brain. *IEEE Trans Nucl Sci* 42:1174–1179.
41. Stearns, C. W. 1995. Scatter correction method for 3D PET using 2D fitted Gaussian functions. *J Nucl Med* 36:105P.
42. Ogawa, K., Y. Harata, T. Ichihara, A. Kubo, and S. Hashimoto. 1991. A practical method for position-dependent Compton-scatter correction in single photon-emission CT. *IEEE Trans Med Imaging* 10:408–412.
43. Frey, E. C., and B. Tsui. 1996. A new method for modeling the spatially-variant, object-dependent scatter response function in SPECT. *IEEE* 1082:1082–1086.
44. Beekman, F. J., H. W. de Jong, and S. van Geloven. 2002. Efficient fully 3-D iterative SPECT reconstruction with Monte Carlo-based scatter compensation. *IEEE Trans Med Imaging* 21:867–877.
45. Dewaraja, Y. K., M. Ljungberg, and J. A. Fessler. 2006. 3-D Monte Carlo-based scatter compensation in quantitative I-131 SPECT reconstruction. *IEEE Trans Nucl Sci* 53:181–188.
46. Ouyang, J., G. El Fakhri, and S. C. Moore. 2008. Improved activity estimation with MC-JOSEM versus TEW-JOSEM in 111In SPECT. *Med Phys* 35:2029–2040.
47. Shcherbinin, S., A. Celler, T. Belhocine, R. Vanderwerf, and A. Driedger. 2008. Accuracy of quantitative reconstructions in SPECT/CT imaging. *Phys Med Biol* 53:4595.
48. Chang, L. T. 1978. A method for attenuation correction in radionuclide computed tomography. *IEEE Trans Nucl Sci* 25:638–643.
49. Sorenson, J. A. 1974. Quantitative measurement of radioactivity in whole-body counting. In *Instrumentation of Nuclear Medicine*, G. J. Hine, and J. A. Soresnon, editors. Academic Press, Waltham, MA, 311–348.
50. Israel, O., and S. J. Goldsmith, editors. 2006. *Hybrid SPECT/CT: Imaging in Clinical Practice*. Taylor & Francis, New York.
51. Defrise, M., and P. Kinahan. 1998. Data acquisition and image reconstruction for 3D PET. In *The Theory and Practice of 3D PET*, B. Bendriem, and D. W. Townsend, editors. Kluwer Academic Publishers, Dordrecht, The Netherlands, 11–53.
52. Colsher, J. G. 1980. Fully thee-dimensional positron emission tomography. *Phys Med Biol* 20:103–115.
53. Kinahan, P., and J. G. Rogers. 1989. Analytic three-dimensional image reconstruction using all detected events. *IEEE Trans Nucl Sci* NS-36:964–968.
54. Defrise, M., P. E. Kinahan, and C. J. Michel. 2005. Image reconstruction algorithms in PET. In *Positron Emission Tomography: Basic Sciences*, D. L. Bailey, D. W. Townsend, P. E. Valk, and M. N. Maisey, editors. Springer, London, 63–91.
55. Miller, T. R., and J. W. Wallis. 1992. Fast maximum-likelihood reconstruction. *J Nucl Med* 33:1710–1711.
56. Miller, T. R., and J. W. Wallis. 1992. Clinically important characteristics of maximum-likelihood reconstruction. *J Nucl Med* 33:1678–1684.
57. Hudson, H. M., and R. S. Larkin. 1994. Accelerated image reconstruction using ordered subsets of projection data. *IEEE Trans Med Imaging* 13:601–609.
58. Tarantola, G., F. Zito, and P. Gerundini. 2003. PET instrumentation and reconstruction algorithms in whole-body applications. *J Nucl Med* 44:756–769.

59. Matej, S., and R. M. Lewitt. 1996. Efficient 3D grids for image reconstruction using speherical-symmetric volume elements. *IEEE Trans Nucl Sci* 42:1361–1370.
60. Matej, S., and R. M. Lewitt. 1996. Practical considerations for 3-D image reconstruction using spherically symemetric volume elements. *IEEE Trans Med Imaging* 15:68–78.
61. Daube-Witherspoon, M. E., S. Matej, J. S. Karp, and R. M. Lewitt. 2001. Application of the 3D row action maximum likelihood algorithm to clinical PET imaging. *IEEE Trans Nucl Sci* 48:24–30.
62. Daube-Witherspoon, M. E., and G. Muehllehner. 1987. Treatment of axial data in three-dimensional PET. *J Nucl Med* 28:1717–1724.
63. Lewitt, R. M., G. Muehllehner, and J. S. Karp. 1994. Three-dimensional image reconstruction for PET by multi-slice rebinning and axial image filtering. *Phys Med Biol* 39:321–339.
64. Defrise, M., P. E. Kinahan, D. W. Townsend, C. Michel, M. Sibomana, and D. F. Newport. 1997. Exact and approximate rebinning algorithms for 3-D PET data. *IEEE Trans Med Imaging* 16:145–158.
65. Hoffman, E. J., S. C. Huang, and M. E. Phelps. 1979. Quantitation in positron emission computed tomography: 1. Effect of object size. *J Comput Assist Tomogr* 3:299–308.
66. Erlandsson, K., I. Buvat, P. H. Pretorius, B. A. Thomas, and B. F. Hutton. 2012. A review of partial volume correction techniques for emission tomography and their applications in neurology, cardiology and oncology. *Phys Med Biol* 57:R119–R159.
67. Nehmeh, S. A., and Y. E. Erdi. 2008. Respiratory motion in positron emission tomography/computed tomography: A review. *Semin Nucl Med* 38:167–176.
68. Nehmeh, S. A., Y. E. Erdi, C. C. Ling, K. E. Rosenzweig, H. Schoder, S. M. Larson, H. A. Macapinlac, O. D. Squire, and J. L. Humm. 2002. Effect of respiratory gating on quantifying PET images of lung cancer. *J Nucl Med* 43:876–881.
69. Nehmeh, S. A., Y. E. Erdi, C. C. Ling, K. E. Rosenzweig, O. D. Squire, L. E. Braban, E. Ford, K. Sidhu, G. S. Mageras, S. M. Larson, and J. L. Humm. 2002. Effect of respiratory gating on reducing lung motion artifacts in PET imaging of lung cancer. *Med Phys* 29:366–371.
70. Nehmeh, S. A., Y. E. Erdi, K. E. Rosenzweig, H. Schoder, S. M. Larson, O. D. Squire, and J. L. Humm. 2003. Reduction of respiratory motion artifacts in PET imaging of lung cancer by respiratory correlated dynamic PET: Methodology and comparison with respiratory gated PET. *J Nucl Med* 44:1644–1648.
71. Sgouros, G., E. Frey, R. Wahl, B. He, A. Prideaux, and R. Hobbs. 2008. Three-dimensional imaging-based radiobiological dosimetry. *Semin Nucl Med* 38:321–334.
72. Flux, G., M. Bardies, M. Monsieurs, S. Savolainen, S. E. Strands, and M. Lassmann. 2006. The impact of PET and SPECT on dosimetry for targeted radionuclide therapy. *Z Med Phys* 16:47–59.
73. Cherry, S. R., J. A. Sorenson, and M. E. Phelps. 2012. *Physics in Nuclear Medicine*. Saunders, Philadelphia, PA.
74. Townsend, D. W. 2001. A combined PET/CT scanner: The choices. *J Nucl Med* 42:533–534.
75. Townsend, D. W., and T. Beyer. 2002. A combined PET/CT scanner: The path to true image fusion. *Br J Radiol* 75 Spec No:S24–S30.
76. Townsend, D. W., T. Beyer, and T. M. Blodgett. 2003. PET/CT scanners: A hardware approach to image fusion. *Semin Nucl Med* 33:193–204.
77. Townsend, D. W. 2008. Positron emission tomography/computed tomography. *Semin Nucl Med* 38:152–166.
78. Horii, S. 1993. ACR-NEMA DICOM support for exchange media: A report on the activity of Working Group V. *Adm Radiol* 12:68–69.
79. Bidgood, W. D., Jr., and S. C. Horii. 1992. Introduction to the ACR-NEMA DICOM standard. *Radiographics* 12:345–355.
80. Kiessling, F., B. J. Pichler, and P. Hauff, editors. 2011. *Small Animal Imaging*. Springer-Verlag, Berlin.
81. Moadel, R. M. 2011. Breast cancer imaging devices. *Semin Nucl Med* 41:229–241.
82. MacDonald, L., J. Edwards, T. Lewellen, D. Haseley, J. Rogers, and P. Kinahan. 2009. Clinical imaging characteristics of the positron emission mammography camera: PEM Flex Solo II. *J Nucl Med* 50:1666–1675.
83. Travin, M. I. 2011. Cardiac cameras. *Semin Nucl Med* 41:182–201.

84. Silinder, M., and A. Y. Ozer. 2008. Recently developed radiopharmaceuticals for positron emission tomography (PET). *FABAD J Pharm Sci* 33:153–162.
85. Guhlke, S., A. M. Verbruggen, and S. Vallabhajosula. 2007. Radiochemistry and radiopharmacy. In *Clinical Nuclear Medicine*, H.-J. Biersack, and L. M. Freeman, editors. Springer, Heidelberg, 34–76.
86. Vallabhajosula, S. 2009. *Molecular Imaging: Radiopharmaceuticals for PET and SPECT.* Springer-Verlag, Berlin.
87. Biersack, H.-J., and L. M. Freeman, editors. 2007. *Clinical Nuclear Medicine.* Springer, Heidelberg.
88. Mathis, C. A., Y. Wang, D. P. Holt, G. F. Huang, M. L. Debnath, and W. E. Klunk. 2003. Synthesis and evaluation of 11C-labeled 6-substituted 2-arylbenzothiazoles as amyloid imaging agents. *J Med Chem* 46:2740–2754.
89. Klunk, W. E., H. Engler, A. Nordberg, Y. Wang, G. Blomqvist, D. P. Holt, M. Bergstrom, I. Savitcheva, G. F. Huang, S. Estrada, B. Ausen, M. L. Debnath, J. Barletta, J. C. Price, J. Sandell, B. J. Lopresti, A. Wall, P. Koivisto, G. Antoni, C. A. Mathis, and B. Langstrom. 2004. Imaging brain amyloid in Alzheimer's disease with Pittsburgh Compound-B. *Ann Neurol* 55:306–319.
90. Bengel, F. M. 2009. Integrated assessment of myocardial perfusion and coronary anatomy by PET-CT. In *PET Imaging: Status Quo and Outlook*, Review articles based on a corporate Bayer Schering Pharma Symposium at EANM 2008. Bayer Healthcare, Bayer Schering Pharma, Berlin, 18–25.
91. Schoder, H., Y. E. Erdi, S. M. Larson, and H. W. Yeung. 2003. PET/CT: A new imaging technology in nuclear medicine. *Eur J Nucl Med Mol Imaging* 30:1419–1437.
92. Smith-Jones, P. M., D. Solit, F. Afroze, N. Rosen, and S. M. Larson. 2006. Early tumor response to Hsp90 therapy using HER2 PET: Comparison with 18F-FDG PET. *J Nucl Med* 47:793–796.
93. Winkelmann, C. T., S. D. Figueroa, G. L. Sieckman, T. L. Rold, and T. J. Hoffman. 2012. Non-invasive microCT imaging characterization and in vivo targeting of BB2 receptor expression of a PC-3 bone metastasis model. *Mol Imaging Biol* 14:667–675.
94. Amis, E. S., Jr., P. F. Butler, K. E. Applegate, S. B. Birnbaum, L. F. Brateman, J. M. Hevezi, F. A. Mettler, R. L. Morin, M. J. Pentecost, G. G. Smith, K. J. Strauss, and R. K. Zeman. 2007. American College of Radiology white paper on radiation dose in medicine. *J Am Coll Radiol* 4:272–284.
95. Ha, S., M. J. Hamamura, W. W. Roeck, L. T. Muftuler, and O. Nalcioglu. 2010. Development of a new RF coil and gamma-ray radiation shielding assembly for improved MR image quality in SPECT/MRI. *Phys Med Biol* 55:2495–2504.
96. Hamamura, M. J., S. Ha, W. W. Roeck, L. T. Muftuler, D. J. Wagenaar, D. Meier, B. E. Patt, and O. Nalcioglu. 2010. Development of an MR-compatible SPECT system (MRSPECT) for simultaneous data acquisition. *Phys Med Biol* 55:1563–1575.
97. Hamamura, M. J., S. Ha, W. W. Roeck, D. J. Wagenaar, D. Meier, B. E. Patt, and O. Nalcioglu. 2010. Initial Investigation of preclinical integrated SPECT and MR imaging. *Technol Cancer Res Treat* 9:21–28.
98. Judenhofer, M. S., H. F. Wehrl, D. F. Newport, C. Catana, S. B. Siegel, M. Becker, A. Thielscher, M. Kneilling, M. P. Lichy, M. Eichner, K. Klingel, G. Reischl, S. Widmaier, M. Rocken, R. E. Nutt, H. J. Machulla, K. Uludag, S. R. Cherry, C. D. Claussen, and B. J. Pichler. 2008. Simultaneous PET-MRI: A new approach for functional and morphological imaging. *Nat Med* 14:459–465.
99. Pichler, B. J., M. S. Judenhofer, and H. F. Wehrl. 2008. PET/MRI hybrid imaging: Devices and initial results. *Eur Radiol* 18:1077–1086.
100. Pichler, B. J., A. Kolb, T. Nagele, and H. P. Schlemmer. 2010. PET/MRI: Paving the way for the next generation of clinical multimodality imaging applications. *J Nucl Med* 51:333–336.
101. Pichler, B. J., H. F. Wehrl, A. Kolb, and M. S. Judenhofer. 2008. Positron emission tomography/magnetic resonance imaging: The next generation of multimodality imaging? *Semin Nucl Med* 38:199–208.
102. Beyer, T., L. S. Freudenberg, J. Czernin, and D. W. Townsend. 2011. The future of hybrid imaging-part 3: PET/MR, small-animal imaging and beyond. *Insights Imaging* 2:235–246.

7

Quantitative Image Analysis

Hsiao-Ming Wu and Wen-Yih I. Tseng

7.1 INTRODUCTION

Although image quality can be characterized by using physical measures such as spatial resolution, contrast, and noise, these measures, do not reflect directly the value of an image in terms of how well clinically or scientifically relevant information can be extracted. Quantitative image analysis is widely used in various imaging modalities, ranging from simple quantification of region-of-interest (ROI) values to full kinetic analysis in PET and dynamic contrast-enhanced (DCE) MRI studies. The analysis can typically be separated into two different types of tasks: a detection task, which has the goal of detecting some abnormality or change in signal, or a quantification task, which measures the concentration of an endogenous signal or an administered contrast agent in a given image voxel. Some other image processing techniques, such as segmentation and registration, are used to facilitate the definition of regions of interest and data analysis.

7.2 LESION DETECTION

One common clinical application of biomedical imaging is to detect lesions, which usually appear as a change in image intensity in a relatively small area with a distinct boundary. To detect a lesion in an image, an observer must be able to distinguish the lesion from the image background.

7.2.1 Rose Model and Contrast-Detail Curves

The study of low-contrast object detectability, based on the *Rose model*, was among the first pieces of research to assess the human visual system in signal detection in the presence of noise [1]. The Rose model, outlined by Dr. Albert Rose, is a probabilistic model of low-contrast threshold detection originally applied to radiographic (x-ray) images, where images are formed by x-ray photons (see Chapter 2) and the number of photons is directly related to the signal to noise (S/N). The model states that an observer can differentiate two regions of the image, called the *target* and *background*, only if there is sufficient information presented in the image. Specifically, if the *signal* is defined to be the difference in the number of detected photons between the regions, and the *noise* is the statistical uncertainty in each of these regions, the observer needs a certain S/N, k, to distinguish the target from its background. Equation 7.1 is a mathematical statement of the Rose model.

$$k = \frac{\text{signal}}{\text{noise}} = C\sqrt{\Phi \cdot A}, \tag{7.1}$$

where C is the contrast of the signal with respect to the background, Φ is the photon density (e.g., photons per unit area) used to form the image, and A is the size (area) of the lesion. Rose found that human observers generally require a k value of 5 to 7 to separate a low-contrast target from its background.

A number of investigators have developed experimental techniques based on the Rose model to evaluate object detectability at the threshold of human visibility in a medical image. One commonly used method, known as *contrast-detail (C-D) curve* analysis (Figure 7.1) is based on a graph that relates the threshold contrast necessary to perceive an object in an image as a function of object diameter. Larger objects can be visualized at a lower contrast, while smaller objects require a higher contrast to be visualized. Therefore, if the threshold contrast is displayed on the vertical axis and detail (or object size) on the horizontal axis, the C-D curve starts at the upper left corner (high contrast, small detail) and declines asymptotically toward the lower right corner (low contrast, large detail) in the shape of a hyperbola.

C-D studies are useful for comparing different imaging systems or methods. For example, a C-D curve can be generated for one set of x-ray imaging parameters (i.e., kVp, mA, and exposure time) and compared against another curve generated for a different set of parameters (Figure 7.1b). The relative positions of the C-D curves would identify which technique is better at producing images for detecting noise-limited objects at low contrast levels. A C-D study, however, may not reflect directly how well clinical information can be extracted from real patient images. Furthermore, modern imaging systems frequently rely on complicated hardware and sophisticated algorithms to produce images. It is essential that system hardware and computer software are optimized and produce useful images to enable physicians to make a reliable clinical interpretation.

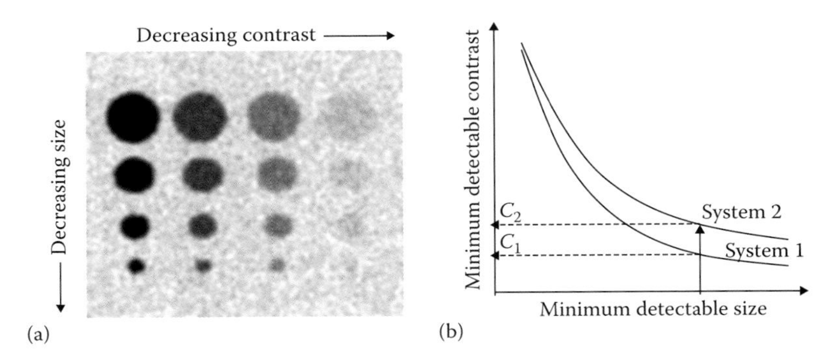

FIGURE 7.1 (a) Simulation image of a Rose phantom. (b) Hypothetical results of a contrast-detail study comparing two imaging systems. The contrast-detail curves of the two systems are obtained by having observers detect circular objects in the corresponding images of a Rose phantom similar to the one shown in (a). The results show that the observer can see an object of a given size at a lower contrast ($C_1 < C_2$) with system 1 than with system 2.

An *observer performance study* that evaluates the ability of an imaging technique to produce useful images or an observer to detect the abnormality is a more objective approach for assessment of image quality in terms of lesion detectability. In such a task, an *observer* is responsible for the detection task and uses images to produce a decision. Observers can be human, model-human, or mathematically ideal (computer) observers. The human observers are usually trained professionals such as radiologists or cardiologists. A model-human observer is a computer program designed to mimic the average performance of human observers. When the quality of an imaging system is to be quantified, it is sometimes advantageous to use a mathematically ideal observer.

7.2.2 Receiver Operating Characteristic Curves

In observer performance studies, we aim to evaluate the performance of observer(s) in detecting abnormalities in images produced from a single or multiple imaging techniques. *Receiver operating characteristic* (ROC) curves are often used in binary classification, such a yes-or-no question (e.g., is there a lesion in the liver?), and the *area under the curve* (AUC) is computed to quantify the performance. The underlying model for ROC analysis is the probability distribution of an observer's confidence in a positive diagnosis on a lesion detection task (Figure 7.2). Based on a *confidence threshold* (i.e., a particular level of confidence in a positive diagnosis), a diagnosis is considered to be positive if it exceeds this threshold, and a diagnosis is considered to be negative if it falls below the threshold. The true-positive fraction (TPF) and false-positive fraction (FPF) are then calculated from the probability density distributions delimited by the confidence threshold.

To perform an ROC analysis, phantom images containing simulated lesions or actual clinical images can be used. Here, we use a simple example to describe the procedure of ROC analysis. A set (e.g., hundreds) of images, each containing either one (positive result) or no (negative result) lesion, are simulated. The images are given to an observer, who is asked to indicate his or her confidence that the lesion actually is present. The confidence levels are numbered in four different levels: 1 = definitely present, 2 = probably present, 3 = probably not present, and 4 = definitely not present. As shown in Figure 7.2, for every possible confidence threshold level, there will be some images with lesions correctly classified as positive (true positive [TP]), but some images will be incorrectly classified as negative

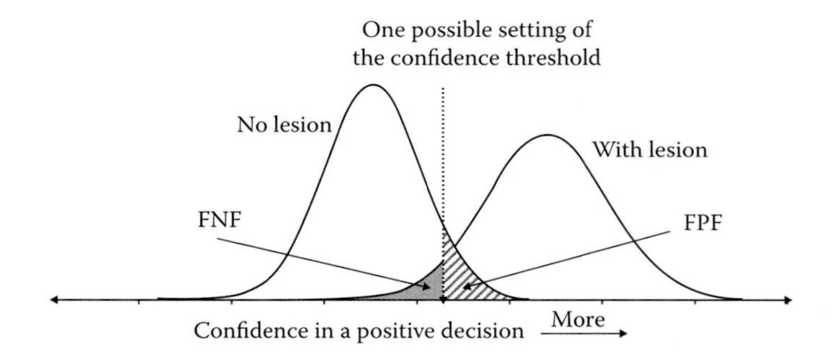

FIGURE 7.2 Schematic example of the model that underlies an ROC analysis. The curves represent the probability density distributions of an observer's confidence in a positive decision. A confidence threshold, represented by the dotted line, separates "positive" decisions from "negative" decisions. The corresponding false-positive fraction (FPF) and false-negative fraction (FNF = 1 − TPF) are indicated by the shaded areas.

(false negative [FN]). On the other hand, some images without a lesion will be correctly classified as negative (true negative [TN]), but some images will be incorrectly classified as positive (false positive [FP]). Two parameters, *sensitivity* and *specificity*, are then calculated at each progressively relaxed degree of confidence, that is, highest confidence of level 1 only, then confidence levels 1 + 2, then confidence levels 1 + 2 + 3, and so forth. *Sensitivity* is calculated as the probability that the lesion is correctly identified when a lesion is present in the image (TPF). *Specificity* is calculated as the probability that "no lesion" is correctly classified when the lesion is not present in the image (TN fraction) (Table 7.1). An ROC curve is then generated by plotting the TPF (sensitivity) as a function of the FPF (equal to 1 – specificity) for each of the progressively relaxed degrees of confidence. Figure 7.3 shows examples of ROC plots that are used to evaluate either a single or multiple techniques. A perfect performance of the observer will have an ROC curve that passes through the upper left corner (sensitivity = 1.0 and specificity = 1.0), while an observer who purely guesses will have an ROC curve that lies on the ascending 45° diagonal. Normally, the ROC should lie above the ascending 45° diagonal.

When two or more imaging devices or image processing techniques are compared, an ROC plot with multiple ROC curves (e.g., imaginary ROC curves, shown in Figure 7.3b) is used. Each ROC curve is plotted against the results of one device (or technique) performed by the same observer. The farther the curve lies above the 45° line, the better the overall performance of the imaging device (or technique). Sometimes, the interpretation of ROC

TABLE 7.1
Definitions of Sensitivity and Specificity in an ROC Study

Test Results	Lesion Present in Image	Number of Occurrences	Lesion Absent in Image	Number of Occurrences
Positive	True positive (TP)	a	False positive (FP)	c
Negative	False negative (FN)	b	True negative (TN)	d
Total		$a + b$		$c + d$
	Sensitivity = $a/(a + b)$		**Specificity = $d/(c + d)$**	

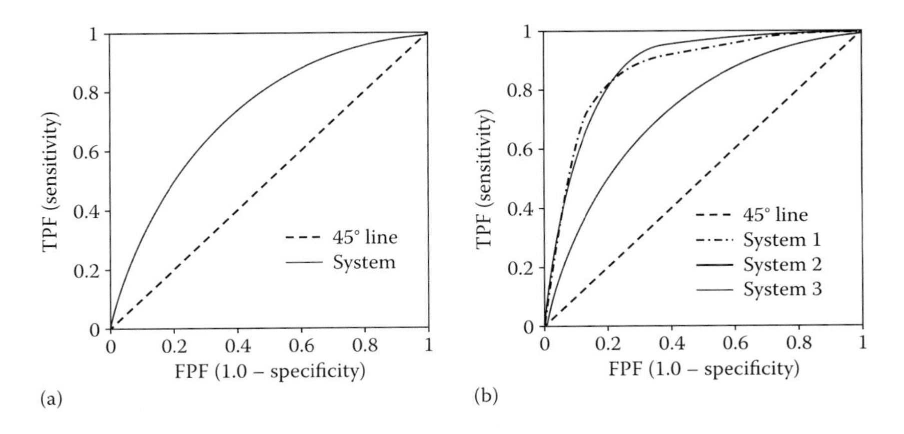

FIGURE 7.3 Examples of an ROC plot that is used to evaluate (a) a single technique or (b) multiple techniques. The ascending 45° diagonal line represents the result an observer would achieve by guessing.

results can be challenging. For example, the ROC curves for systems 1 and 2 in Figure 7.3b cause some ambiguity in interpretation of the results because the curves cross each other. One approach to simplifying the interpretation of an ROC analysis is to calculate the AUC and report the ROC curve as a single number [2]. The AUC, usually denoted by A_z, of an ROC curve is widely accepted as a measure of performance for a lesion detection task. This number can range from 0 (all readings wrong) to 1.0 (all readings correct). A value of 0.5 indicates an overall accuracy of 50%, which is equivalent to "guessing." To compare the performances of two or more systems, statistical differences among their A_z numbers are evaluated.

In general, an ROC curve can be generated from any model that predicts binary status with a certain probability. The choice of a probability threshold is crucial in balancing FP and FN errors. The threshold allows TPs to be traded off against FPs. A particular threshold, corresponding to a particular point on the ROC curve, can be chosen according to the relative cost and benefit of FPs and FNs for a given clinical scenario.

7.2.3 Model Observers

An observer performance study usually involves viewing a large number of images and requires tremendous effort and patience from human observers. Therefore, mathematical models or computer algorithms (*model observers*) that correlate with human performance have been developed to expedite the objective assessment of image quality. A model observer is a mathematical equation that computes a scalar test statistic based on the pixel values of an image and renders a decision by comparing the statistic to a threshold value. Current research on model observers as predictors of human performance has centered on linear observers, such as the *channelized Hotelling observer* (CHO) [3]. Evidence has suggested that there are multiple channels or "analyzers" in the human visual system, with different channels sensitive to different ranges of spatial frequency. CHO uses frequency-selective channels to mimic the human visual response and applies the Hotelling observer to the channel outputs. It has shown good correlation with human observer performance and has been used to evaluate various image reconstruction and image processing methods.

The computational time required to objectively assess image quality is still far greater than those use measures not related to a performance task. However, as these techniques become more mature and more efficient, task-based approaches will play a vital role in understanding the strengths and limitations of various imaging systems and algorithms.

7.3 IMAGE REGISTRATION

Image registration, sometimes called *coregistration*, is an image processing technique in which image volumes from different sources are realigned into a common anatomical coordinate space [4]. Multiple tomographic images are often acquired from the same subject at different times and/or with multiple imaging modalities. Because tomographic images are usually viewed on a 2-D screen, it is beneficial if images acquired at different times or from different modalities are realigned to the same position or orientation. Although manual adjustment to match two image volumes to a common space is feasible, the process can be labor intensive. Many computer algorithms have been developed over the years to automate and improve the accuracy of image volume alignment. These techniques have been most successfully employed in brain imaging because the brain is constrained within the skull and can reasonably be assumed in the absence of major pathologies to be nondeformable over time. In this case, *rigid body registration,* which uses only translations and rotations, can be applied for *intrasubject registration* provided the images are not distorted. It has also been used in motion correction for interscan or during dynamic sequences to correct for patient movement. Very often, serial scans are registered and then subtracted from each other to look for subtle changes over time. One particular interest is in coregistration of images of the same subject obtained from different modalities such as CT, MRI, and PET. Here, image registration allows functional and biological signals (such as those obtained by functional MRI [fMRI] and PET) to be mapped onto the high-resolution anatomy provided by CT or MRI (Figure 7.4).

Another type of registration, *spatial normalization,* is more sophisticated and involves *nonrigid transformations.* It is sometimes called *elastic mapping* or *elastic deformation* and is usually used for intersubject registration. Although human brains are similar in anatomy and functionality, there are broad deviations in size and shape. Significant variations are seen in the cortex at the gyral level. Therefore, voxel-based group mapping, which applies a special normalization technique, is popular in fMRI. Spatial normalization allows the use of a standardized atlas and eliminates the need for manual drawing of regions for an individual image. It is also useful in enabling image averaging and reducing intersubject variability. Neuroimaging processing tools, including the widely used *statistical parametric mapping* (SPM) software, have been developed to register fMRI or PET images of different subjects into a standardized atlas, such as the Montreal Neurological Institute (MNI) template or Talairach space [5].

SPM is commonly used to identify functionally specialized brain responses and is the most prevalent approach to characterize functional anatomy and disease-related changes. Figure 7.5 shows an example that uses SPM analysis to study the longitudinal metabolic changes in glucose utilization in nondemented subjects with genetic risk for developing Alzheimer's disease. Figure 7.6 shows a further application in which fMRI activation of a group of subjects during word generation (left) and rhyming (right) tasks was normalized and mapped to an MRI-derived cortical surface template. The fMRI studies reveal unique

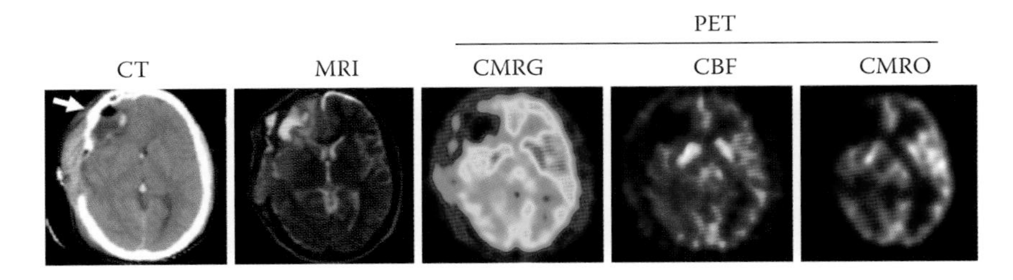

FIGURE 7.4 Representative cross-sectional images from coregistered tomographic images of a patient with traumatic brain injury. The patient had a cerebral contusion (indicated by the white arrow in the CT image) in the right frontal cortex. The detailed anatomical information showing contusion and edema, provided by CT and MRI, help provide context to interpret the changes in glucose metabolism, blood flow, and oxygen utilization shown in the PET scans. The cerebral metabolic rate of glucose (CMRG), cerebral blood flow (CBF), and cerebral metabolic rate of oxygen (CMRO) are parametric images obtained from the sequential dynamic PET studies (see Section 7.5.2).

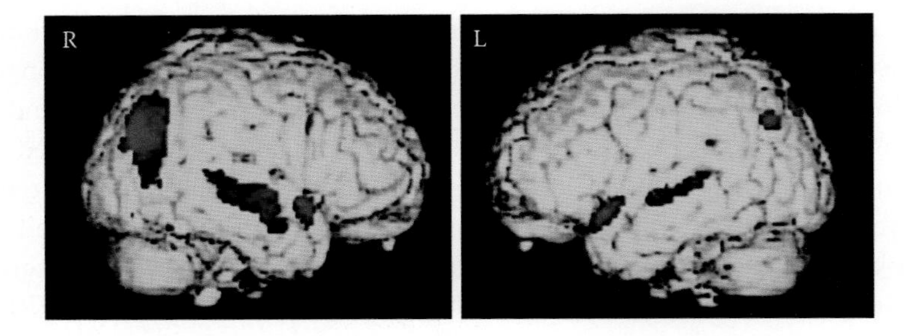

FIGURE 7.5 SPM analysis on spatially normalized brains. (From Small GW et al., *PNAS*, 37, 2000. Copyright 2000 National Academy of Sciences, U.S.A.) [^{18}F]FDG-PET images from a baseline scan and a second scan 2 years later at follow-up, from 20 nondemented subjects with the apolipoprotein E-4 (APOE-4) allele, were pooled and analyzed. The SPM analysis comparing the two time points showed that the regions (in red) with greatest metabolic decline (t-statistic; $P < .001$) after 2 years were in the right lateral temporal and inferior parietal cortex.

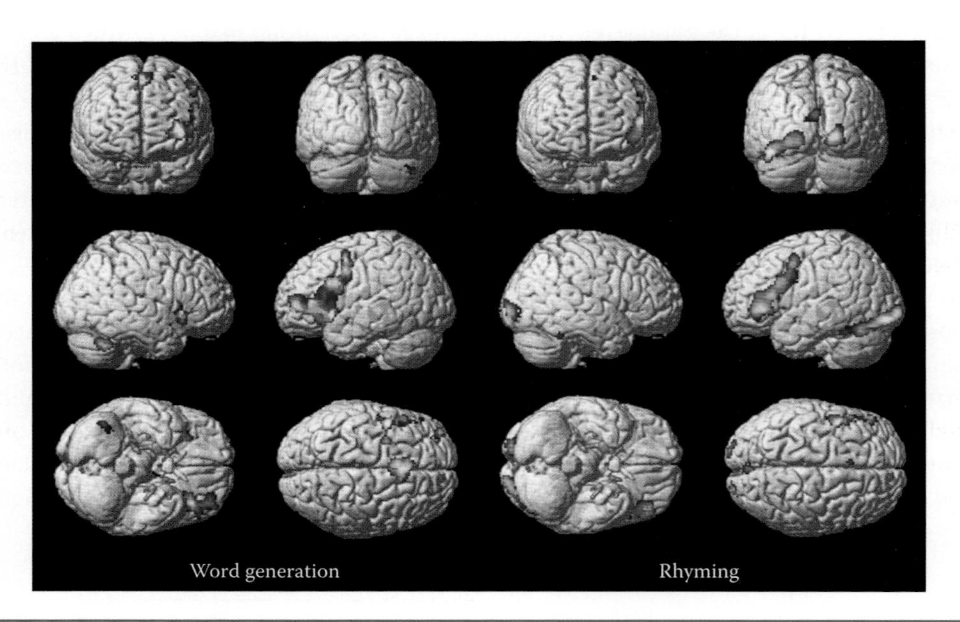

FIGURE 7.6 Hemispheric surface maps that show regions of activation (orange to red color) obtained using functional MRI during a word generation task (left) and rhyming task (right). (Courtesy of Dr. Shen-Hsing Annabel Chen. Studies performed at National Taiwan University, Taipei.)

spatial patterns of blood oxygen level–dependent (BOLD) activation, consistent with the current knowledge of left lateralization of language functions.

Nonrigid transformations are generally required for both intersubject and intrasubject registration tasks in organs other than the brain. In the rest of the body, organs and tissues can easily change their shape between or even during imaging sessions, depending on exactly how the subject is positioned on the scanner bed and many other factors, such as the movement of food through the digestive tract, filling of the bladder, and. Physiologic motion from the beating heart and respiration also pose significant registration challenges if the imaging modality cannot acquire data quickly enough to "freeze" such motion.

7.4 IMAGE ANALYSIS

To analyze an image, an operator needs to identify the locations to be analyzed and the methods of data analysis. There are three types of analysis generally used on biomedical images, ROI analysis, histogram analysis, and voxel-based group mapping.

7.4.1 ROI ANALYSIS

A simple method to quantitatively analyze an image is to quantify the mean intensity in an *ROI*. The ROI can cover a specific organ, such as the kidney or liver, or a subarea inside an organ, such as a lesion. The size, shape, and location of an ROI can be defined by the user chosen from a selection of predefined geometric shapes (e.g., rectangles, circles) or hand-drawn with an irregular shape. An ROI is commonly defined in a single image plane and can be extended to multiple contiguous planes to form a *volume of interest* (VOI). Statistics, such as the mean value, the standard deviation of the voxel intensities, the total number of voxels in the ROI (or VOI), and the area (ROI) or volume (VOI) are then derived and reported.

7.4.1.1 Segmentation

Manually drawn ROIs have the drawback of being labor intensive and are subject to *intra-observer* and *interobserver* variations. Alternatively, the definition of an ROI or VOI can be automated using computer programs with edge detection or segmentation techniques to minimize variation among operators. Widespread structural changes not evident on visual inspection of high-resolution MRI have been identified in patients with brain disease, such as Alzheimer's disease or cerebral dysgenesis and epilepsy, by applying these techniques to study the dimensions and volumes of brain structures from the images [6].

One of the most commonly used algorithms for *edge detection* is the Laplacian. The *Laplacian*, or *Laplace operator*, ∇^2, is given by sum of second partial derivatives of the function with respect to each variable. The Laplace operation in two dimensions of a function f is given by

$$\nabla^2 f = \frac{\partial^2 f}{\partial x^2} + \frac{\partial^2 f}{\partial y^2}, \tag{7.2}$$

where x and y are the standard Cartesian coordinates of the xy-plane. The Laplacian reaches a local minimum value where a high rate of change between neighboring pixel values (i.e., edges) is present. In practice, the operator specifies a starting point; the algorithm then searches all possible directions and constructs a boundary line where the Laplacian is minimized. *Image segmentation* is a process that sorts and groups voxel elements based on their intensities (or sometimes, their intensity as a function of time) with the goal of identifying regions with similar morphological or functional characteristics. *Cluster analysis* is often used by segmentation techniques to classify and group the voxels in 2-D or 3-D into regions. Sometimes, the regions can be defined on a coregistered high-resolution anatomical image (e.g., CT or MRI) and then are transferred onto a functional image (e.g., PET or SPECT) to determine the levels of radiotracer activity in these structural regions. One such example is shown in Figure 7.7. The edge detection technique was used to delineate a whole-brain ROI in MR images and excluded the skull and the extracerebral tissues. The remaining gray

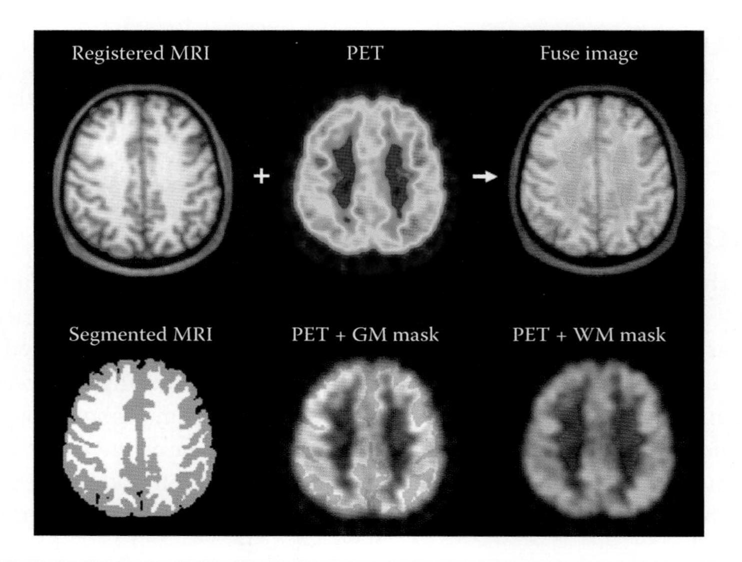

FIGURE 7.7 (Upper row, from left to right) The coregistered MRI, PET image, and the fusion of the two. Images are from a normal volunteer. The red line shown in the MRI is the ROI for the whole brain defined by an edge detection technique. (Bottom row, from left to right) The GM/WM segmented MRI image (GM in gray, WM in white) obtained by using cluster analysis and the fused images showing the PET image (in gray scale) with restricted GM and WM masks (in red). The PET images shown are the parametric CMRG images. The GM and WM restricted masks are generated from the segmented MRI. Some voxels at the interface of GM and WM are eliminated in the corresponding GM and WM masks in order to avoid underestimation of PET signals due to partial volumes effects. (From Wu H.M. et al., *Neurosurgery*, 55, 2004. With permission.)

matter (GM), white matter (WM), and cerebrospinal fluid (CSF) are then separated by the segmentation technique to generate GM and WM masks. The method allowed the cerebral metabolic rates of glucose (CMRGs) in GM and WM to be quantified, without user-defined ROIs, from a PET scan acquired following the intravenous administration of 2-deoxy-2-[^{18}F]fluoro-D-glucose, also known as ^{18}F-fluorodeoxyglucose or [^{18}F]FDG.

To perform an ROI analysis of brains with improved objectivity and reproducibility, fully automated ROI-based analysis software has been developed. A brain atlas is first prepared by experts who define and label the brain ROIs on high-resolution MRI images. Prior to data analysis, the 3-D atlas image is registered to the target image; the resulting transformation is then used to deform the atlas ROIs to the coordinate system of the target images. In this approach, registration using a nonlinear deformation algorithm is often used to account for the possible anatomical variation between the atlas and target images [7].

7.4.1.2 Percent Injected Dose per Gram and Standardized Uptake Value

ROI analysis can be used with many different imaging modalities. For example, in nuclear medicine studies, the images commonly have voxel values in units of count density (counts/voxel/s) that can be converted by an appropriate calibration factor into the average concentration (in kBq/mL) of an administered radiotracer in a given volume over a given time interval. In addition to the ROI statistics, two derived parameters, the *percent injected dose per gram of tissue* (%ID/g tissue) and the *standardized uptake value* (SUV), are commonly calculated when the relative amounts of a tracer accumulated in different tissues are compared or the distribution of a tracer throughout the body is studied.

The %ID/g tissue is the percentage of the injected dose of activity that is in 1 g of a target tissue. It is a way of normalizing the radioactive tracer taken by a given tissue to the total amount of tracer injected into the subject.

$$\%\text{ID/g} = C_\text{t} \cdot \frac{V_\text{t}}{W_\text{t}} \cdot \frac{1}{D_0} \cdot 100\% \tag{7.3}$$

where C_t is the mean radioactivity concentration (kBq/mL) obtained from an ROI placed over the tissue/organ of interest. W_t and V_t are the weight (g) and volume (mL) of the tissue, respectively. The density term V_t/W_t is often assumed to be 1, a reasonable assumption for soft tissues. D_0 is the dose of the radiotracer (kBq) injected.

The SUV is related to %ID/g tissue but also normalizes for the mass or surface area (W_S) of the subject.

$$\text{SUV} = C_\text{t} \cdot \frac{V_\text{t}}{W_\text{t}} \cdot \frac{1}{D_0} \cdot W_\text{S} = \%\text{ID/g} \cdot \frac{W_\text{S}}{100} \tag{7.4}$$

SUVs are widely used to measure [^{18}F]FDG uptake in various tumors. It has been reported that normalization of [^{18}F]FDG uptake with body weight (SUV_bw) overestimates [^{18}F]FDG uptake in heavy patients as their body fat does not take up [^{18}F]FDG. The SUV normalized with body surface area (SUV_bsa) appears to be more preferable since it is minimally affected by the body size [8].

ROI quantification is essential for measuring clinical changes, such as tumor growth and the efficacy of therapeutic interventions. The accuracy of ROI quantification, however, is significantly affected by corrections to the data (e.g., for attenuation and scatter) and also the image reconstruction algorithms used. For example, in PET and SPECT studies, the relatively low S/N of the data means that all practical image reconstruction algorithms employ operations that result in a trade-off between minimizing the bias and minimizing the variance of the reconstructed image. Therefore, for all modalities, the bias and the variance errors inherent in images should be considered to ascertain the significance of any changes seen by ROI analysis.

7.4.2 Histogram Analysis

Histogram analysis has become increasingly popular, especially in the study of neurological diseases where the biological effects are diffuse and widespread. Unlike a focal lesion, such as a cancerous mass, a diffuse change may extend over a larger area, may have no distinct boundary, and is harder to detect by visual observation. Histogram analysis uses all the voxels within the image or organ of interest and plots a frequency distribution of the parameter of interest (typically the voxel intensity). It avoids the need to define ROIs and has the advantage of being sensitive to subtle changes in the intensity distribution within the tissue/organs of interest. A particular application of this technique has been the study of the normal-appearing WM in multiple sclerosis. Quantitative MR measures including magnetization transfer ratio, diffusion, and T_1 relaxation time (see Chapter 3) are sensitive to this subtle abnormality and are extensively studied using histogram analysis. Subregions of the brain, such as GM and WM, can also be segmented and studied. Generation of the histogram usually starts with segmentation to extract the brain VOI. An *absolute histogram* is plotted with a range of image values showing the number of voxels per interval (or bin width) on the y-axis. Since brain volumes vary among individuals, the histogram is usually normalized by dividing the voxel number on the y-axis by the total number of voxels to generate a histogram with frequency distribution. This *normalized histogram* has the property that the area under the histogram is unity and therefore is easier to interpret when making comparisons.

Features of these histograms are studied or compared by extracting parameters that are intended to capture important information contained in the histogram. Typical features are peak height, peak location, and mean parameter value. An extensive discussion of histogram analysis in MR brain imaging can be found in the literature [9].

7.4.3 Voxel-Based Group Mapping

When ROI analysis is carried out, an observer may intentionally place the boundary of an ROI around a distinct structure or abnormality (e.g., a lesion). Therefore, ROI analysis can be biased and subject to large intraobserver variations. Histogram analysis, on the other hand, uses a large volume of voxels with no subjective ROI definition but has the disadvantage that spatial information is lost. A third method, *voxel-based group mapping*, provides an alternative to overcome the pitfalls of these two methods. It is popular and particularly widely used in brain imaging. The techniques involve several key steps: (1) The images of all subjects are spatially *registered* to the same stereotactic space; (2) the images are segmented into WM, GM, and CSF to restrict the data analysis to a subregion of interest (e.g., cortical GM); (3) the images are smoothed with Gaussian filters; and (4) a variety of statistical tests are carried out on the entire brain on a voxel-by-voxel basis [5]. As shown in Figure 7.5, with spatially normalized brain images using SPM software, this voxel-based analysis clearly identifies a significant metabolic decline in specific brain regions, although clinical symptoms have not yet developed in these subjects.

Another normalization technique, *cortical surface mapping*, has been used in fMRI to visualize and assess the functional activities on the entire cortical surface [10]. Other applications, such as building a large database of normal uptake of a specific radiotracer (e.g., [^{18}F]FDG), have been developed. These allow an individual's image to be compared against the normal database to determine areas with abnormal uptake [11].

7.5 PARAMETER ESTIMATION

Dynamic imaging, with PET, SPECT, or contrast-enhanced MRI or CT, enables the distribution of a radiotracer or a contrast agent (e.g., typically gadolinium [Gd] based for MRI and iodine based for CT) to be measured as a function of time and can allow a range of functional (physiology, metabolism, molecular targets/pathways) parameters to be assessed. When radioactive compounds are dealt with as contrast agents, they usually are referred to as tracers because only a small or "trace" mass of the imaging probe is administered. The recorded signals are reconstructed into either a single image (static image integrated over a given time interval) or a time series of images (dynamic images) (Figure 7.8). With an understanding of the biologic fate of the tracer in the body, it is possible to extract biologically meaningful parameters from the images. Unlike semiquantitative analysis, quantitative analysis, in principle, requires a second source of data—the time course of

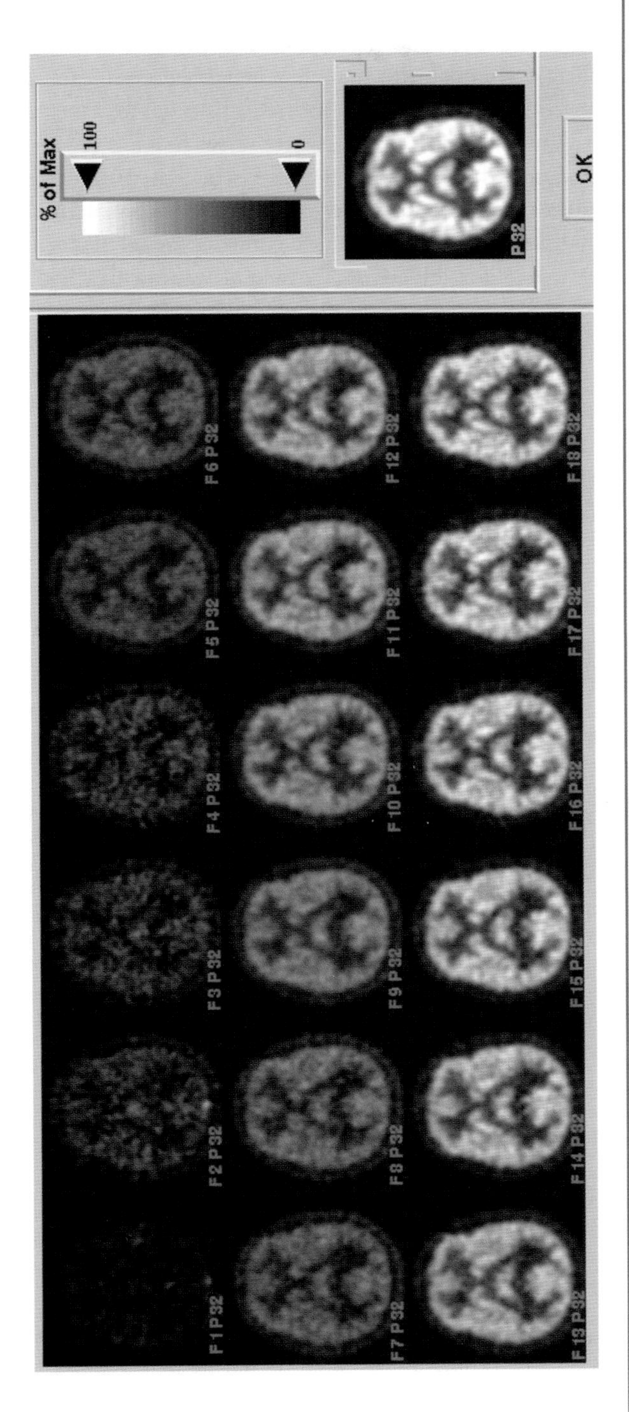

FIGURE 7.8 Activity distribution in a single transverse plane through the midbrain (plane 32, frames 1–18) as a function of time. Data taken from a set of 3-D dynamic brain images (63 planes) of a normal subject acquired at different time intervals (total scan time, 60 min; durations, [F1–F4] 0.5 min, [F5–F8] 2 min, [F11–F18] 5 min) after bolus administration of [^{18}F]FDG. The image at the lower right reflects a single summed image (integrated between 10–60 min) reconstructed from the same PET data.

the administered tracer or contrast agent in whole blood or plasma. The choice of analysis depends on the medical or biologic information desired, as well as on equipment availability and technical complexity.

In this section, examples of nuclear medicine studies are primarily used due to their long history and their broad application in quantification of biological process. This has been facilitated by the wide range of radiotracers that have been used in studies and the high sensitivity of radioactive assays, which allows imaging with injected masses that are orders of magnitude below typical pharmacologic doses. The general concepts illustrated here, however, are applicable to other imaging techniques as well. For example, the principle of compartmental modeling will be discussed and demonstrated with the widely used compartmental model for quantifying the metabolic rate for glucose using the [^{18}F]FDG-PET model. The same principle applies to MR studies using T_2^*-weighted dynamic imaging of Gd to enable blood perfusion measurements using a one-compartment model, or renal filtration and vascular parameter measurements using DCE MRI and a two-compartment model. An excellent text that captures the essentials of quantitative MR measures and how these measures are translated to their biological correlates is recommended for further reading [9]. In perfusion CT, images are dynamically acquired to trace the signal evolution following the passage of iodine contrast material. Perfusion parameters can be derived semiquantitatively from the signal dynamics (e.g., maximum slope and time to peak) as well as quantitatively by employing the principles of tracer kinetics for non-diffusible tracers. Perfusion CT has been shown to improve stroke detection and has proven clinically feasible in several different organs [12]. For more details, principles, and applications, please refer to the review in Ref. [13].

In MRI studies, signal intensity (SI) time curves of dynamic enhancement following contrast injection are first obtained from the tissue of interest (i.e., myocardium) and from a region that represents the arterial input to tissue (i.e., blood pool in the left ventricular cavity). Semiquantitative parameters can be determined from the characteristics of the SI-time curves. For example, the maximal slope of the enhancement curve (up-slope) is determined from the first derivative of the SI-time curves (Figure 7.9a). Other semiquantitative parameters, such as peak value of the SI (peak SI), time to peak enhancement (tPeak), and AUC, can be included to estimate different characteristics of perfusion

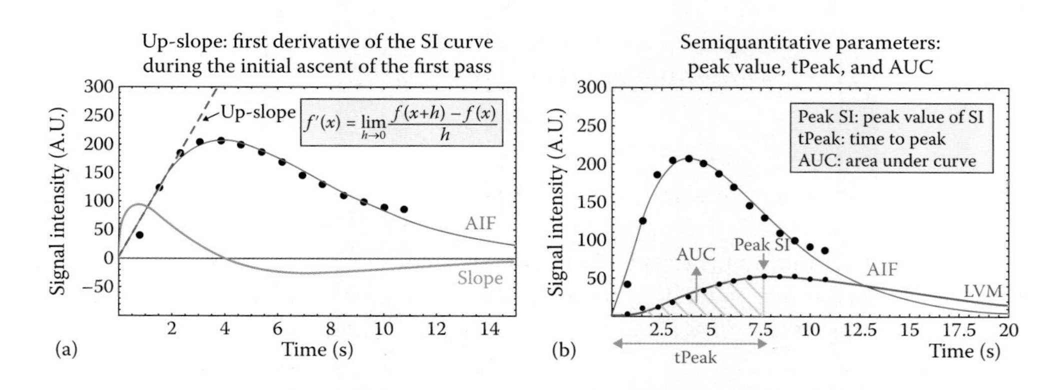

FIGURE 7.9 Signal intensity (SI)–time curves measured from the tissue of interest (blue curve) and the region of the arterial input function (AIF, red curves). Up-slope is determined by the maximal slope of the initial ascent of the first pass of the SI-time curves (a, green curve). Other parameters such as peak value of SI (peak SI), time to peak (tPeak), and area under the curve (AUC) can be estimated from SI-time curves of either the left ventricular myocardium (LVM) or the AIF (b).

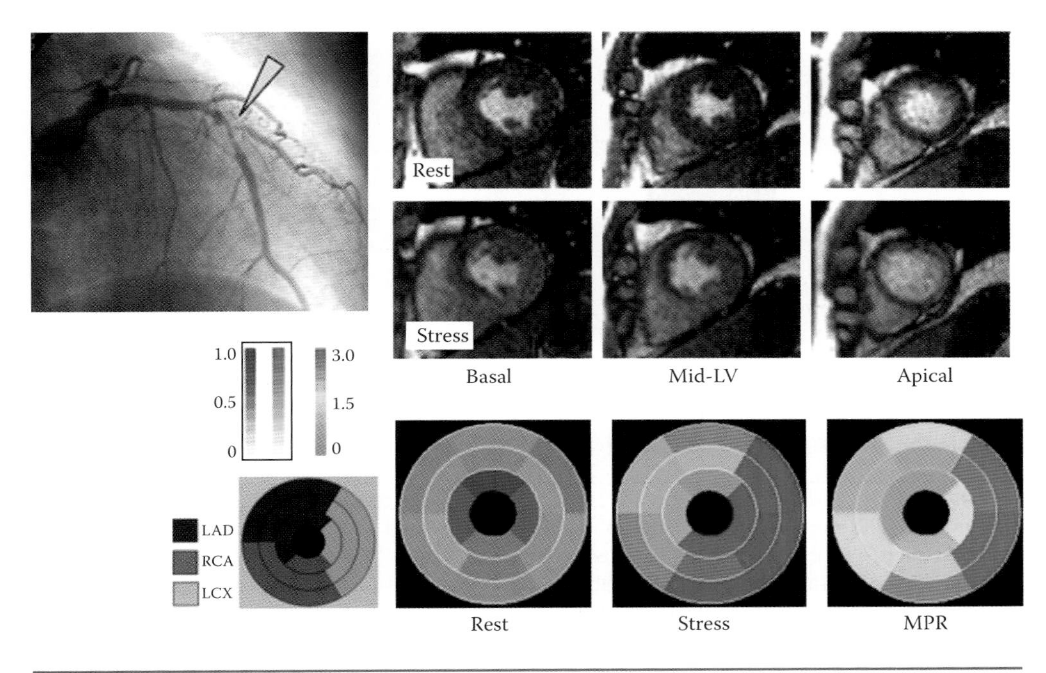

FIGURE 7.10 Perfusion MRI of a patient with significant stenosis of the left anterior descending artery (LAD) as documented by coronary angiography (blue arrow). Two perfusion MRI examinations were performed on the patient, one at rest and one at stress. Using the maps of up-slope at rest and at stress, a map of myocardial perfusion ratio (MPR) can be obtained. The myocardial segments corresponding to the LAD territory show markedly decreased MPR, indicating the presence of ischemia. The RCA (right coronary artery) and LC (left circumflex artery) territories are relatively normal.

(Figure 7.9b). These parameters, which are estimated from the tissue of interest, are often divided by the corresponding parameters estimated from the arterial input region to normalize the variation of the arterial input function. In this example, the SI-time curves were acquired from the same subject in two different conditions, one at rest and one under pharmacologically induced stress. The myocardial perfusion ratio (MPR) is estimated by dividing the up-slope at stress by the up-slope at rest. Ischemic change in the myocardium is reflected when the MPR value is lower than a certain normative value (Figure 7.10). Thus, this semiquantitative analysis has direct clinical applicability.

7.5.1 Quantitative Analysis

Quantitative analysis in medical imaging often uses either *tracer kinetic modeling* or *multiple-time graphical analysis*. These analyses can be viewed as extensions of clinical pharmacokinetic analysis. Imaging tracers and drugs often share common concepts in structural design and principle of action if they have the same target, such as an enzyme, receptor, or neurotransmitter system. Through sampling of the blood (the accessible pool), pharmacologists developed methods to make predictions of pharmacokinetic parameters of a drug based on blood concentration curves. However, these approaches have limited capability to determine the drug distribution in a tissue of interest (the inaccessible pool) because it is inferred through a model and not actually measured. Acquisition of dynamic images makes the time course of tissue concentrations available. With both blood and tissue data available, biological processes, such as the utilization of glucose or other metabolic substrates,

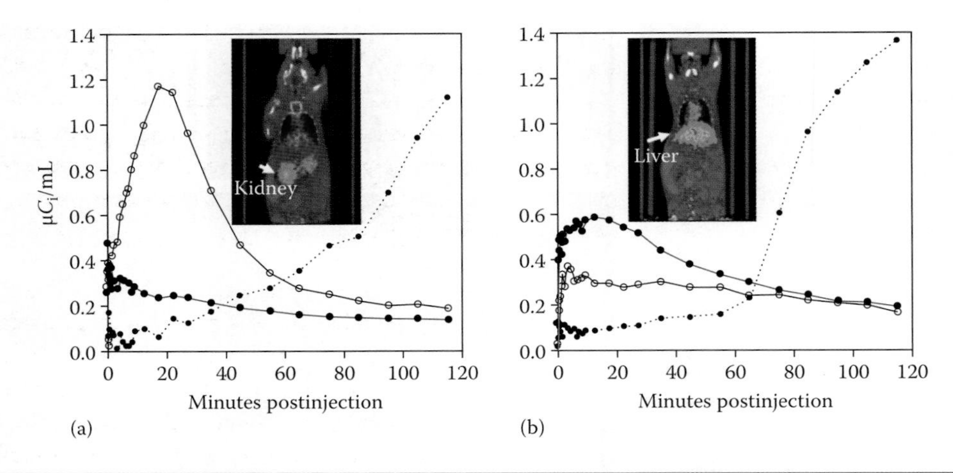

FIGURE 7.11 (a and b) Graphs show the time course of tissue concentrations (open circles with line, kidney; solid circles with line, liver; solid circles with dashed line, bladder) derived from the dynamic micro-PET images of mice injected with two different ^{124}I-labeled antibody fragments. The insert in each graph shows a static whole-body PET image (summed 20–40 min postinjection) in color overlaid in a CT scan (gray scale) of the mouse. Notice that the clearance routes for the two radiotracers are different. In (b) a substantial amount of the radiotracer was metabolized and cleared by the liver. In (a), on the other hand, the radiotracer was mostly excreted by the kidneys. The data suggest that the inclusion of a liver compartment may be necessary for modeling a radiotracer (b) in a quantitative analysis. (Courtesy of Dr. Anna Wu. The studies were performed at the University of California, Los Angeles.)

tissue perfusion, cell proliferation, receptor-ligand binding, and other biological events, can be accurately quantified through the use of appropriate imaging tracers and mathematical models. Small animal imaging, in particular, provides unprecedented opportunities for studying biological processes involved in the pharmacokinetics of a new drug, including absorption, distribution, metabolism, and excretion. A comprehensive model can be configured to quantify biological processes in different tissues or organs using dynamic whole-body images. An example is shown in Figure 7.11.

7.5.2 Tracer Kinetic Modeling

Quantitative analysis using tracer kinetic modeling requires (1) the time course of the injected radiotracer in plasma (the *input function*), (2) the time course of the tissue response curve (the tissue *time-activity curve* [TAC]), and (3) a mathematical model that describes the fate of the tracer after administration. Mathematical models can be classified as *noncompartmental, compartmental,* or *distributive.* Distributive models are the most complex and have the ability to describe not only the time dependence of drug (or imaging tracer) distribution but also the spatial dependence. They require mathematical equations that involve partial derivatives to incorporate spatial dependence. Compartmental models, on the other hand, deal with the time domain only and are simpler to model mathematically. They are easier to implement and often provide adequate parameter estimates. Therefore, compartmental models are widely used for medical imaging. Furthermore, only a trace amount (i.e., nonpharmacological dose) of an imaging tracer is usually administered into the subject, and under these conditions, the model is linear and is referred to as a *tracer kinetic model.* In addition, an imaging probe is often designed and developed so that only a few biological processes (or limiting steps) are involved once it enters the body. As a result, models with a relatively small number of compartments (typically less than four) are required to describe the observed imaging data. This is important because the number of compartments and number of parameters that can be estimated are typically severely constrained by the S/N of the available imaging data. The precise number of compartments and the definition and the relationship among compartments in a tracer kinetic model need to be determined based on existing knowledge and biochemical principles.

7.5.2.1 Input Function

By definition, the input function is the time course of the injected radiotracer or contrast agent in the arterial plasma. This represents the delivery or "input" of the tracer to the tissues of the body. Although the method of placement of an arterial catheter and manual sampling to determine plasma concentration is still considered the gold standard for deriving the input function, the challenges associated with manual and quite-invasive blood sampling impede the routine use of tracer kinetic modeling. Alternatively, the input function may be derived in human studies by hand-warmed venous blood samples. Hand-warming dilates the arterial and venous vessels, therefore arterializing the venous blood effectively. Another alternative is using the whole blood TAC derived from dynamic images of the heart using an ROI placed over the large volume of blood in the left ventricle (LV) or left atrium. This approach has the advantage of being noninvasive but samples tracer activities in whole blood rather than in plasma. Correction with plasma–to–whole blood concentration ratios and correction for any metabolites of the original tracer, if present, are necessary. In some cases, correction of spillover signal from any tracer in the adjacent heart muscle is also required.

7.5.2.2 Tissue TAC

The tissue TAC is usually derived using the ROI method. An ROI is first defined on an image with good S/N, for example, a summed image from a sequence of late-phase images or from a coregistered structural image. The mean values of the ROI from sequential dynamic images, showing the change of tracer concentration as a function of time, are then obtained to generate the tissue TAC. Examples of tissue TAC's were shown in (Figure 7.11). Typically, the signal measured from the ROIs used to obtain a tissue TAC does not come from a single tissue compartment but is the sum of several sources, including tracer in the blood, as well as the extravascular and intracellular spaces. This is inevitable due to the limited spatial resolution of the imaging system. Thus, contributions of individual components in a tissue TAC need to be included in the mathematical equations that formulate the model. In addition, calibration of the tissue TAC to the device used to measure the input function (if measured using non-image-based methods) must be performed. In nuclear medicine, this is commonly achieved by cross-calibrating a well counter used to count blood samples to the PET or SPECT scanner by imaging a cylinder containing a uniform and known concentration of a radionuclide and then taking a sample from that cylinder and counting it in the well counter.

7.5.2.3 Compartmental Model

A *compartmental model* consists of a finite number of compartments with specified interconnections, inputs, and losses (Figure 7.12). A *compartment* is defined as a volume or space within which the tracer rapidly becomes uniformly distributed. In some cases, a compartment has an obvious physical interpretation, such as tracer molecules in the intravascular blood pool and those entering the cells. For other cases, the physical interpretation may be less obvious, such as for a tracer that may be metabolized and trapped; thus, the free and the

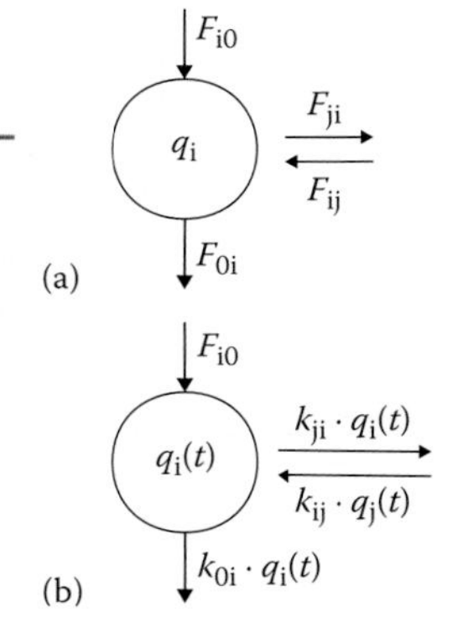

FIGURE 7.12 (a) Diagram of a compartmental model that shows the *i*th compartment (i = 1, 2, ..., *n*) of an *n*-compartment model. The flux, *F*, mathematically describes the mass transfer of substance among compartments interacting with the *i*th compartment. F_{ji} is the transfer of the substance from compartment i to compartment j, and F_{ij} is the transfer of substance from compartment j to compartment i. F_{i0} and F_{0i} represent new input to and the loss from compartment i, respectively. (b) For a linear compartmental model with constant coefficients, the flux *F* can be replaced by the corresponding rate constant, *k*, times the amount of substance, *q*, in the compartment.

bound forms of the tracer are defined as separate compartments. The volume of a compartment is usually defined in terms of whether it is open or closed. An open compartment is one from which the tracer can escape to other compartments, whereas a closed compartment is one from which the tracer cannot escape. If a nondiffusible tracer is injected into a closed compartment, conservation of mass requires that after the distribution of the tracer reaches equilibrium or steady-state condition,

$$V_d = \frac{A}{C_b}, \tag{7.5}$$

where A is the amount of tracer injected, C_b is the steady-state concentration of the tracer in the blood, compartment and V_d is the *distribution volume* of the compartment. In most cases, tracers are exchanged between blood and tissue and are able to escape from the compartments. In this case, the concentration of the tracer in tissue will typically be different from that in blood after the tracer reaches its equilibrium distribution. The ratio of tissue concentration (C_t, in units of Bq/g) to blood concentration (C_b, in units of Bq/mL) at equilibrium is called the *partition coefficient* (λ).

$$\lambda = \frac{C_t}{C_b} \tag{7.6}$$

If one assumes that the concentration of tracer in tissue is the same as the concentration in blood, this leads to an *apparent distribution volume* (V_a) in tissue.

$$V_a = \frac{A_t}{C_b}, \tag{7.7}$$

where A_t is the activity in tissue. Since $A_t = C_t V_t$, where V_t is the volume of tissue, λ can also be calculated as

$$\lambda = \frac{V_a}{V_t}. \tag{7.8}$$

Therefore, another interpretation of λ is that it is the distribution volume per unit mass of tissue for a diffusible tracer in the model.

The equations that govern a linear compartmental model are based on the physical principles of mass balance. That is, the rate of change of mass for a given compartment must equal the net mass coming into a compartment per unit time minus the net mass leaving per unit time. *Flux* refers to the amount (mg/min) of a substance transported between two compartments per unit time. *Rate constants* (min^{-1}) describe the relationships between the concentrations and fluxes of a substance in two compartments. For a first-order process,

$$F_{\mathrm{ji}}(t) = k_{\mathrm{ji}} \cdot q_{\mathrm{i}}(t), \tag{7.9}$$

where F_{ji} is the amount of the tracer that is transported from compartment i to compartment j, and k_{ji} is a constant and is the transfer rate from compartment i to compartment j. $q_{\mathrm{i}}(t)$ is the mass in compartment i at time t. In imaging studies, the model is developed and describes only the biological processes within the field of view of an imaging system. The first compartment is usually the intravascular space and is represented by the input function. Based on the principles of mass balance, the mathematical expression describing the rate of change of mass (dq_{i}/dt) in the other compartments i (i ≠ 1) is

$$\frac{dq_{\mathrm{i}}}{dt} = \sum_{\substack{\mathrm{j}=1 \\ \mathrm{j}\neq \mathrm{i}}}^{n} k_{\mathrm{ij}} q_{\mathrm{j}}(t) - \left[\sum_{\substack{\mathrm{j}=1 \\ \mathrm{j}\neq \mathrm{i}}}^{n} k_{\mathrm{ji}} \right] q_{\mathrm{i}}(t), \tag{7.10}$$

where n is the total number of compartments.

7.5.2.4 Model for Measuring Rate of Glucose Utilization Using [^{18}F]FDG

Here, we use a brain PET study using the radiotracer [^{18}F]FDG as an example. The study is designed to measure the rate of glucose utilization by the brain. FDG is an analog of glucose that is similar to glucose in terms of transport and initial phosphorylation. Like glucose, it is transported from the blood to the brain by a carrier-mediated diffusion mechanism. Once it has entered the brain cells, FDG molecules are catalyzed by the enzyme hexokinase to form FDG-6-PO_4 similar to the phosphorylation of glucose to form glucose-6-PO_4. FDG-6-PO_4, however, is a poor substrate for further metabolism. Unlike glucose-6-PO_4, FDG-6-PO_4 is not converted into glycogen to any significant extent and is metabolically trapped and accumulates in tissues. Therefore, [^{18}F]FDG is an excellent imaging probe to study glucose utilization, and it is extensively used in the study of brain diseases, cardiovascular diseases, cancer, and so forth. A compartmental model configured for FDG and corresponding to comparable distributions of glucose is shown in Figure 7.13b, although glucose itself continues on through further metabolic steps. This three-compartment model consists of [^{18}F]FDG in plasma, [^{18}F]FDG in tissue, and [^{18}F]FDG-6-PO_4 trapped in tissue. The first-order rate constants K_1^* (mL/min/g) and k_2^* (min^{-1}) describe the transport of [^{18}F]FDG from blood to brain and brain to blood, respectively; the first-order rate constants k_3^* (min^{-1}) and k_4^* (min^{-1}) describe the phosphorylation of [^{18}F]FDG and dephosphorylation of [^{18}F]FDG-6-PO_4, respectively. The asterisk symbol in each rate constant is added to distinguish it from the corresponding rate constants for endogenous glucose. The rate constant K_1^* is written with an uppercase letter K and has different units because it also includes

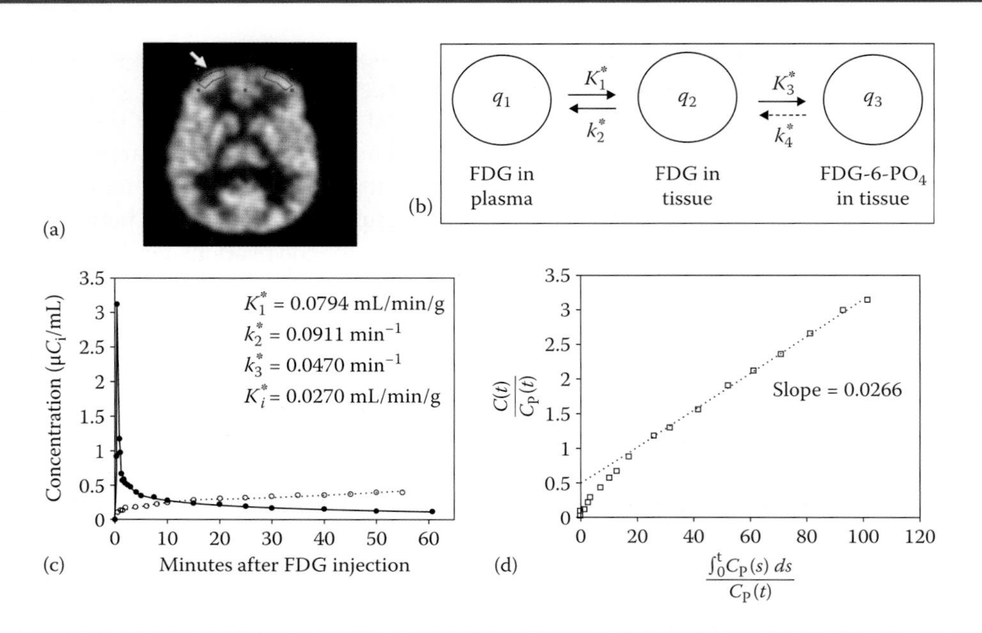

FIGURE 7.13 Estimation of [18F]FDG uptake rate constant (K_i^*) of human brain tissue from a dynamic FDG-PET brain scan. (a) Brain image at level of midbrain acquired after a bolus intravenous injection of [18F]FDG. The image represents the counts summed over the period 40–60 min after injection. A pair of ROIs (in red) was drawn on the frontal cortex of the brain to define tissue TACs. (b) Diagram of a three-compartment FDG model for glucose kinetic analysis. The concentration in compartment 1 [$C_p(t)$, the input function] is measured by taking arterial blood samples during the 60 min duration of the PET scan. The PET image reflects the sum of the concentrations in compartments 2 and 3 [$C(t)$]. (c) The concentration of arterial plasma samples (solid dots) and the derived input function (solid line) as well as the tissue TAC (open dots) derived from the dynamic PET images using the frontal-cortex ROI in (a). The dashed line shows the results of the fit to the three-compartment model shown in (b). The parameters estimated by model fitting (in this case, it was assumed that $k_4^* = 0$) are shown in the plot. (d) The Patlak plot generated using the graphical analysis method (Equation 7.20) with the same PET data and input function as shown in (c). Note that the slope of the Patlak plot ($t > 20$ min) is close to the value of K_i estimated by the three-compartment model fitting.

the perfusion and extraction of [18F]FDG in the capillary bed. The activity of phosphatase (which dephosphorylates [18F]FDG-6-PO$_4$) in the brain is low and sometimes ignored (i.e., by assuming $k_4^* = 0$). Based on mass balance, the mathematical equations describing the three-compartment FDG model are

$$\begin{aligned}
\frac{dq_1(t)}{dt} &= -K_1^* q_1(t) + k_2^* q_2(t) \\
\frac{dq_2(t)}{dt} &= K_1^* q_1(t) - k_2^* q_2(t) - k_3^* q_2(t) + k_4^* q_3(t) \\
\frac{dq_3(t)}{dt} &= -k_4^* q_3(t) + k_3^* q_2(t),
\end{aligned} \tag{7.11}$$

where $q_i(t)$ is the mass of compartment i at time t.

In practice, the differential equations often are expressed in terms of concentrations rather than mass. Using the FDG model (Figure 7.13b) as an example, Equation 7.12 can be rewritten as

$$\begin{aligned}\frac{dC_2(t)}{dt} &= K_1^* C_p^*(t) - k_2^* C_2(t) - k_3^* C_2(t) + k_4^* C_3(t)\\ \frac{dC_3(t)}{dt} &= -k_4^* C_3(t) + k_3^* C_2(t),\end{aligned} \tag{7.12}$$

where C_i is the concentration in compartment i. $C_p^*(t)$ is the [^{18}F]FDG concentration curve in plasma (i.e., compartment 1). In an [^{18}F]FDG study, the TAC [ROI(t) in Equation 7.11] derived from a tissue region combines the activities [$C_2(t) + C_3(t)$] from free and bound tracer in tissue. To account for the radioactivity in tissue vasculature, a blood term, $V_B C_B(t)$, is included:

$$\text{ROI}(t) = C_2(t) + C_3(t) + V_B C_B(t), \tag{7.13}$$

where $C_B(t)$ is the total radioactivity concentration in vasculature, and V_B is the volume fraction of blood in tissue. The [^{18}F]FDG uptake rate constant (K_i^*, mL/min/g) is calculated by using the equation $K_i^* = (K_1^* \cdot k_3^*)/(k_2^* + k_3^*)$ [14,15]. If the steady-state concentration of glucose (*Glu*, μmol/mL) in the blood is measured and the lumped constant (LC), which accounts for the difference in FDG and glucose kinetics, is known, then the metabolic rate of glucose (MRG, μmol/min/g) in the brain can be calculated by

$$\text{MRG} = \frac{\text{Glu} \cdot K_i^*}{\text{LC}}. \tag{7.14}$$

LC is a constant that equals the ratio of the net extraction of [^{18}F]FDG to that of endogenous glucose at steady state [14].

Solutions to the differential equations of a linear compartmental model can be derived using a variety of mathematical techniques. A common approach is to use weighted non-linear regression for parameter estimation. This is an iterative process in which starting estimates (or initial guesses) of all parameters along with the equations that govern a particular model are used to generate a refined estimate of the parameters. This approach of minimizing the square of the difference between the observations (tracer concentration over time derived from a dynamic sequence of images) and model prediction is repeated until some criteria are met for achieving convergence. An example of a brain PET study using the three-compartment FDG model is shown in Figure 7.13a through c. The detailed methods used to solve differential equations and the techniques to estimate model parameters are beyond the scope of this chapter. However, appropriate models used for a range of imaging studies have been defined and validated. Many user-friendly software packages are available that allow users to select the compartmental model for a particular tracer or contrast agent and provide routines to estimate the model parameters [16,17].

7.5.2.5 Modeling Dynamic MRI Data

In MRI studies, two techniques, *dynamic susceptibility contrast* (DSC) and DCE, are often used to assess tissue perfusion based on two different contrast mechanisms of the Gd-based contrast agent (see Chapter 3, Section 3.5.6). DSC MRI is used to assess perfusion in the brain in which the contrast agent stays within the blood vessels due to the tight junction of the blood-brain barrier. In this case, the SI of the brain tissue decreases with increasing concentration of the contrast agent within the vessels due to the magnetic susceptibility effect.

The SI behavior of such a nondiffusible agent can be modeled based on the central volume principle. DSC MRI assumes a linear relationship between the concentration of the contrast agent $C(t)$ and the change of transverse relaxation rate $\Delta R_2^*(t)$:

$$C(t) \propto \Delta R_2^*(t) = -\frac{1}{\mathrm{TE}} \ln\left(\frac{\mathrm{S}(t)}{\mathrm{S}_0}\right), \tag{7.15}$$

where TE is the echo time, S_0 is the SI before the arrival of the contrast agent, and $\mathrm{S}(t)$ is the SI at time t. For cerebral perfusion measurements, the Gd-based contrast material is mainly employed as an intravascular agent unless the blood-brain barrier is disrupted as a consequence of disease. The contrast concentration in brain tissue $C_\mathrm{B}(t)$ is thus given by

$$C_\mathrm{B}(t) = CBF \cdot C_\mathrm{A}(t) \otimes R(t), \tag{7.16}$$

where $C_\mathrm{A}(t)$ is the contrast concentration in the feeding artery (also referred to as the arterial input function), CBF is the cerebral blood flow, $R(t)$ is the residue function or impulse response function (monotonically decreasing from 1), and ⊗ denotes convolution (Figure 7.14).

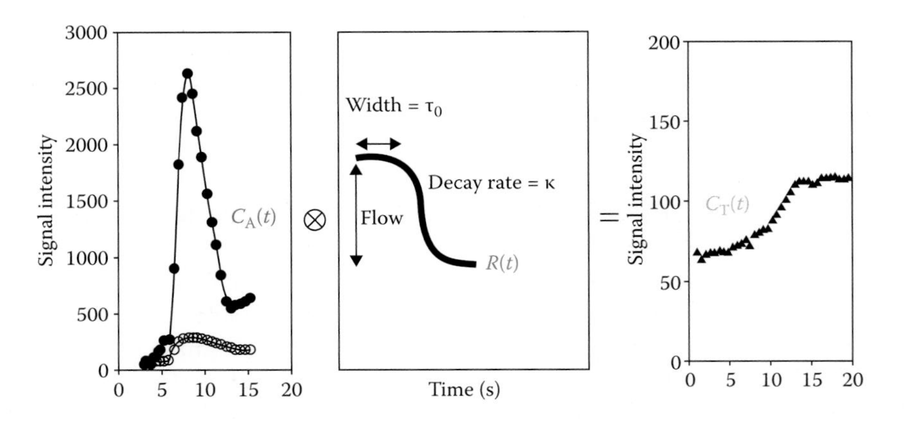

FIGURE 7.14 Tissue signal intensity–time curve $C_\mathrm{T}(t)$ is the convolution of the blood pool arterial input function $C_\mathrm{A}(t)$ and the residue function or impulse response function $R(t)$ scaled by the blood flow to brain tissue.

Perfusion parameters, including CBF, cerebral blood volume (CBV), and mean transit time (MTT), can then be calculated as follows.

$$\mathrm{CBF} = \left[C_{\mathrm{B}}(t) \otimes^{-1} C_{\mathrm{A}}(t)\right]_{t=0}$$

$$\mathrm{CBV} = \frac{\int_0^{\infty} C_{\mathrm{B}}(\tau)\,d\tau}{\int_0^{\infty} C_{\mathrm{A}}(\tau)\,d\tau} \tag{7.17}$$

$$\mathrm{MTT} = \frac{\mathrm{CBV}}{\mathrm{CBF}},$$

where $\otimes^{-1}$ denotes deconvolution.

DCE MRI is another technique used to assess perfusion in tissues in which contrast agent leaks from the intravascular space to the extravascular space due to permeability of the vascular wall. In this case, the SI of the tissue increases with the concentration of the contrast agent in the tissue due to shortening of the T_1 relaxation time. Since Gd-based contrast agents are water soluble, they do not enter the cells, and they exist only in two compartments, the intravascular space and the extravascular extracellular space. The SI behavior of such a diffusible water-soluble agent can be modeled based on a bicompartment model.

The master equation of this bicompartment model can be described as follows.

$$\frac{dC_{\mathrm{t}}}{dt} = K^{\mathrm{trans}}(C_{\mathrm{p}} - C_{\mathrm{e}})$$

$$\frac{dC_{\mathrm{t}}}{dt} = K^{\mathrm{trans}}(C_{\mathrm{p}} - C_{\mathrm{t}}/v_{\mathrm{e}}) = K^{\mathrm{trans}}C_{\mathrm{p}} - k_{\mathrm{ep}}C_{\mathrm{t}}, \tag{7.18}$$

where C_{t} is the tissue concentration, C_{p} is the plasma concentration, C_{e} is the extravascular extracellular concentration, K^{trans} is the exchange rate between the two compartments, v_{e} is the ratio of extravascular extracellular volume to the volume of tissue, and $k_{\mathrm{ep}} = K^{\mathrm{trans}}/v_{\mathrm{e}}$.

Given initial conditions that $C_{\mathrm{p}}(0) = 0$ and $C_{\mathrm{t}}(0) = 0$, we have the following solution for the differential equation (Equation 7.18).

$$C_{\mathrm{t}}(t) = K^{\mathrm{trans}} \int C_{\mathrm{p}}(\tau) e^{-k_{\mathrm{ep}}(t-\tau)}\,d\tau = C_{\mathrm{p}}(t) \otimes K^{\mathrm{trans}} e^{-k_{\mathrm{ep}}t} = C_{\mathrm{p}}(t) \otimes K^{\mathrm{trans}} R(t) \tag{7.19}$$

The solution states that the tissue concentration is the convolution of the plasma concentration and the residue function scaled by K^{trans}. This equation is very similar to the equation for nondiffusible contrast agent in brain perfusion; the only difference is that CBF is replaced with K^{trans}. The constituents of K^{trans} vary depending on different conditions. Under flow-limited conditions, where permeability is large, K^{trans} equals the blood plasma flow per unit volume of tissue. Under permeability-limited condition, where the flow is large, K^{trans} equals the permeability surface area product per unit volume of tissue.

7.5.2.6 Parametric Images

The images (i.e., raw images) produced by an imaging scanner are typically displayed with voxel values either in an intensity unit (e.g., Hounsfield units, SI, activity, dB, etc.) or in count rates (counts/voxel/s) that directly derive from the parameter that the scanner measures. Alternatively, one may desire to present images in terms of values that reflect biological parameters (e.g., perfusion, vascular permeability, glucose metabolism, receptor binding potential [BP], etc.). Such an image, in which the voxel values are derived from the raw images following mathematical manipulation, is called a *parametric image*. For example, consider two T_2-weighted images of the same piece of tissue obtained by an MR scanner with two different echo times. A third image, a T_2 map, can be obtained by fitting the voxel intensities in the two images to an exponential decay function. The T_2 map is a parametric image that is independent of scanner parameters, such as transmitter or receiver settings.

In more sophisticated cases, parametric images can be generated by applying tracer kinetic analysis, as described in Section 7.5.2.4. Figure 7.15 shows some examples of parametric images where the image values are the CMRG, the CBF, and the cerebral metabolic rate of oxygen (CMRO). The images were obtained by performing the sequential [^{15}O]H_2O, ^{15}O-O_2, and [^{18}F]FDG-PET brain scans on the same subject with arterial blood sampling. Sometimes, artifacts may appear in a parametric image because the same tracer kinetic model may not apply to all the tissue types present within the image. Furthermore, generation of a parametric image requires a sequence of dynamic images. The low S/N commonly found in dynamic images may introduce noise-related artifacts in estimating the parameter of interest on a voxel-by-voxel basis.

7.5.3 Multiple-Time Graphical Analysis

Unlike compartmental analysis, methods using multiple-time graphical analysis are independent of a particular model structure. They were developed based on compartmental model theory but with certain conditions or assumptions. The plasma input function and tissue TAC are transformed and combined into a single curve that approaches linearity when these conditions are met. Graphical methods are easy to implement and are generally

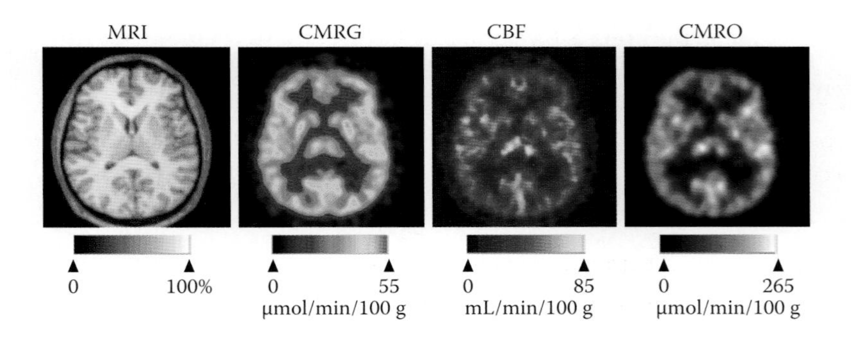

FIGURE 7.15 An example that demonstrates the quantitative capability of PET imaging in assessing a range of biological parameters by using different radiotracers. (From left to right) Structural MRI scan and quantitative parametric PET images with voxel values corresponding to the cerebral metabolic rate of glucose (CMRG), cerebral blood flow (CBF), and the cerebral metabolic rate of oxygen (CMRO), respectively, in a human brain. Only a single midbrain section from each 3-D data set is shown.

considered more robust than full kinetic modeling. If a macroparameter, such as the uptake rate constant (K_i^*) of [^{18}F]FDG or the BP of dopamine transporter in the brain, is the parameter of interest, the laborious procedure of compartmental modeling can be greatly simplified by performing graphical analysis. The most frequently used graphical analysis methods for irreversibly and reversibly binding tracers are the *Patlak plot* and the *Logan plot*, respectively. Due to its simplicity and computational efficiency, multiple-time graphical analysis is often the choice when a parametric image is desired.

7.5.3.1 Patlak Graphical Analysis

Patlak graphical analysis is also known as the Gjedde-Patlak plot, the Patlak-Rutland plot, or the Patlak plot [18,19]. It is based on a compartmental model but with no limitation on the number of reversible compartments. It was developed under the assumption that there must be at least one irreversible reaction in the system where the tracer or its labeled metabolites cannot escape that compartment and are trapped. The Gjedde-Patlak plot with plasma input, $C_P(t)$, is described by

$$\frac{C(t)}{C_P(t)} = K_i \frac{\int_0^t C_P(s)\,ds}{C_P(t)} + int, \tag{7.20}$$

where $C(t)$ is the ROI-derived or voxel-wise tissue TAC. The Gjedde-Patlak plot is generated by plotting $C(t)/C_p(t)$ (y-axis) against $\int_0^t C_p(s)ds/C_p(t)$ (x-axis), and the curve becomes linear after the tracer concentrations in the reversible compartments and in plasma are in equilibrium ($t > t^*$). The slope of the linear phase of the plot is the net uptake rate constant (K_i^*, in units of mL/min/g) of the tracer with an extrapolated intercept denoted as *int*. This type of analysis is feasible if the kinetics of a tracer can be approximated by a "central" compartment that is in rapid equilibrium with plasma and a "peripheral" compartment where the tracer enters and is irreversibly trapped during the time over which measurements are taken. Patlak analysis is widely used for [^{18}F]FDG brain PET studies, and an example is shown in Figure 7.13d. Patlak analysis is appropriate in this case because the enzyme activity of phosphatase in the brain is low (or negligible) and [^{18}F]FDG-6-PO_4 can reasonably be assumed to be irreversibly trapped in brain tissue during the time of the scan. Using Patlak analysis and a linear regression analysis, a 3-D parametric image of the uptake rate constant K_i in the entire brain can be generated in less than a minute using current-generation desktop computers.

7.5.3.2 Logan Plot

The most commonly used graphical analysis method for a tracer that binds reversibly is the *Logan plot* [20]. The Logan plot with plasma input, $C_P(t)$, is described by

$$\frac{\int_0^t C(s)\,ds}{C(t)} = DV_T \frac{\int_0^t C_P(s)\,ds}{C(t)} + int, \tag{7.21}$$

where $C(t)$ is the ROI-derived or voxel-wise tissue TAC. The Logan plot is generated by plotting $\int_0^t C(s)ds/C(t)$ (y-axis) against $\int_0^t C_p(s)ds/C(t)$ (x-axis). At some t ($t > t^*$), the intercept (int) of the linear phase of the plot effectively reaches a constant value. The slope of the line is the total distribution volume (DV_T) of the tracer in the target tissue.

In some instances, a TAC from a *reference region* can be used in place of the arterial plasma input. A good reference region should not have any significant specific binding of the tracer under study. The Logan plot with reference input, $C_{REF}(t)$, is described by

$$\frac{\int_0^t C(s)ds}{C(t)} = \text{DVR}\frac{\int_0^t C_{REF}(s)ds}{C(t)} + int, \tag{7.22}$$

where $C(t)$ and $C_{REF}(t)$ are the ROI-derived or voxel-wise TAC of the target tissue and the reference tissue, respectively. When $t > t^*$, and $C(t)/C_{REF}(t)$ becomes constant, the slope of the Logan plot, DVR, corresponds to the ratio of the distribution volume DV_T of the target to the reference tissues [21]. In receptor-ligand kinetics, BP is a combined measure of available receptor density and affinity of the ligand to the receptor. If a reference region without specific binding is available, BP can be calculated from DVR as

$$\text{BP} = \text{DVR} - 1. \tag{7.23}$$

The Logan plot has been widely used to quantify the DV_T and BP of reversible receptor binding in the brain. The method makes certain assumptions, for example, that system equilibrium is reached, where the plot shows a linear phase. If these assumptions or conditions are not met, the resulting parameter estimates may be biased. Different approaches have been proposed to improve the estimates of DV_T and BP, for example, in cases where

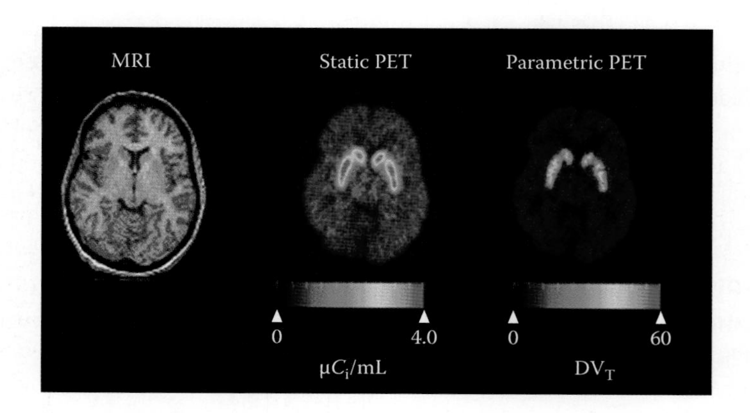

FIGURE 7.16 Coregistered MRI (gray scale) and the DV_T parametric PET images (in color) generated from dynamic human PET studies using the radiotracer [^{11}C]WIN35,428. A summed PET image (integrated from 40–90 min after tracer injection) is also displayed for reference. (Reprinted from *NeuroImage*, 49, Zhou Y. et al., "Multi-graphical analysis of dynamic PET," 2947–2957, Copyright 2010, with permission from Elsevier.)

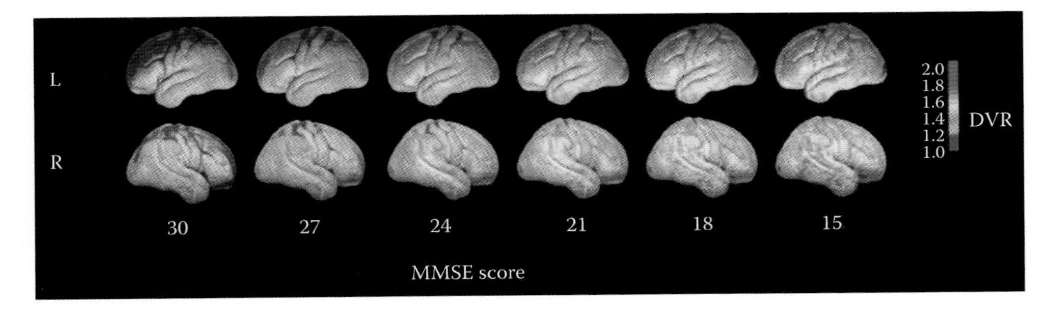

FIGURE 7.17 Hemispheric surface maps that show the progression of 2-(1-{6-[(2-[F-18]fluoroethyl)(methyl)amino]-2-naphthyl}ethylidene) malononitrile ([^{18}F]FDDNP) binding in subjects with different mini-mental state examination (MMSE) scores. The parametric DVR images (see Equation 7.22 for definition) were generated from the dynamic [^{18}F]FDDNP PET studies of the subjects and normalized to a common cortical surface map. There is little signal in the left temporal lobe at a normal MMSE of 30 that increases in the temporal lobe and spreads to the parietal and frontal areas as the MMSE score drops. This pattern of DVR spreading mimics the pathologic progression of beta-amyloid plaque and neurofibrillary tangle accumulation in Alzheimer's disease. (Reprinted from *NeuroImage*, 49, Protas H.D. et al., "FDDNP binding using MR derived cortical surface maps," 240–248, Copyright 2010, with permission from Elsevier.)

the time available for imaging is limited, or to reduce the effects of high noise levels typically present in fast dynamic images.

Figure 7.16 shows a dynamic [^{11}C]WIN35,428 PET study and the use of Logan plots to quantify dopamine transporter density in the human brain. The parametric images of DV_T were generated using the Logon plot. Applying parameter estimation and voxel-based analysis, Figure 7.17 shows another application of the Logan plot (this time using a reference tissue) in which [^{18}F]FDDNP PET images of a group of subjects with different mental status were normalized and mapped to an MR-derived cortical surface template. The DVR calculated from the [^{18}F]FDDNP PET studies reveals a spatial pattern of beta-amyloid plaque and neurofibrillary tangle accumulation that is consistent with the known pathological progression of Alzheimer's disease.

7.6 SUMMARY

This chapter gives an overview of common approaches used in quantitative imaging. The basic principles of lesion detection, image analysis, and quantitative data analysis have been introduced. The discipline of radiology started with simple visual discerning of shadows or abnormalities on x-ray films viewed on a light box. Today, modern cross-sectional imaging scanners, including CT, MR, SPECT, and PET, are equipped with powerful workstations and sophisticated software for image display, image manipulation, and data analysis. With these tools, radiologists and physicians can read the images and make diagnoses more efficiently and with more confidence. Image registration, for example, is now a standard procedure in many institutions for studies of brain diseases, whether it is within subject for motion correction or between subjects for group-mapping voxel analysis. Semiquantitative metrics, such as SUV, are routinely used in clinical FDG-PET studies for diagnosis and staging of tumors. DCE MRI uses K_{trans} to determine vascular permeability and evaluate the microvasculature within tumors. It has been reported that for interpretation of mammograms, double reading, either by radiologists or with application of a computer-aided detection algorithm, can increase the sensitivity of breast cancer detection. In breast ultrasound, application of computer-aided detection algorithms has also resulted in improved sensitivity and/or specificity.

More sophisticated quantitative analyses, such as compartmental modeling and ROC studies, are widely used both in clinical and preclinical research. These quantitative tools help correct for errors or imperfections in the data (e.g., motion correction). They can be used to quantitatively evaluate whether a new instrument or protocol is better at a specific task, can aid in providing objective measures of image SI in structures/tissue of interest, and can be used with appropriate models to relate image intensity to biologically relevant parameters. These techniques therefore play a critical role in modern biomedical imaging science.

REFERENCES

1. Rose, A. 1953. Quantum and noise limitations of the visual process. *J Opt Soc Am* 43:715–716.
2. Metz, C. E. 1978. Basic principles of ROC analysis. *Semin Nucl Med* 8:283–298.
3. Barrett, H. H., J. Yao, J. P. Rolland, and K. J. Myers. 1993. Model observers for assessment of image quality. *Proc Natl Acad Sci U S A* 90:9758–9765.
4. Hill, D. L., P. G. Batchelor, M. Holden, and D. J. Hawkes. 2001. Medical image registration. *Phys Med Biol* 46:R1–R45.
5. Lancaster, J. L., M. G. Woldorff, L. M. Parsons, M. Liotti, C. S. Freitas, L. Rainey, P. V. Kochunov, D. Nickerson, S. A. Mikiten, and P. T. Fox. 2000. Automated Talairach atlas labels for functional brain mapping. *Hum Brain Mapp* 10:120–131.
6. Smith, S. M., N. De Stefano, M. Jenkinson, and P. M. Matthews. 2001. Normalized accurate measurement of longitudinal brain change. *J Comput Assist Tomogr* 25:466–475.
7. Klein, A., J. Andersson, B. A. Ardekani, J. Ashburner, B. Avants, M. C. Chiang, G. E. Christensen, D. L. Collins, J. Gee, P. Hellier, J. H. Song, M. Jenkinson, C. Lepage, D. Rueckert, P. Thompson, T. Vercauteren, R. P. Woods, J. J. Mann, and R. V. Parsey. 2009. Evaluation of 14 nonlinear deformation algorithms applied to human brain MRI registration. *Neuroimage* 46:786–802.
8. Kim, C. K., N. C. Gupta, B. Chandramouli, and A. Alavi. 1994. Standardized uptake values of FDG: Body surface area correction is preferable to body weight correction. *J Nucl Med* 35:164–167.
9. Tofts, P. S. 2003. *Quantitative MRI of the Brain: Measuring Changes Caused by Disease.* John Wiley & Sons, Inc., Hoboken, NJ.
10. Fischl, B., and A. M. Dale. 2000. Measuring the thickness of the human cerebral cortex from magnetic resonance images. *Proc Natl Acad Sci U S A* 97:11050–11055.
11. Silverman, D. 2009. *PET in the Evaluation of Alzheimer's Disease and Related Disorders.* Springer Publishing, New York.
12. Hopyan, J., A. Ciarallo, D. Dowlatshahi, P. Howard, V. John, R. Yeung, L. Zhang, J. Kim, G. MacFarlane, T. Y. Lee, and R. I. Aviv. 2010. Certainty of stroke diagnosis: Incremental benefit with CT perfusion over noncontrast CT and CT angiography. *Radiology* 255:142–153.
13. Miles, K. A. 2006. Perfusion imaging with computed tomography: Brain and beyond. *Eur Radiol* 16:M37–M43.
14. Sokoloff, L., M. Reivich, C. Kennedy, M. H. Des Rosiers, C. S. Patlak, K. D. Pettigrew, O. Sakurada, and M. Shinohara. 1977. The [14C]deoxyglucose method for the measurement of local cerebral glucose utilization: Theory, procedure, and normal values in the conscious and anesthetized albino rat. *J Neurochem* 28:897–916.
15. Phelps, M. E., S. C. Huang, E. J. Hoffman, C. Selin, L. Sokoloff, and D. E. Kuhl. 1979. Tomographic measurement of local cerebral glucose metabolic rate in humans with (F-18)2-fluoro-2-deoxy-D-glucose: Validation of method. *Ann Neurol* 6:371–388.
16. Muzic, R. F., Jr., and S. Cornelius. 2001. COMKAT: Compartment model kinetic analysis tool. *J Nucl Med* 42:636–645.

17. Huang, S. C., D. Truong, H. M. Wu, A. F. Chatziioannou, W. Shao, A. M. Wu, and M. E. Phelps. 2005. An internet-based kinetic imaging system (KIS) for MicroPET. *Mol Imaging Biol* 7:330–341.
18. Gjedde, A. 1982. Calculation of cerebral glucose phosphorylation from brain uptake of glucose analogs in vivo: A re-examination. *Brain Res* 257:237–274.
19. Patlak, C. S., R. G. Blasberg, and J. D. Fenstermacher. 1983. Graphical evaluation of blood-to-brain transfer constants from multiple-time uptake data. *J Cereb Blood Flow Metab* 3:1–7.
20. Logan, J. 2000. Graphical analysis of PET data applied to reversible and irreversible tracers. *Nucl Med Biol* 27:661–670.
21. Logan, J. 2003. A review of graphical methods for tracer studies and strategies to reduce bias. *Nucl Med Biol* 30:833–844.

Appendix

Constants, Units, Conversions, and Useful Relations

CONSTANTS

		Value	Units
c	speed of light in vacuum	2.998×10^{8}	m/s
h	Planck's constant	6.626×10^{-34}	J s
k	Boltzmann's constant	1.381×10^{-23}	J/K

UNITS

Name	Typical Symbol	Unit
	Fundamental Units	
Length	x, y, z	meter (m)
Mass	m	kilogram (kg)
Time	t or τ	second (s)
Current	I	amp (A)
Temperature	T	kelvin (K)
Amount of substance	—	mole (mol)
Luminous intensity	I_v	candela (cd)

Name	Typical Symbol	SI Unit
	Derived Units	
Absorbed dose	D	gray (Gy)
Acoustic impedance	Z	Rayl
Acoustic intensity	I	watt/cm^2 (W/cm^2)
Activity	A	becquerel (Bq)
Air kerma	K	gray (Gy)
Angular frequency	ω	radian/s
Area	A	m^2
Density (mass)	ρ	kg/m^3
Distance	d	meter (m)
Equivalent dose	H	sievert (Sv)
Energy	E	joule (J)
Force	F	newton (N)
Frequency	ν or f	hertz (Hz) or s^{-1}
Gyromagnetic ratio	γ	radian/s/tesla
Magnetic flux density	B	tesla (T)
Magnetization	M	amp/meter (A/m)
Period	T	s
Pressure	p or P	pascal (Pa)
Solid angle	Ω	steradian (sr)
Velocity	c or v	m/s
Voltage	V	volt (V)
Volume	V	m^3
Wavelength	λ	m
X-ray fluence	Φ	photons/mm^2

USEFUL CONVERSIONS

Energy	1 electron volt (eV) = 1.6×10^{-19} joules (J)
	1 J = 6.24×10^{18} eV
Magnetic flux density	1 gauss (G) = 0.0001 T
	1 T = 10,000 G
Activity	1 Curie (Ci) = 3.7×10^{10} Bq
	1 Bq = 2.7×10^{-11} Ci
Temperature	Kelvin (K) to centigrade (°C)
	T(°C) = T(K) − 273.15
	T(K) = T(°C) + 273.15

UNIT PREFIXES

Zepto	z	10^{-21}
Atto	a	10^{-18}
Femto	f	10^{-15}
Pico	p	10^{-12}
Nano	n	10^{-9}
Micro	μ	10^{-6}
Milli	m	10^{-3}
Centi	c	10^{-2}
Kilo	k	10^{3}
Mega	M	10^{6}
Giga	G	10^{9}
Tera	T	10^{12}

RATIO BETWEEN TWO POWER OR AMPLITUDE LEVELS

The ratio R_P between two *power levels*, P_1 and P_0 is often described in terms of decibels (dB), where

$$R_P\ (\text{dB}) = 10 \log_{10} (P_1/P_0).$$

P_1 and P_0 may be power quantities such as acoustic intensity or luminous intensity. 10 dB corresponds to a ratio of 10. 1 dB corresponds to a ratio of ~1.26.

The ratio R_A between *two amplitudes* A_1 and A_0 (assuming that the power is proportional to the square of the amplitude) in dB is given by

$$R_A\ (\text{dB}) = 10 \log_{10} (A_1^2/A_0^2) = 20 \log_{10} (A_1/A_0).$$

A_1 and A_0 could represent quantities such as pressure, current, or voltage.

In both cases, the quantities and units must be the same in the numerator and the denominator before the ratio is calculated.

Index

Page numbers followed by f and t indicate figures and tables, respectively.

D

E

F

G

H

M

P

S

T

U

V

W

X

Y